Graduate Texts in Mathematics **173**

Springer

New York
Berlin
Heidelberg
Barcelona
Budapest
Hong Kong
London
Milan
Paris
Santa Clara
Singapore
Tokyo

Graduate Texts in Mathematics

continued after index

Reinhard Diestel

Graph Theory

With 103 Illustrations

 Springer

Reinhard Diestel
Faculty of Mathematics
Chemnitz University
D-09107 Chemnitz
Germany

The Königsberg bridge etching (Figure 1.8.1 on page 18) is reprinted by permission of Oxford University Press.

Mathematics Subject Classifications (1991): 05-01, 05Cxx

Library of Congress Cataloging-in-Publication Data
Diestel, Reinhard.
    [Graphentheorie. English]
    Graph theory / Reinhard Diestel.
      p.   cm. − (Graduate texts in mathematics ; 173)
    Includes index.
    ISBN 0-387-98210-8 (hardcover : alk. paper). − ISBN 0-387-98211-6
    (softcover : alk. paper)
      1. Graph theory.   I. Title.   II. Series.
    QA166.D51413   1997
    511′.5 − dc21   97-6932

Printed on acid-free paper.

This book is a translation of *Graphentheorie*, Springer-Verlag, 1996.

Production managed by Karina Mikhli; manufacturing supervised by Joe Quatela.
Photocomposed copy prepared from the author's TeX file.
Printed and bound by R.R. Donnelley & Sons, Harrisonburg, VA.
Printed in the United States of America.

9 8 7 6 5 4 3 2 1

ISBN 0-387-98210-8 Springer-Verlag New York Berlin Heidelberg (Hardcover) SPIN 10572041
ISBN 0-387-98211-6 Springer-Verlag New York Berlin Heidelberg (Softcover) SPIN 10572059

To Dagmar

Preface

Almost two decades have passed since the appearance of those graph theory texts that still set the agenda for most introductory courses taught today. The canon created by those books has helped to identify some main fields of study and research, and will doubtless continue to influence the development of the discipline for some time to come.

Yet much has happened in those 20 years, in graph theory no less than elsewhere: deep new theorems have been found, seemingly disparate methods and results have become interrelated, entire new branches have arisen. To name just a few such developments, one may think of how the new notion of list colouring has bridged the gulf between invariants such as average degree and chromatic number, how probabilistic methods and the regularity lemma have pervaded extremal graph theory and Ramsey theory, or how the entirely new field of graph minors and tree-decompositions has brought standard methods of surface topology to bear on long-standing algorithmic graph problems.

Clearly, then, the time has come for a reappraisal: *what are, today, the essential areas, methods and results that should form the centre of an introductory graph theory course aiming to equip its audience for the most likely developments ahead?*

I have tried in this book to offer material for such a course. In view of the increasing complexity and maturity of the subject, I have broken with the tradition of attempting to cover both theory and applications: this book offers an introduction to the theory of graphs as part of (pure) mathematics; it contains neither explicit algorithms nor 'real world' applications. My hope is that the potential for depth gained by this restriction in scope will serve students of computer science as much as their peers in mathematics: assuming that they prefer algorithms but will benefit from an encounter with pure mathematics of *some* kind, it seems an ideal opportunity to look for this close to where their heart lies!

In the selection and presentation of material, I have tried to accommodate two conflicting goals. On the one hand, I believe that an introductory text should be lean and concentrate on the essential, so as to offer guidance to those new to the field. As a graduate text, moreover, it should get to the heart of the matter quickly: after all, the idea is to convey at least an impression of the depth and methods of the subject. On the other hand, it has been my particular concern to write with sufficient detail to make the text enjoyable and easy to read: guiding questions and ideas will be discussed explicitly, and all proofs presented will be rigorous and complete.

A typical chapter, therefore, begins with a brief discussion of what are the guiding questions in the area it covers, continues with a succinct account of its classic results (often with simplified proofs), and then presents one or two deeper theorems that bring out the full flavour of that area. The proofs of these latter results are typically preceded by (or interspersed with) an informal account of their main ideas, but are then presented formally at the same level of detail as their simpler counter-parts. I soon noticed that, as a consequence, some of those proofs came out rather longer in print than seemed fair to their often beautifully simple conception. I would hope, however, that even for the professional reader the relatively detailed account of those proofs will at least help to minimize reading time...

If desired, this text can be used for a lecture course with little or no further preparation. The simplest way to do this would be to follow the order of presentation, chapter by chapter: apart from two clearly marked exceptions, any results used in the proof of others precede them in the text.

Alternatively, a lecturer may wish to divide the material into an easy basic course for one semester, and a more challenging follow-up course for another. To help with the preparation of courses deviating from the order of presentation, I have listed in the margin next to each proof the reference numbers of those results that are used in that proof. These references are given in round brackets: for example, a reference (4.1.2) in the margin next to the proof of Theorem 4.3.2 indicates that Lemma 4.1.2 will be used in this proof. Correspondingly, in the margin next to Lemma 4.1.2 there is a reference [4.3.2] (in square brackets) informing the reader that this lemma will be used in the proof of Theorem 4.3.2. Note that this system applies between different sections only (of the same or of different chapters): the sections themselves are written as units and best read in their order of presentation.

The mathematical prerequisites for this book, as for most graph theory texts, are minimal: a first grounding in linear algebra is assumed for Chapter 1.9 and once in Chapter 5.5, some basic topological concepts about the Euclidean plane and 3-space are used in Chapter 4, and a previous first encounter with elementary probability will help with Chapter 11. (Even here, all that is assumed formally is the knowledge of basic definitions: the few probabilistic tools used are developed in the text.) There are two areas of graph theory which I find both fascinating and important, especially from the perspective of pure mathematics adopted here, but which are not covered in this book: these are algebraic graph theory and infinite graphs.

At the end of each chapter, there is a section with exercises and another with bibliographical and historical notes. Many of the exercises were chosen to complement the main narrative of the text: they illustrate new concepts, show how a new invariant relates to earlier ones, or indicate ways in which a result stated in the text is best possible. Particularly easy exercises are identified by the superscript $^-$, the more challenging ones carry a $^+$. The notes are intended to guide the reader on to further reading, in particular to any monographs or survey articles on the theme of that chapter. They also offer some historical and other remarks on the material presented in the text.

Ends of proofs are marked by the symbol $\Box$. Where this symbol is found directly below a formal assertion, it means that the proof should be clear after what has been said—a claim waiting to be verified! There are also some deeper theorems which are stated, without proof, as background information: these can be identified by the absence of both proof and $\Box$.

Almost every book contains errors, and this one will hardly be an exception. I shall try to post on the Web any corrections that become necessary. The relevant site may change in time, but will always be accessible via the following two addresses:

http://www.springer-ny.com/supplements/diestel/
http://www.springer.de/catalog/html-files/deutsch/math/3540609180.html

Please let me know about any errors you find.

Little in a textbook is truly original: even the style of writing and of presentation will invariably be influenced by examples. The book that no doubt influenced me most is the classic GTM graph theory text by Bollobás: it was in the course recorded by this text that I learnt my first graph theory as a student. Anyone who knows this book well will feel its influence here, despite all differences in contents and presentation.

I should like to thank all who gave so generously of their time, knowledge and advice in connection with this book. I have benefited particularly from the help of N. Alon, G. Brightwell, R. Gillett, R. Halin, M. Hintz, A. Huck, I. Leader, T. Łuczak, W. Mader, V. Rödl, A.D. Scott, P.D. Seymour, G. Simonyi, M. Škoviera, R. Thomas, C. Thomassen and P. Valtr. I am particularly grateful also to Tommy R. Jensen, who taught me much about colouring and all I know about k-flows, and who invested immense amounts of diligence and energy in his proofreading of the preliminary German version of this book.

March 1997   *RD*

Contents

1   The Basics

_____

This chapter gives a gentle yet concise introduction to most of the terminology used later in the book. Fortunately, much of standard graph theoretic terminology is so intuitive that it is easy to remember; the few terms better understood in their proper setting will be introduced later, when their time has come.

Section 1.1 offers a brief but self-contained summary of the most basic definitions in graph theory, those centred round the notion of a graph. Most readers will have met these definitions before, or will have them explained to them as they begin to read this book. For this reason, Section 1.1 does not dwell on these definitions more than clarity requires: its main purpose is to collect the most basic terms in one place, for easy reference later.

From Section 1.2 onwards, all new definitions will be brought to life almost immediately by a number of simple yet fundamental propositions. Often, these will relate the newly defined terms to one another: the question of how the value of one invariant influences that of another underlies much of graph theory, and it will be good to become familiar with this line of thinking early.

By $\mathbb{N}$ we denote the set of natural numbers, including zero. The set $\mathbb{Z}/n\mathbb{Z}$ of integers modulo n is denoted by $\mathbb{Z}_n$; its elements are written as $\bar{i} := i + n\mathbb{Z}$. For a real number x we denote by $\lfloor x \rfloor$ the greatest integer $\leqslant x$, and by $\lceil x \rceil$ the least integer $\geqslant x$. Logarithms written as 'log' are taken at base 2; the natural logarithm will be denoted by 'ln'. A set $\mathcal{A} = \{A_1, \ldots, A_k\}$ of disjoint subsets of a set A is a _partition_ of A if $A = \bigcup_{i=1}^{k} A_i$ and $A_i \neq \emptyset$ for every i. Another partition $\{A'_1, \ldots, A'_\ell\}$ of A _refines_ the partition $\mathcal{A}$ if each A'_i is contained in some A_j. By $[A]^k$ we denote the set of all k-element subsets of A. Sets with k elements will be called k-_sets_; subsets with k elements are k-_subsets_.

$\mathbb{Z}_n$

$\lfloor x \rfloor, \lceil x \rceil$
log, ln
partition

$[A]^k$

k-set

1.1 Graphs

graph

A *graph* is a pair $G = (V, E)$ of sets satisfying $E \subseteq [V]^2$; thus, the elements of E are 2-element subsets of V. To avoid notational ambiguities, we shall always assume tacitly that $V \cap E = \emptyset$. The elements of V are the

vertex

vertices (or *nodes*, or *points*) of the graph G, the elements of E are its

edge

edges (or *lines*). The usual way to picture a graph is by drawing a dot for each vertex and joining two of these dots by a line if the corresponding two vertices form an edge. Just how these dots and lines are drawn is considered irrelevant: all that matters is the information which pairs of vertices form an edge and which do not.

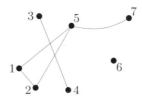

Fig. 1.1.1. The graph on $V = \{1, \ldots, 7\}$ with edge set
$$E = \{\{1,2\}, \{1,5\}, \{2,5\}, \{3,4\}, \{5,7\}\}$$

on

A graph with vertex set V is said to be a graph *on* V. The vertex

$V(G), E(G)$

set of a graph G is referred to as $V(G)$, its edge set as $E(G)$. These conventions are independent of any actual names of these two sets: the vertex set W of a graph $H = (W, F)$ is still referred to as $V(H)$, not as $W(H)$. We shall not always distinguish strictly between a graph and its vertex or edge set. For example, we may speak of a vertex $v \in G$ (rather than $v \in V(G)$), an edge $e \in G$, and so on.

order

The number of vertices of a graph G is its *order*, written as $|G|$;

$|G|, \|G\|$

its number of edges is denoted by $\|G\|$. Graphs are *finite* or *infinite* according to their order; unless otherwise stated, the graphs we consider are all finite.

$\emptyset$
trivial
graph

For the *empty graph* $(\emptyset, \emptyset)$ we simply write $\emptyset$. A graph of order 0 or 1 is called *trivial*. Sometimes, e.g. to start an induction, trivial graphs can be useful; at other times they form silly counterexamples and become a nuisance. To avoid cluttering the text with non-triviality conditions, we shall mostly treat the trivial graphs, and particularly the empty graph $\emptyset$, with generous disregard.

incident

A vertex v is *incident* with an edge e if $v \in e$; then e is an edge *at* v.

ends

The two vertices incident with an edge are its *endvertices* or *ends*, and an edge *joins* its ends. An edge $\{x, y\}$ is usually written as xy (or yx). If $x \in X$ and $y \in Y$, then xy is an *X–Y edge*. The set of all *X–Y* edges

$E(X, Y)$

in a set E is denoted by $E(X, Y)$; instead of $E(\{x\}, Y)$ and $E(X, \{y\})$ we simply write $E(x, Y)$ and $E(X, y)$. The set of all the edges in E at a

$E(v)$

vertex v is denoted by $E(v)$.

Two vertices x, y of G are *adjacent*, or *neighbours*, if xy is an edge   *adjacent*
of G. Two edges $e \neq f$ are *adjacent* if they have an end in common. If all   *neighbour*
the vertices of G are pairwise adjacent, then G is *complete*. A complete   *complete*
graph on n vertices is a K^n; a K^3 is called a *triangle*. K^n

Pairwise non-adjacent vertices or edges are called *independent*.
More formally, a set of vertices or of edges is *independent* (or *stable*)   *inde-*
if no two of its elements are adjacent.   *pendent*

Let $G = (V, E)$ and $G' = (V', E')$ be two graphs. We call G and
G' *isomorphic*, and write $G \simeq G'$, if there exists a bijection $\varphi : V \to V'$
with $xy \in E \Leftrightarrow \varphi(x)\varphi(y) \in E'$ for all $x, y \in V$. Such a map φ is called $\simeq$
an *isomorphism*; if $G = G'$, it is called an *automorphism*. We do not
normally distinguish between isomorphic graphs. Thus, we usually write   *isomor-*
$G = G'$ rather than $G \simeq G'$, speak of *the* complete graph on 17 vertices,   *phism*
and so on.

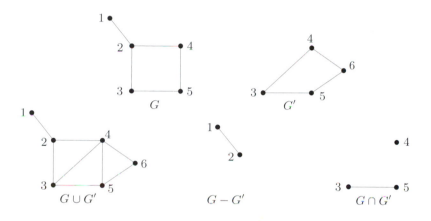

Fig. 1.1.2. Union, difference and intersection; the vertices 2,3,4
induce (or span) a triangle in $G \cup G'$ but not in G

We set $G \cup G' := (V \cup V', E \cup E')$ and $G \cap G' := (V \cap V', E \cap E')$. $G \cap G'$
If $G \cap G' = \emptyset$, then G and G' are *disjoint*. If $V' \subseteq V$ and $E' \subseteq E$, then   *subgraph*
G' is a *subgraph* of G (and G a *supergraph* of G'), written as $G' \subseteq G$. $G' \subseteq G$
Less formally, we say that G *contains* G'.

If $G' \subseteq G$ and G' contains *all* the edges $xy \in E$ with $x, y \in V'$, then
G' is an *induced subgraph* of G; we say that V' *induces* or *spans* G' in G,   *induced*
   subgraph

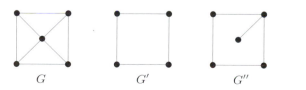

Fig. 1.1.3. A graph G with subgraphs G' and G'':
G' is an induced subgraph of G, but G'' is not

G[U]   and write $G' =: G[V']$. Thus if $U \subseteq V$ is any set of vertices, then $G[U]$
         denotes the graph on U whose edges are precisely the edges of G with
         both ends in U. If H is a subgraph of G, not necessarily induced, we
spanning abbreviate $G[V(H)]$ to $G[H]$. Finally, $G' \subseteq G$ is a *spanning* subgraph
         of G if V' spans all of G, i.e. if $V' = V$.

                 If U is any set of vertices (usually of G), we write $G - U$ for
         $G[V \smallsetminus U]$. In other words, $G - U$ is obtained from G by *deleting* all the
-        vertices in $U \cap V$ and their incident edges. If $U = \{v\}$ is a singleton,
         we write $G - v$ rather than $G - \{v\}$. Instead of $G - V(G')$ we simply
+   write $G - G'$. For a subset F of $[V]^2$ we write $G - F := (V, E \smallsetminus F)$ and
         $G + F := (V, E \cup F)$; as above, $G - \{e\}$ and $G + \{e\}$ are abbreviated to
edge- $G - e$ and $G + e$. We call G *edge-maximal* with a given graph property
maximal if G itself has the property but no graph $G + xy$ does, for non-adjacent
         vertices $x, y \in G$.

minimal   More generally, when we call a graph *minimal* or *maximal* with some
maximal property but have not specified any particular ordering, we are referring
         to the subgraph relation. When we speak of minimal or maximal sets of
         vertices or edges, the reference is simply to set inclusion.

$G * G'$ If G and G' are disjoint, we denote by $G * G'$ the graph obtained
         from $G \cup G'$ by joining all the vertices of G to all the vertices of G'. For
comple- example, $K^2 * K^3 = K^5$. The *complement* $\overline{G}$ of G is the graph on V
ment $\overline{G}$
         with edge set $[V]^2 \smallsetminus E$. The *line graph* $L(G)$ of G is the graph on E in
line graph which $x, y \in E$ are adjacent as vertices if and only if they are adjacent
$L(G)$ as edges in G.

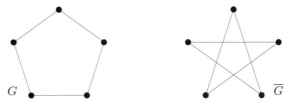

Fig. 1.1.4. A graph isomorphic to its complement

1.2 The degree of a vertex

         Let $G = (V, E)$ be a (non-empty) graph. The set of neighbours of a
$N(v)$ vertex v in G is denoted by $N_G(v)$, or briefly by $N(v)$.[1] More generally
         for $U \subseteq V$, the neighbours in $V \smallsetminus U$ of vertices in U are called *neighbours*
         of U; their set is denoted by $N(U)$.

---

[1] Here, as elsewhere, we drop the index referring to the underlying graph if the
reference is clear.

The *degree* (or *valency*) $d_G(v) = d(v)$ of a vertex v is the number   degree $d(v)$
$|E(v)|$ of edges at v; by our definition of a graph,[2] this is equal to the
number of neighbours of v. A vertex of degree 0 is *isolated*. The number   isolated
$\delta(G) := \min \{ d(v) \mid v \in V \}$ is the *minimum degree* of G, the number $\delta(G)$
$\Delta(G) := \max \{ d(v) \mid v \in V \}$ its *maximum degree*. If all the vertices $\Delta(G)$
of G have the same degree k, then G is *k-regular*, or simply *regular*. A   regular
3-regular graph is called *cubic*.   cubic
    The number

$$d(G) := \frac{1}{|V|} \sum_{v \in V} d(v)$$

    $d(G)$

is the *average degree* of G. Clearly,   average
    degree

$$\delta(G) \leqslant d(G) \leqslant \Delta(G).$$

The average degree quantifies globally what is measured locally by the
vertex degrees: the number of edges of G per vertex. Sometimes it will
be convenient to express this ratio directly, as $\varepsilon(G) := |E|/|V|$. $\varepsilon(G)$
    The quantities d and ε are, of course, intimately related. Indeed,
if we sum up all the vertex degrees in G, we count every edge exactly
twice: once from each of its ends. Thus

$$|E| = \tfrac{1}{2} \sum_{v \in V} d(v) = \tfrac{1}{2} d(G) \cdot |V|,$$

and therefore

$$\varepsilon(G) = \tfrac{1}{2} d(G).$$

Proposition 1.2.1. *The number of vertices of odd degree in a graph is*   [10.3.3]
always even.

Proof. A graph on V has $\tfrac{1}{2} \sum_{v \in V} d(v)$ edges, so $\sum d(v)$ is an even
number.   □

If a graph has large minimum degree, i.e. everywhere, locally, many
edges per vertex, it also has many edges per vertex globally: $\varepsilon(G) = \tfrac{1}{2} d(G) \geqslant \tfrac{1}{2} \delta(G)$. Conversely, of course, its average degree may be large
even when its minimum degree is small. However, the vertices of large
degree cannot be scattered completely among vertices of small degree: as
the next proposition shows, every graph G has a subgraph whose average
degree is no less than the average degree of G, and whose minimum
degree is more than half its average degree:

Proposition 1.2.2. *Every graph G with at least one edge has a subgraph H with $\delta(H) > \varepsilon(H) \geqslant \varepsilon(G)$.*

---

[2] but not for multigraphs; see Section 1.10

Proof. To construct H from G, let us try to delete vertices of small degree one by one, until only vertices of large degree remain. Up to which degree $d(v)$ can we afford to delete a vertex v, without lowering ε? Clearly, up to $d(v) = \varepsilon$: then the number of vertices decreases by 1 and the number of edges by at most ε, so the overall ratio ε of edges to vertices will not decrease.

Formally, we construct a sequence $G = G_0 \supseteq G_1 \supseteq \ldots$ of induced subgraphs of G as follows. If G_i has a vertex v_i of degree $d(v_i) \leqslant \varepsilon(G_i)$, we let $G_{i+1} := G_i - v_i$; if not, we terminate our sequence and set $H := G_i$. By the choices of v_i we have $\varepsilon(G_{i+1}) \geqslant \varepsilon(G_i)$ for all i, and hence $\varepsilon(H) \geqslant \varepsilon(G)$.

What else can we say about the graph H? Since $\varepsilon(K^1) = 0 < \varepsilon(G)$, none of the graphs in our sequence is trivial, so in particular $H \neq \emptyset$. The fact that H has no vertex suitable for deletion thus implies $\delta(H) > \varepsilon(H)$, as claimed. $\square$

1.3 Paths and cycles

path

A *path* is a non-empty graph $P = (V, E)$ of the form

$$V = \{ x_0, x_1, \ldots, x_k \} \qquad E = \{ x_0x_1, x_1x_2, \ldots, x_{k-1}x_k \},$$

where the x_i are all distinct. The vertices x_0 and x_k are *linked* by P and are called its *ends*; the vertices $x_1, \ldots, x_{k-1}$ are the *inner* vertices of P.

length
P^k

The number of edges of a path is its *length*, and the path of length k is denoted by P^k. Note that k is allowed to be zero; thus, $P^0 = K^1$.

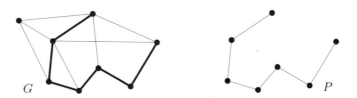

Fig. 1.3.1. A path $P = P^6$ in G

We often refer to a path by the natural sequence of its vertices,[3] writing, say, $P = x_0x_1 \ldots x_k$ and calling P a path *from* x_0 *to* x_k (as well as *between* x_0 and x_k).

---

[3]   More precisely, by one of the two natural sequences: $x_0 \ldots x_k$ and $x_k \ldots x_0$ denote the same path. Still, it often helps to fix one of these two orderings of $V(P)$ notationally: we may then speak of things like the 'first' vertex on P with a certain property, etc.

For $0 \leqslant i \leqslant j \leqslant k$ we write
xPy, P̊

$$Px_i := x_0 \ldots x_i$$
$$x_i P := x_i \ldots x_k$$
$$x_i P x_j := x_i \ldots x_j$$

and

$$\mathring{P} := x_1 \ldots x_{k-1}$$
$$P\mathring{x}_i := x_0 \ldots x_{i-1}$$
$$\mathring{x}_i P := x_{i+1} \ldots x_k$$
$$\mathring{x}_i P \mathring{x}_j := x_{i+1} \ldots x_{j-1}$$

for the appropriate subpaths of P. We use similar intuitive notation for the concatenation of paths; for example, if the union $Px \cup xQy \cup yR$ of three paths is again a path, we may simply denote it by $PxQyR$.
PxQyR

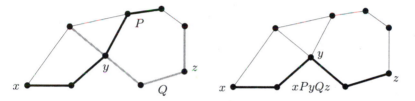

Fig. 1.3.2. Paths P, Q and $xPyQz$

Given sets A, B of vertices, we call $P = x_0 \ldots x_k$ an *A–B path* if
A–B path
$V(P) \cap A = \{x_0\}$ and $V(P) \cap B = \{x_k\}$. As before, we write a–B path rather than $\{a\}$–B path, etc. Two or more paths are *independent*
inde-
pendent
if none of them contains an inner vertex of another. Two a–b paths, for instance, are independent if and only if a and b are their only common vertices.

Given a graph H, we call P an *H-path* if P is non-trivial and meets
H-path
H exactly in its ends. In particular, the edge of any H-path of length 1 is never an edge of H.

If $P = x_0 \ldots x_{k-1}$ is a path and $k \geqslant 3$, then the graph $C :=$ $P + x_{k-1}x_0$ is called a *cycle*. As with paths, we often denote a cycle
cycle
by its (cyclic) sequence of vertices; the above cycle C might be written as $x_0 \ldots x_{k-1}x_0$. The *length* of a cycle is its number of edges (or vertices);
length
C^k
the cycle of length k is called a *k-cycle* and denoted by C^k.

The minimum length of a cycle (contained) in a graph G is the *girth*
girth g(G)
$g(G)$ of G; the maximum length of a cycle in G is its *circumference*. (If
circum-
ference
G does not contain a cycle, we set the former to ∞, the latter to zero.) An edge which joins two vertices of a cycle but is not itself an edge of
chord
the cycle is a *chord* of that cycle. Thus, an *induced cycle* in G, a cycle in
induced
cycle
G forming an induced subgraph, is one that has no chords (Fig. 1.3.3).

Fig. 1.3.3. A cycle C^8 with chord xy, and induced cycles C^6, C^4

If a graph has large minimum degree, it contains long paths and cycles:

Proposition 1.3.1. *Every graph G contains a path of length $\delta(G)$ and a cycle of length at least $\delta(G) + 1$ (provided that $\delta(G) \geqslant 2$).*

Proof. Let $x_0 \ldots x_k$ be a longest path in G. Then all the neighbours of x_k lie on this path (Fig. 1.3.4). Hence $k \geqslant d(x_k) \geqslant \delta(G)$. If $i < k$ is minimum with $x_i x_k \in E(G)$, then $x_i \ldots x_k x_i$ is a cycle of length at least $\delta(G) + 1$.   □

Fig. 1.3.4. A longest path $x_0 \ldots x_k$, and the neighbours of x_k

Minimum degree and girth, on the other hand, are not related (unless we fix the number of vertices): as we shall see in Chapter 11, there are graphs combining arbitrarily large minimum degree with arbitrarily large girth.

distance
$d_G(x,y)$

The *distance* $d_G(x,y)$ in G of two vertices x, y is the length of a shortest x–y path in G; if no such path exists, we set $d(x,y) := \infty$. The greatest distance between any two vertices in G is the *diameter* of G, denoted by $\mathrm{diam}(G)$. Diameter and girth are, of course, related:

diameter
$\mathrm{diam}(G)$

Proposition 1.3.2. *Every graph G containing a cycle satisfies $g(G) \leqslant 2\,\mathrm{diam}(G) + 1$.*

Proof. Let C be a shortest cycle in G. If $g(G) \geqslant 2\,\mathrm{diam}(G) + 2$, then C has two vertices whose distance in C is at least $\mathrm{diam}(G) + 1$. In G, these vertices have a lesser distance; any shortest path P between them is therefore not a subgraph of C. Thus, P contains a C-path xPy. Together with the shorter of the two x–y paths in C, this path xPy forms a shorter cycle than C, a contradiction.   □

A vertex is *central* in G if its greatest distance from any other vertex is as small as possible. This distance is the *radius* of G, denoted by $\mathrm{rad}(G)$. Thus, formally, $\mathrm{rad}(G) = \min_{x \in V(G)} \max_{y \in V(G)} d_G(x, y)$. As one easily checks (exercise), we have

$$\mathrm{rad}(G) \leqslant \mathrm{diam}(G) \leqslant 2\,\mathrm{rad}(G)\,.$$

central

radius

$\mathrm{rad}(G)$

Diameter and radius are not directly related to the minimum or average degree: a graph can combine large minimum degree with large diameter, or small average degree with small diameter (examples?).

The maximum degree behaves differently here: a graph of large order can only have small radius and diameter if its maximum degree is large. This connection is quantified very roughly in the following proposition:

Proposition 1.3.3. *A graph G of radius at most k and maximum degree at most d has no more than $1 + kd^k$ vertices.*

[9.4.1]
[9.4.2]

Proof. Let z be a central vertex in G, and let D_i denote the set of vertices of G at distance i from z. Then $V(G) = \bigcup_{i=0}^{k} D_i$, and $|D_0| = 1$. Since $\Delta(G) \leqslant d$, we have $|D_i| \leqslant d\,|D_{i-1}|$ for $i = 1, \ldots, k$, and thus $|D_i| \leqslant d^i$ by induction. Adding up these inequalities we obtain

$$|G| \leqslant 1 + \sum_{i=1}^{k} d^i \leqslant 1 + kd^k.$$

$\square$

A *walk* (of *length k*) in a graph G is a non-empty alternating sequence $v_0 e_0 v_1 e_1 \ldots e_{k-1} v_k$ of vertices and edges in G such that $e_i = \{v_i, v_{i+1}\}$ for all $i < k$. If $v_0 = v_k$, the walk is *closed*. If the vertices in a walk are all distinct, it defines an obvious path in G. In general, every walk between two vertices contains[4] a path between these vertices (proof?).

walk

1.4 Connectivity

A non-empty graph G is called *connected* if any two of its vertices are linked by a path in G. If $U \subseteq V(G)$ and $G[U]$ is connected, we also call U itself connected (in G).

connected

Proposition 1.4.1. *The vertices of a connected graph G can always be enumerated, say as $v_1, \ldots, v_n$, so that $G_i := G[v_1, \ldots, v_i]$ is connected for every i.*

[1.5.2]
[1.6.1]

---

[4] We shall often use terms defined for graphs also for walks, as long as their meaning is obvious.

Proof. Pick any vertex as v_1, and assume inductively that $v_1, \ldots, v_i$ have been chosen for some $i < |G|$. Now pick a vertex $v \in G - G_i$. As G is connected, it contains a v–v_1 path P. Choose as v_{i+1} the last vertex of P in $G - G_i$; then v_{i+1} has a neighbour in G_i. The connectedness of every G_i follows by induction on i.   □

component   Let $G = (V, E)$ be a graph. A maximal connected subgraph of G is called a *component* of G. Note that a component, being connected, is always non-empty; the empty graph, therefore, has no components.

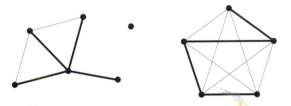

Fig. 1.4.1. A graph with three components, and a minimal spanning connected subgraph in each component

If $A, B \subseteq V$ and $X \subseteq V \cup E$ are such that every A–B path in *separate* G contains a vertex or an edge from X, we say that X *separates* the sets A and B in G. This implies in particular that $A \cap B \subseteq X$. More generally we say that X *separates* G, and call X a *separating set* in G, if X separates two vertices of $G - X$ in G. A vertex which separates *cutvertex*   two other vertices of the same component is a *cutvertex*, and an edge *bridge*   separating its ends is a *bridge*. Thus, the bridges in a graph are precisely those edges that do not lie on any cycle.

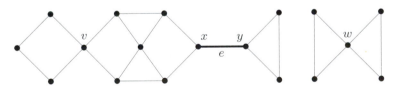

Fig. 1.4.2. A graph with cutvertices v, x, y, w and bridge $e = xy$

k-connected G is called *k-connected* (for $k \in \mathbb{N}$) if $|G| > k$ and $G - X$ is connected for every set $X \subseteq V$ with $|X| < k$. In other words, no two vertices of G are separated by fewer than k other vertices. Every (non-empty) graph is 0-connected, and the 1-connected graphs are precisely the non-trivial connected graphs. The greatest integer k such that G is k-connected *connectivity*   is the *connectivity* $\kappa(G)$ of G. Thus, $\kappa(G) = 0$ if and only if G is $\kappa(G)$ disconnected or a K^1, and $\kappa(K^n) = n - 1$ for all $n \geqslant 1$.

If $|G| > 1$ and $G - F$ is connected for every set $F \subseteq E$ of fewer *ℓ-edge-*   than ℓ edges, then G is called *ℓ-edge-connected*. The greatest integer ℓ *connected*

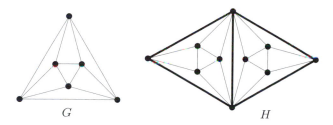

Fig. 1.4.3. The *octahedron* G (left) with $\kappa(G) = \lambda(G) = 4$,
and a graph H with $\kappa(H) = 2$ but $\lambda(H) = 4$

such that G is ℓ-edge-connected is the *edge-connectivity* $\lambda(G)$ of G. In
particular, we have $\lambda(G) = 0$ if G is disconnected.

edge-connectivity $\lambda(G)$

For every non-trivial graph G we have

$$\kappa(G) \leqslant \lambda(G) \leqslant \delta(G)$$

(exercise), so in particular high connectivity requires a large minimum
degree. Conversely, large minimum degree does not ensure high connectivity, not even high edge-connectivity (examples?). It does, however,
imply the existence of a highly connected subgraph:

Theorem 1.4.2. (Mader 1972)
Every graph of average degree at least $4k$ has a k-connected subgraph.

[8.1.1]
[11.2.3]

Proof. For $k \in \{0, 1\}$ the assertion is trivial; we consider $k \geqslant 2$ and a
graph $G = (V, E)$ with $|V| =: n$ and $|E| =: m$. For inductive reasons it
will be easier to prove the stronger assertion that G has a k-connected
subgraph whenever

(i) $n \geqslant 2k - 1$ and

(ii) $m \geqslant (2k - 3)(n - k + 1) + 1$.

(This assertion is indeed stronger, i.e. (i) and (ii) follow from our assumption of $d(G) \geqslant 4k$: (i) follows, since otherwise $2k > n + 1$ and
hence $m = \frac{1}{2}d(G)n \geqslant 2kn > n(n + 1)$; (ii) follows directly from
$m = \frac{1}{2}d(G)n \geqslant 2kn$.)
We apply induction on n. If $n = 2k - 1$, then $k = \frac{1}{2}(n + 1)$, and
hence $m \geqslant \frac{1}{2}n(n - 1)$ by (ii). Thus $G = K^n \supseteq K^{k+1}$, proving our
claim. We now assume that $n \geqslant 2k$. If v is a vertex with $d(v) \leqslant 2k - 3$,
we can apply the induction hypothesis to $G - v$ and are done. So we
assume that $\delta(G) \geqslant 2k - 2$. If G is k-connected, there is nothing to
show. We may therefore assume that G has the form $G = G_1 \cup G_2$ with
$|G_1 \cap G_2| = k - 1$ and $|G_1|, |G_2| < n$. For every vertex v in $G_1 - G_2$ or
$G_2 - G_1$ we have $d(v) \geqslant 2k - 2$ by assumption (and such vertices exist),
so $|G_1|, |G_2| \geqslant 2k - 1$. But then at least one of the graphs G_1, G_2 must

satisfy the induction hypothesis (completing the proof): if neither does, we have

$$\|G_i\| \leqslant (2k-3)(|G_i| - k + 1)$$

for $i = 1, 2$, and hence

$$
\begin{aligned}
m & \leqslant \|G_1\| + \|G_2\| \\
& \leqslant (2k-3)\big(|G_1| + |G_2| - 2k + 2\big) \\
& = (2k-3)(n - k + 1) \qquad (\text{by } |G_1 \cap G_2| = k - 1)
\end{aligned}
$$

contradicting (ii).   □

1.5 Trees and forests

forest
tree
leaf

An *acyclic* graph, one not containing any cycles, is called a *forest*. A connected forest is called a *tree*. (Thus, a forest is a graph whose components are trees.) The vertices of degree 1 in a tree are its *leaves*. Every nontrivial tree has at least two leaves—take, for example, the ends of a longest path. This little fact often comes in handy, especially in induction proofs about trees: if we remove a leaf from a tree, what remains is still a tree.

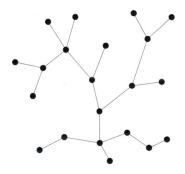

Fig. 1.5.1. A tree

[1.6.1]
[1.9.6]
[4.2.7]

Theorem 1.5.1. *The following assertions are equivalent for a graph T:*

(i) T *is a tree;*

(ii) *any two vertices of T are linked by a unique path in T;*

(iii) T *is minimally connected, i.e. T is connected but $T - e$ is disconnected for every edge $e \in T$;*

(iv) T *is maximally acyclic, i.e. T contains no cycle but $T + xy$ does, for any two non-adjacent vertices $x, y \in T$.*   □

The proof of Theorem 1.5.1 is straightforward, and a good exercise for anyone not yet familiar with all the notions it relates. Extending our notation for paths from Section 1.3, we write xTy for the unique path in a tree T between two vertices x, y (see (ii) above). xTy

A frequently used application of Theorem 1.5.1 is that every connected graph contains a spanning tree: by the equivalence of (i) and (iii), any minimal connected spanning subgraph will be a tree. Figure 1.4.1 shows a spanning tree in each of the three components of the graph depicted.

Corollary 1.5.2. *The vertices of a tree can always be enumerated, say*   [12.4.5]
as $v_1, \ldots, v_n$, so that every v_i with $i \geqslant 2$ has a unique neighbour in
$\{v_1, \ldots, v_{i-1}\}$.

Proof. Use the enumeration from Proposition 1.4.1.   □   (1.4.1)

Corollary 1.5.3. *A connected graph with n vertices is a tree if and*   [1.9.6]
only if it has $n - 1$ edges.   [3.5.1]
    [3.5.4]
Proof. Induction on i shows that the subgraph spanned by the first   [4.2.7]
i vertices in Corollary 1.5.2 has $i - 1$ edges; for $i = n$ this proves the   [8.2.2]
forward implication. Conversely, let G be any connected graph with n
vertices and $n - 1$ edges. Let G' be a spanning tree in G. Since G' has
$n - 1$ edges by the first implication, it follows that $G = G'$.   □

Corollary 1.5.4. *If T is a tree and G is any graph with $\delta(G) \geqslant |T| - 1$,*   [9.2.1]
then $T \subseteq G$, i.e. G has a subgraph isomorphic to T.   [9.2.3]

Proof. Find a copy of T in G inductively along its vertex enumeration
from Corollary 1.5.2.   □

1.6 Bipartite graphs

Let $r \geqslant 2$ be an integer. A graph $G = (V, E)$ is called *r-partite* if   *r-partite*
V admits a partition into r classes such that every edge has its ends
in different classes: vertices in the same partition class must not be
adjacent. Instead of '2-partite' one usually says *bipartite*.   *bipartite*

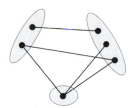 

$$K_{2,2,2} = K_2^3$$

Fig. 1.6.1. Two 3-partite graphs

An *r*-partite graph in which *every* two vertices from different partition classes are adjacent is called *complete*; the complete *r*-partite graphs for all *r* together are the *complete multipartite* graphs. The complete *r*-partite graph $\overline{K^{n_1}} * \ldots * \overline{K^{n_r}}$ is denoted by $K_{n_1,\ldots,n_r}$; if $n_1 = \ldots = n_r =: s$, we abbreviate this to K_s^r. Thus, K_s^r is the complete *r*-partite graph in which every partition class contains exactly *s* vertices.[5] (Figure 1.6.1 shows the example of the octahedron K_2^3; compare its drawing with that in Figure 1.4.3.) Graphs of the form $K_{1,n}$ are called *stars*.

complete r-partite

$K_{n_1,\ldots,n_r}$
K_s^r

star

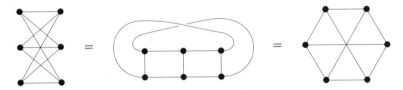

Fig. 1.6.2. Three drawings of the bipartite graph $K_{3,3} = K_3^2$

odd cycle

Clearly, a bipartite graph cannot contain an *odd cycle*, a cycle of odd length. In fact, the bipartite graphs are characterized by this property:

[5.3.1]
[6.4.2]

Proposition 1.6.1. *A graph is bipartite if and only if it contains no odd cycle.*

(1.4.1)
(1.5.1)

Proof. Let $G = (V, E)$ be a graph without odd cycles; we show that G is bipartite. Clearly a graph is bipartite if all its components are bipartite or trivial, so we may assume that G is connected. Let T be a spanning tree in G, and pick a vertex $r \in T$.[6] For each $v \in V$, the unique path rTv has odd or even length. This defines a bipartition of V; we show that G is bipartite with this partition.

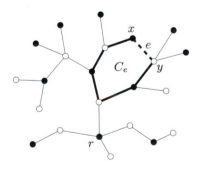

Fig. 1.6.3. The cycle C_e in $T + e$

---

[5] Note that we obtain a K_s^r if we replace each vertex of a K^r by an independent *s*-set; our notation of K_s^r is intended to hint at this connection.

[6] Such a 'special' vertex of a tree is sometimes called a *root*.

Let $e = xy$ be an edge of G. If $e \in T$, we may assume that $x \in rTy$; otherwise $rTyx = rTx$ by Theorem 1.5.1 (ii), and we rename x, y as y, x. Then $rTy = rTxy$, so x and y lie in different partition classes. If $e \notin T$, then $C_e := xTy + e$ is a cycle (Fig. 1.6.3), and by the case treated already, the vertices along xTy alternate between the two classes. Since C_e is even by assumption, x and y again lie in different classes. □

1.7 Contraction and minors

In Section 1.1 we saw two fundamental containment relations between graphs: the subgraph relation, and the 'induced subgraph' relation. In this section we meet another: the minor relation.

Let $e = xy$ be an edge of a graph $G = (V, E)$. By G/e we denote the graph obtained from G by *contracting* the edge e into a new vertex v_e, which becomes adjacent to all the former neighbours of x and of y. Formally, G/e is a graph (V', E') with vertex set $V' := (V \smallsetminus \{x, y\}) \cup \{v_e\}$ (where v_e is the 'new' vertex, i.e. $v_e \notin V \cup E$) and edge set

G/e

contraction

v_e

$$E' := \left\{ vw \in E \mid \{v, w\} \cap \{x, y\} = \emptyset \right\}$$
$$\cup \left\{ v_e w \mid xw \in E \smallsetminus \{e\} \text{ or } yw \in E \smallsetminus \{e\} \right\}.$$

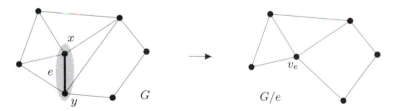

Fig. 1.7.1. Contracting the edge $e = xy$

More generally, if X is another graph and $\{V_x \mid x \in V(X)\}$ is a partition of V into connected subsets such that, for any two vertices $x, y \in X$, there is a V_x–V_y edge in G if and only if $xy \in E(X)$, we call G an MX and write[7] $G = MX$ (Fig. 1.7.2). The sets V_x are the *branch sets* of this MX. Intuitively, we obtain X from G by contracting every

MX

branch sets

---

[7] Thus formally, the expression MX—where M stands for 'minor'; see below— refers to a whole class of graphs, and $G = MX$ means (with slight abuse of notation) that G belongs to this class.

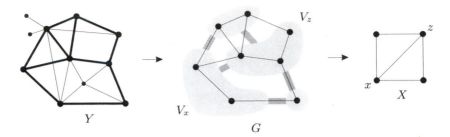

Fig. 1.7.2. $Y \supseteq G = MX$, so X is a minor of Y

branch set to a single vertex and deleting any 'parallel edges' or 'loops' that may arise.

G/U
v_U

If $V_x = U \subseteq V$ is one of the branch sets above and every other branch set consists just of a single vertex, we also write G/U for the graph X and v_U for the vertex $x \in X$ to which U contracts, and think of the rest of X as an induced subgraph of G. The contraction of a single edge uu' defined earlier can then be viewed as the special case of $U = \{\, u, u' \,\}$.

[12.4.2]

Proposition 1.7.1. *G is an MX if and only if X can be obtained from G by a series of edge contractions, i.e. if and only if there are graphs $G_0, \dots, G_n$ and edges $e_i \in G_i$ such that $G_0 = G$, $G_n \simeq X$, and $G_{i+1} = G_i / e_i$ for all $i < n$.*

Proof. Induction on $|G| - |X|$.   □

minor; $\preccurlyeq$

If $G = MX$ is a subgraph of another graph Y, we call X a *minor* of Y and write $X \preccurlyeq Y$. Note that every subgraph of a graph is also its minor; in particular, every graph is its own minor. By Proposition 1.7.1, any minor of a graph can be obtained from it by first deleting some vertices and edges, and then contracting some further edges. Conversely, any graph obtained from another by repeated deletions and contractions (in any order) is its minor: this is clear for one deletion or contraction, and follows for several from the transitivity of the minor relation (Proposition 1.7.3).

subdivision
TX

If we replace the edges of X with independent paths between their ends (so that none of these paths has an inner vertex on another path or in X), we call the graph G obtained a *subdivision* of X and write $G = TX$.[8] If $G = TX$ is the subgraph of another graph Y, then X is a *topological minor* of Y (Fig. 1.7.3).

topological
minor

branch
vertices

If $G = TX$, we view $V(X)$ as a subset of $V(G)$ and call these vertices the *branch vertices* of G; the other vertices of G are its *subdividing*

---

[8] So again TX denotes an entire class of graphs: all those which, viewed as a topological space in the obvious way, are homeomorphic to X. The T in TX stands for 'topological'.

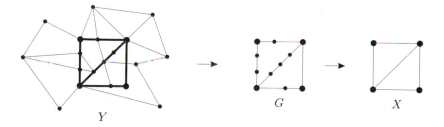

Fig. 1.7.3. $Y \supseteq G = TX$, so X is a topological minor of Y

vertices. Thus, all subdividing vertices have degree 2, while the branch vertices retain their degree from X.

Proposition 1.7.2.

(i) *Every TX is also an MX (Fig. 1.7.4); thus, every topological minor of a graph is also its (ordinary) minor.*

(ii) *If $\Delta(X) \leqslant 3$, then every MX contains a TX; thus, every minor with maximum degree at most 3 of a graph is also its topological minor.* □

[4.4.2]
[8.3.1]

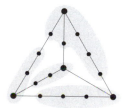

Fig. 1.7.4. A subdivision of K^4 viewed as an MK^4

Proposition 1.7.3. *The minor relation $\preccurlyeq$ and the topological-minor relation are partial orderings on the class of finite graphs, i.e. they are reflexive, antisymmetric and transitive.* □

[12.4.1]

1.8 Euler tours

Any mathematician who happens to find himself in the East Prussian city of Königsberg (and in the 18th century) will lose no time to follow the great Leonhard Euler's example and inquire about a round trip through the old city that traverses each of the bridges shown in Figure 1.8.1 exactly once.

Thus inspired,[9] let us call a closed walk in a graph an *Euler tour* if

---

[9] Anyone to whom such inspiration seems far-fetched, even after contemplating Figure 1.8.2, may seek consolation in the *multigraph* of Figure 1.10.1.

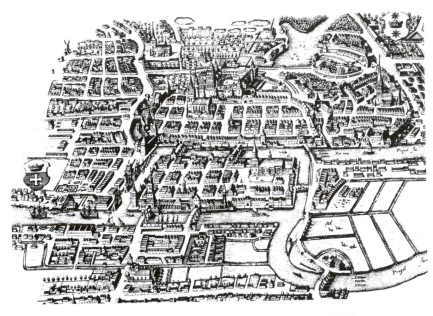

Fig. 1.8.1. The bridges of Königsberg (anno 1736)

Eulerian   it traverses every edge of the graph exactly once. A graph is *Eulerian* if
it admits an Euler tour.

Fig. 1.8.2. A graph formalizing the bridge problem

Theorem 1.8.1. (Euler 1736)

[2.1.5]
[10.3.3]   *A connected graph is Eulerian if and only if every vertex has even degree.*

Proof. The degree condition is clearly necessary: a vertex appearing k
times in an Euler tour (or $k+1$ times, if it is the starting and finishing
vertex and as such counted twice) must have degree $2k$.

Conversely, let G be a connected graph with all degrees even, and
let

$$W = v_0 e_0 \dots e_{\ell-1} v_\ell$$

be a longest walk in G using no edge more than once. Since W cannot
be extended, it already contains all the edges at v_ℓ. By assumption, the

number of such edges is even. Hence $v_\ell = v_0$, so W is a closed walk.

Suppose W is not an Euler tour. Then G has an edge e outside W but incident with a vertex of W, say $e = uv_i$. (Here we use the connectedness of G, as in the proof of Proposition 1.4.1.) Then the walk

$$uev_ie_i \ldots e_{\ell-1}v_\ell e_0 \ldots e_{i-1}v_i$$

is longer than W, a contradiction. $\qquad\qquad\qquad\qquad\qquad\qquad\square$

1.9 Some linear algebra

Let $G = (V, E)$ be a graph with n vertices and m edges, say $V = \{v_1, \ldots, v_n\}$ and $E = \{e_1, \ldots, e_m\}$. The *vertex space* $\mathcal{V}(G)$ of G is the vector space over the 2-element field $\mathbb{F}_2 = \{0, 1\}$ of all functions $V \to \mathbb{F}_2$. Every element of $\mathcal{V}(G)$ corresponds naturally to a subset of V, the set of those vertices to which it assigns a 1, and every subset of V is uniquely represented in $\mathcal{V}(G)$ by its indicator function. We may thus think of $\mathcal{V}(G)$ as the power set of V made into a vector space: the sum $U + U'$ of two vertex sets $U, U' \subseteq V$ is their symmetric difference (why?), and $U = -U$ for all $U \subseteq V$. The zero in $\mathcal{V}(G)$, viewed in this way, is the empty (vertex) set $\emptyset$. Since $\{\{v_1\}, \ldots, \{v_n\}\}$ is a basis of $\mathcal{V}(G)$, its *standard basis*, we have $\dim \mathcal{V}(G) = n$.

In the same way as above, the functions $E \to \mathbb{F}_2$ form the *edge space* $\mathcal{E}(G)$ of G: its elements are the subsets of E, vector addition amounts to symmetric difference, $\emptyset \subseteq E$ is the zero, and $F = -F$ for all $F \subseteq E$. As before, $\{\{e_1\}, \ldots, \{e_m\}\}$ is the *standard basis* of $\mathcal{E}(G)$, and $\dim \mathcal{E}(G) = m$.

Since the edges of a graph carry its essential structure, we shall mostly be concerned with the edge space. Given two edge sets $F, F' \in \mathcal{E}(G)$ and their coefficients $\lambda_1, \ldots, \lambda_m$ and $\lambda'_1, \ldots, \lambda'_m$ with respect to the standard basis, we write

$$\langle F, F' \rangle := \lambda_1 \lambda'_1 + \ldots + \lambda_m \lambda'_m \in \mathbb{F}_2 .$$

Note that $\langle F, F' \rangle = 0$ may hold even when $F = F' \neq \emptyset$: indeed, $\langle F, F' \rangle = 0$ if and only if F and F' have an even number of edges in common. Given a subspace $\mathcal{F}$ of $\mathcal{E}(G)$, we write

$$\mathcal{F}^\perp := \{D \in \mathcal{E}(G) \mid \langle F, D \rangle = 0 \text{ for all } F \in \mathcal{F}\} .$$

This is again a subspace of $\mathcal{E}(G)$ (the space of all vectors solving a certain set of linear equations—which?), and we have

$$\dim \mathcal{F} + \dim \mathcal{F}^\perp = m .$$

The *cycle space* $\mathcal{C} = \mathcal{C}(G)$ is the subspace of $\mathcal{E}(G)$ spanned by all the cycles in G—more precisely, by their edge sets.[10] The dimension of $\mathcal{C}(G)$ is the *cyclomatic number* of G.

Proposition 1.9.1. *The induced cycles in G generate its entire cycle space.*

Proof. By definition of $\mathcal{C}(G)$ it suffices to show that the induced cycles in G generate every cycle $C \subseteq G$ with a chord e. This follows at once by induction on $|C|$: the two cycles in $C + e$ with e but no other edge in common are shorter than C, and their symmetric difference is precisely C. □

Proposition 1.9.2. *An edge set $F \subseteq E$ lies in $\mathcal{C}(G)$ if and only if every vertex of (V, F) has even degree.*

Proof. The forward implication holds by induction on the number of cycles needed to generate F, the backward implication by induction on the number of cycles in (V, F). □

If $\{V_1, V_2\}$ is a partition of V, the set $E(V_1, V_2)$ of all the edges of G *crossing* this partition is called a *cut*. Recall that for $V_1 = \{v\}$ this cut is denoted by $E(v)$.

Proposition 1.9.3. *Together with $\emptyset$, the cuts in G form a subspace $\mathcal{C}^*$ of $\mathcal{E}(G)$. This space is generated by cuts of the form $E(v)$.*

Proof. Let $\mathcal{C}^*$ denote the set of all cuts in G, together with $\emptyset$. To prove that $\mathcal{C}^*$ is a subspace, we show that for all $D, D' \in \mathcal{C}^*$ also $D + D'$ $(= D - D')$ lies in $\mathcal{C}^*$. Since $D + D = \emptyset \in \mathcal{C}^*$ and $D + \emptyset = D \in \mathcal{C}^*$, we may assume that D and D' are distinct and non-empty. Let $\{V_1, V_2\}$ and $\{V_1', V_2'\}$ be the corresponding partitions of V. Then $D + D'$ consists of all the edges that cross one of these partitions but not the other (Fig. 1.9.1). But these are precisely the edges between $(V_1 \cap V_1') \cup (V_2 \cap V_2')$ and $(V_1 \cap V_2') \cup (V_2 \cap V_1')$, and by $D \neq D'$ these two sets form another partition of V. Hence $D + D' \in \mathcal{C}^*$, and $\mathcal{C}^*$ is indeed a subspace of $\mathcal{E}(G)$.

Our second assertion, that the cuts $E(v)$ generate all of $\mathcal{C}^*$, follows from the fact that every edge $xy \in G$ lies in exactly two such cuts (in $E(x)$ and in $E(y)$); thus every partition $\{V_1, V_2\}$ of V satisfies $E(V_1, V_2) = \sum_{v \in V_1} E(v)$. □

---

[10] For simplicity, we shall not normally distinguish between cycles and their edge sets in connection with the cycle space.

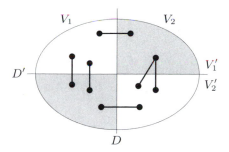

Fig. 1.9.1. Cut edges in $D + D'$

The subspace $\mathcal{C}^* =: \mathcal{C}^*(G)$ of $\mathcal{E}(G)$ from Proposition 1.9.3 will be called the *cut space* of G. It is not difficult to find among the cuts $E(v)$ an explicit basis for $\mathcal{C}^*(G)$, and thus to determine its dimension (exercise); together with Theorem 1.9.5 this yields an independent proof of Theorem 1.9.6.

The following lemma will be useful when we study the duality of plane graphs in Chapter 4.6:

cut space
$\mathcal{C}^*(G)$

Lemma 1.9.4. *The minimal cuts in a connected graph generate its entire cut space.*

[4.6.2]

Proof. Note first that a cut in a connected graph $G = (V, E)$ is minimal if and only if both sets in the corresponding partition of V are connected in G. Now consider any connected subgraph $C \subseteq G$. If D is a component of $G - C$, then also $G - D$ is connected (Fig. 1.9.2); the edges between D and $G - D$ thus form a minimal cut. By choice of D, this cut is precisely the set $E(C, D)$ of all C–D edges in G.

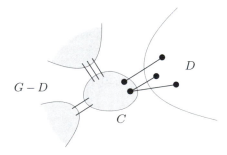

Fig. 1.9.2. $G - D$ is connected, and $E(C, D)$ a minimal cut

To prove the lemma, let a partition $\{V_1, V_2\}$ of V be given, and consider a component C of $G[V_1]$. Then $E(C, V_2) = E(C, G - C)$ is the disjoint union of the edge sets $E(C, D)$ over all components D of $G - C$, and is thus the disjoint union of minimal cuts (see above). Now the disjoint union of all these edge sets $E(C, V_2)$, taken over all the components C of $G[V_1]$, is precisely our cut $E(V_1, V_2)$. So this cut is generated by minimal cuts, as claimed. $\square$

Theorem 1.9.5. *The cycle space $\mathcal{C}$ and the cut space $\mathcal{C}^*$ of any graph satisfy*

$$\mathcal{C} = \mathcal{C}^{*\perp} \quad and \quad \mathcal{C}^* = \mathcal{C}^{\perp}.$$

Proof. Let us consider a graph $G = (V, E)$. Clearly, any cycle in G has an even number of edges in each cut. This implies $\mathcal{C} \subseteq \mathcal{C}^{*\perp}$.

Conversely, recall from Proposition 1.9.2 that for every edge set $F \notin \mathcal{C}$ there exists a vertex v incident with an odd number of edges in F. Then $\langle E(v), F \rangle = 1$, so $E(v) \in \mathcal{C}^*$ implies $F \notin \mathcal{C}^{*\perp}$. This completes the proof of $\mathcal{C} = \mathcal{C}^{*\perp}$.

To prove $\mathcal{C}^* = \mathcal{C}^{\perp}$, it now suffices to show $\mathcal{C}^* = (\mathcal{C}^{*\perp})^{\perp}$. Here $\mathcal{C}^* \subseteq (\mathcal{C}^{*\perp})^{\perp}$ follows directly from the definition of $\perp$. But since

$$\dim \mathcal{C}^* + \dim \mathcal{C}^{*\perp} = m = \dim \mathcal{C}^{*\perp} + \dim (\mathcal{C}^{*\perp})^{\perp},$$

$\mathcal{C}^*$ has the same dimension as $(\mathcal{C}^{*\perp})^{\perp}$, so $\mathcal{C}^* = (\mathcal{C}^{*\perp})^{\perp}$ as claimed. $\square$

[4.5.1]

Theorem 1.9.6. *Every connected graph G with n vertices and m edges satisfies*

$$\dim \mathcal{C}(G) = m - n + 1 \quad and \quad \dim \mathcal{C}^*(G) = n - 1.$$

(1.5.1)
(1.5.3)

Proof. Let $G = (V, E)$. As $\dim \mathcal{C} + \dim \mathcal{C}^* = m$ by Theorem 1.9.5, it suffices to find $m - n + 1$ linearly independent vectors in $\mathcal{C}$ and $n - 1$ linearly independent vectors in $\mathcal{C}^*$: since these numbers add up to m, neither the dimension of $\mathcal{C}$ nor that of $\mathcal{C}^*$ can then be strictly greater.

Let T be a spanning tree in G. By Corollary 1.5.3, T has $n - 1$ edges, so $m - n + 1$ edges of G lie outside T. For each of these $m - n + 1$ edges $e \in E \setminus E(T)$, the graph $T + e$ contains a cycle C_e (see Fig. 1.6.3 and Theorem 1.5.1 (iv)). Since none of the edges e lies on $C_{e'}$ for $e' \neq e$, these $m - n + 1$ cycles are linearly independent.

For each of the $n - 1$ edges $e \in T$, the graph $T - e$ has exactly two components (Theorem 1.5.1 (iii)), and the set D_e of edges in G between these components form a cut (Fig.1.9.3). Since none of the edges $e \in T$ lies in $D_{e'}$ for $e' \neq e$, these $n - 1$ cuts are linearly independent. $\square$

incidence
matrix

The *incidence matrix* $B = (b_{ij})_{n \times m}$ of a graph $G = (V, E)$ with $V = \{v_1, \ldots, v_n\}$ and $E = \{e_1, \ldots, e_m\}$ is defined over $\mathbb{F}_2$ by

$$b_{ij} := \begin{cases} 1 & \text{if } v_i \in e_j \\ 0 & \text{otherwise.} \end{cases}$$

As usual, let B^t denote the transpose of B. Then B and B^t define linear maps $B : \mathcal{E}(G) \to \mathcal{V}(G)$ and $B^t : \mathcal{V}(G) \to \mathcal{E}(G)$ with respect to the standard bases.

Fig. 1.9.3. The cut D_e

Proposition 1.9.7.

(i) *The kernel of B is $\mathcal{C}(G)$.*

(ii) *The image of B^t is $\mathcal{C}^*(G)$.* □

The *adjacency matrix* $A = (a_{ij})_{n \times n}$ of G is defined by

adjacency matrix

$$a_{ij} := \begin{cases} 1 & \text{if } v_i v_j \in E \\ 0 & \text{otherwise.} \end{cases}$$

Our last proposition establishes a simple connection between A and B (now viewed as real matrices). Let D denote the real diagonal matrix $(d_{ij})_{n \times n}$ with $d_{ii} = d(v_i)$ and $d_{ij} = 0$ otherwise.

Proposition 1.9.8. $BB^t = A + D$. □

1.10 Other notions of graphs

For completeness, we now mention a few other notions of graphs which feature less frequently or not at all in this book.

A *hypergraph* is a pair (V, E) of disjoint sets, where the elements of E are non-empty subsets (of any cardinality) of V. Thus, graphs are special hypergraphs.

hypergraph

A *directed graph* (or *digraph*) is a pair (V, E) of disjoint sets (of *vertices* and *edges*) together with two maps init: $E \to V$ and ter: $E \to V$ assigning to every edge e an *initial vertex* init(e) and a *terminal vertex* ter(e). The edge e is said to be *directed from* init(e) *to* ter(e). Note that a directed graph may have several edges between the same two vertices x, y. Such edges are called *multiple edges*; if they have the same direction

directed graph
init(e)
ter(e)

loop

orientation

oriented graph

multigraph

(say from x to y), they are *parallel*. If $\mathrm{init}(e) = \mathrm{ter}(e)$, the edge e is called a *loop*.

A directed graph D is an *orientation* of an (undirected) graph G if $V(D) = V(G)$ and $E(D) = E(G)$, and if $\{\mathrm{init}(e), \mathrm{ter}(e)\} = \{x, y\}$ for every edge $e = xy$. Intuitively, such an *oriented graph* arises from an undirected graph simply by directing every edge from one of its ends to the other. Put differently, oriented graphs are directed graphs without loops or multiple edges.

A *multigraph* is a pair (V, E) of disjoint sets (of *vertices* and *edges*) together with a map $E \to V \cup [V]^2$ assigning to every edge either one or two vertices, its *ends*. Thus, multigraphs too can have loops and multiple edges: we may think of a multigraph as a directed graph whose edge directions have been 'forgotten'. To express that x and y are the ends of an edge e we still write $e = xy$, though this no longer determines e uniquely.

A graph is thus essentially the same as a multigraph without loops or multiple edges. Somewhat surprisingly, proving a graph theorem more generally for multigraphs may, on occasion, simplify the proof. Moreover, there are areas in graph theory (such as plane duality; see Chapters 4.6 and 6.5) where multigraphs arise more naturally than graphs, and where any restriction to the latter would seem artificial and be technically complicated. We shall therefore consider multigraphs in these cases, but without much technical ado: terminology introduced earlier for graphs will be used correspondingly.

Two differences, however, should be pointed out. First, a multigraph may have cycles of length 1 or 2: loops, and pairs of multiple edges (or *double edges*). Second, the notion of edge contraction is simpler in multigraphs than in graphs. If we contract an edge $e = xy$ in a multigraph $G = (V, E)$ to a new vertex v_e, there is no longer a need to delete any edges other than e itself: edges parallel to e become loops at v_e, while edges xv and yv become parallel edges between v_e and v (Fig. 1.10.1). Thus, formally, $E(G/e) = E \smallsetminus \{e\}$, and only the incidence map $e' \mapsto \{\mathrm{init}(e'), \mathrm{ter}(e')\}$ of G has to be adjusted to the new vertex set in G/e. The notion of a minor adapts to multigraphs accordingly.

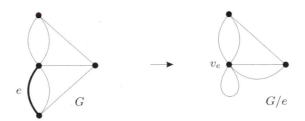

Fig. 1.10.1. Contracting the edge e in the multigraph corresponding to 1.8.1

Finally, it should be pointed out that authors who usually work with multigraphs tend to call them graphs; in their terminology, our graphs would be called simple graphs.

Exercises

1.⁻ What is the number of edges in a K^n?

2. Let $d \in \mathbb{N}$ and $V := \{0,1\}^d$; thus, V is the set of all 0–1 sequences of length d. The graph on V in which two such sequences form an edge if and only if they differ in exactly one position is called the *d-dimensional cube*. Determine the average degree, number of edges, diameter, girth and circumference of this graph.

    (Hint for circumference. Induction on d.)

3.⁺ Let G be a graph containing a cycle C, and assume that G contains a path of length at least k between two vertices of C. Show that G contains a cycle of length at least $\sqrt{2k}$.

4.⁻ Is the bound in Proposition 1.3.2 best possible?

5. Show that $\mathrm{rad}(G) \leqslant \mathrm{diam}(G) \leqslant 2\,\mathrm{rad}(G)$ for every graph G.

6.⁺ Assuming that $d \geqslant 2$ and $k \geqslant 3$, improve the bound in Proposition 1.3.3 to d^k.

7.⁻ Show that the components of a graph partition its vertex set. (In other words, show that every vertex belongs to exactly one component.)

8.⁻ Show that every 2-connected graph contains a cycle.

9. (i)⁻ Determine $\kappa(G)$ and $\lambda(G)$ for $G = P^k, C^k, K^k, K_{m,n}$ $(k, m, n \geqslant 3)$.

    (ii)⁺ Determine the connectivity of the n-dimensional cube (defined in Exercise 2).

    (Hint for (ii). Induction on n.)

10. Show that $\kappa(G) \leqslant \lambda(G) \leqslant \delta(G)$ for every non-trivial graph G.

11.⁻ Is there a function $f : \mathbb{N} \to \mathbb{N}$ such that, for all $k \in \mathbb{N}$, every graph of minimum degree at least $f(k)$ is k-connected?

12. Prove Theorem 1.5.1.

13. Show that any tree T has at least $\Delta(T)$ leaves.

14. Let G be a connected graph, and let $r \in G$ be a vertex. Show that G has a spanning tree T such that every edge $xy \in G$ satisfies either $x \in rTy$ or $y \in rTx$. (Intuitively, if we think of r as the root of T, then all the edges of G run along branches of T, never across.)

    (Hint. T is called a *depth-first search tree* of G with root r.)

15.⁺ Let $\mathcal{T}$ be a set of subtrees of a tree T. Assume that the trees in $\mathcal{T}$ have pairwise non-empty intersection. Show that their overall intersection $\bigcap \mathcal{T}$ is non-empty.

16.   Show that every automorphism of a tree fixes a vertex or an edge.

17.   Are the partition classes of a regular bipartite graph always of the same size?

18.   Show that a graph is bipartite if and only if every *induced* cycle has even length.

19.   Find a function $f: \mathbb{N} \to \mathbb{N}$ such that, for all $k \in \mathbb{N}$, every graph of average degree at least $f(k)$ has a bipartite subgraph of minimum degree at least k.

20.   Show that the minor relation $\preccurlyeq$ defines a partial ordering on any set of (finite) graphs. Is the same true for infinite graphs?

21.⁻   Show that the elements of the cycle space of a graph G are precisely the unions of the edges sets of edge-disjoint cycles in G.

22.   Given a graph G, find among all cuts of the form $E(v)$ a basis for the cut space of G.

23.   Consider the cycle and cut spaces of an odd cycle C. What is the relationship between these spaces, and what is their relationship to the entire edge space $\mathcal{E}(C)$? How do the answers to these questions change when C has even length?

24.   Give a direct proof of the fact that the cycles C_e defined in the proof of Theorem 1.9.6 generate the cycle space of the graph G considered.

       (Hint. Induction on $|F \smallsetminus E(T)|$ for given $F \in \mathcal{C}(G)$.)

25.⁺   Give a direct proof of the fact that the cuts D_e defined in the proof of Theorem 1.9.6 generate the cut space of the graph.

       (Hint. Induction on $|D \cap E(T)|$ for a given cut D.)

26.   What are the dimensions of the cycle and the cut space of a graph with k components?

Notes

The terminology used in this book is mostly standard. Alternatives do exist, of course, and some of these are stated when a concept is first defined. There is one small point where our notation deviates slightly from standard usage. Whereas complete graphs, paths, cycles etc. of given order are mostly denoted by K_n, P_k, C_ℓ and so on, we follow Bollobás in using superscripts instead of subscripts. This has the advantage of leaving the variables K, P, C etc. free for ad-hoc use: we may now enumerate components as $C_1, C_2, \ldots$, speak of paths $P_1, \ldots, P_k$, and so on—without any danger of confusion.

Theorem[11] 1.4.2 is due to W. Mader, Existenz n-fach zusammenhängen-der Teilgraphen in Graphen genügend großer Kantendichte, *Abh. Math. Sem.*

---

[11] In the interest of readability, the end-of-chapter notes in this book give references only for Theorems, and only in cases where these references cannot be found in a monograph or survey cited for that chapter.

Univ. Hamburg **37** (1972) 86–97. Theorem 1.8.1 is from L. Euler, Solutio problematis ad geometriam situs pertinentis, *Comment. Acad. Sci. I. Petropolitanae* **8** (1736), 128–140.

Of the large subject of algebraic methods in graph theory, Section 1.9 does not claim to convey an adequate impression. The standard monograph here is N.L. Biggs, *Algebraic Graph Theory*, Cambridge University Press 1974. A more recent and comprehensive account is given by C.D. Godsil & G.F. Royle, *Algebraic Graph Theory*, in preparation. Surveys on the use of algebraic methods can also be found in the *Handbook of Combinatorics* (R.L. Graham, M. Grötschel & L. Lovász, eds.), North-Holland 1995.

2 Matching

Suppose we are given a graph and are asked to find in it as many independent edges as possible. How should we go about this? Will we be able to pair up all its vertices in this way? If not, how can we be sure that this is indeed impossible? Somewhat surprisingly, this basic problem does not only lie at the heart of numerous applications, it also gives rise to some rather interesting graph theory.

A set M of independent edges in a graph $G = (V, E)$ is called a *matching*. M is a matching *of $U \subseteq V$* if every vertex in U is incident with an edge in M. The vertices in U are then called *matched* (by M); vertices not incident with any edge of M are *unmatched*.

A k-regular spanning subgraph is called a *k-factor*. Thus, a subgraph $H \subseteq G$ is a 1-factor of G if and only if $E(H)$ is a matching of V. The problem of how to characterize the graphs that have a 1-factor, i.e. a matching of their entire vertex set, will be our main theme in this chapter.

matching

matched

factor

2.1 Matching in bipartite graphs

For this whole section, we let $G = (V, E)$ be a fixed bipartite graph with bipartition $\{A, B\}$. Vertices denoted as a, a' etc. will be assumed to lie in A, vertices denoted as b etc. will lie in B.

How can we find a matching in G with as many edges as possible? Let us start by considering an arbitrary matching M in G. A path in G which starts in A at an unmatched vertex and then contains, alternately, edges from $E \setminus M$ and from M, is an *alternating path* with respect to M. An alternating path P that ends in an unmatched vertex of B is called an *augmenting path* (Fig. 2.1.1), because we can use it to turn M into a larger matching: the symmetric difference of M with $E(P)$ is again a

$G = (V, E)$

A, B

a, b etc.

alternating path

augmenting path

Fig. 2.1.1. Augmenting the matching M by the alternating
       path P

matching (consider the edges at a given vertex), and the set of matched
vertices is increased by two, the ends of P.

Alternating paths play an important role in the practical search for
large matchings. In fact, if we start with any matching and keep applying
augmenting paths until no further such improvement is possible, the
matching obtained will always be an optimal one, a matching with the
largest possible number of edges (Exercise 1). The algorithmic problem
of finding such matchings thus reduces to that of finding augmenting
paths—which is an interesting and accessible algorithmic problem.

Our first theorem characterizes the maximal cardinality of a matching
in G by a kind of duality condition. Let us call a set $U \subseteq V$ a *cover* of E
(or a *vertex cover* of G) if every edge of G is incident with a vertex in U.

vertex cover

Theorem 2.1.1. (König 1931)
*The maximum cardinality of a matching in G is equal to the minimum
cardinality of a vertex cover.*

M

Proof. Let M be a matching in G of maximum cardinality. From every
edge in M let us choose one of its ends: its end in B if some alternating
path ends in that vertex, and its end in A otherwise (Fig. 2.1.2). We
shall prove that the set U of these $|M|$ vertices covers G; since any vertex
cover of G must cover M, there can be none with fewer than $|M|$ edges,
and so the theorem will follow.

U

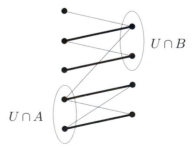

Fig. 2.1.2. The vertex cover U

Let $ab \in E$ be an edge; we show that either a or b lies in U. If $ab \in M$, this holds by definition of U, so we assume that $ab \notin M$. Since M is a maximal matching, it contains an edge $a'b'$ with $a = a'$ or $b = b'$. In fact, we may assume that $a = a'$: for if a is unmatched (and $b = b'$), then ab is an alternating path, and so the end of $a'b' \in M$ chosen for U was the vertex $b' = b$. Now if $a' = a$ is not in U, then $b' \in U$, and some alternating path P ends in b'. But then there is also an alternating path P' ending in b: either $P' := Pb$ (if $b \in P$) or $P' := Pb'a'b$. By the maximality of M, however, P' is not an augmenting path. So b must be matched, and was chosen for U from the edge of M containing it. $\square$

Let us return to our main problem, the search for some necessary and sufficient conditions for the existence of a 1-factor. In our present case of a bipartite graph, we may as well ask more generally when G contains a matching of A; this will define a 1-factor of G if $|A| = |B|$, a condition that has to hold anyhow if G is to have a 1-factor.

A condition clearly necessary for the existence of a matching of A is that every subset of A has enough neighbours in B, i.e. that

$$|N(S)| \geqslant |S| \qquad \text{for all } S \subseteq A.$$

marriage
condition

The following *marriage theorem* says that this obvious necessary condition is in fact sufficient:

Theorem 2.1.2. (Hall 1935)
marriage
theorem
G contains a matching of A if and only if $|N(S)| \geqslant |S|$ for all $S \subseteq A$.

We give three proofs for the non-trivial implication of this theorem, i.e. that the 'marriage condition' implies the existence of a matching of A. The first of these is based on König's theorem; the second is a direct constructive proof by augmenting paths; the third will be an independent proof from first principles.

First proof. Let $U = A' \cup B'$ be a vertex cover of G of minimum cardinality, with $A' \subseteq A$ and $B' \subseteq B$. If G contains no matching of A, Theorem 2.1.1 implies that

$$|A'| + |B'| = |U| < |A|,$$

and hence

$$|B'| < |A| - |A'| = |A \smallsetminus A'|$$

(Fig. 2.1.3). By definition of U, however, G has no edges between $A \smallsetminus A'$ and $B \smallsetminus B'$, so

$$|N(A \smallsetminus A')| \leqslant |B'| < |A \smallsetminus A'|$$

and the marriage condition fails for $S := A \smallsetminus A'$. $\square$

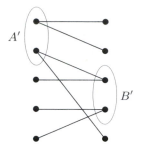

Fig. 2.1.3. A cover by fewer than $|A|$ vertices

Second proof. Consider a matching M of G that leaves a vertex of A unmatched; we shall construct an augmenting path with respect to M. Let $a_0, b_1, a_1, b_2, a_2, \ldots$ be a maximal sequence of distinct vertices $a_i \in A$ and $b_i \in B$ satisfying the following conditions for all $i \geqslant 1$ (Fig. 2.1.4):

 (i) a_0 is unmatched;

 (ii) b_i is adjacent to some vertex $a_{f(i)} \in \{a_0, \ldots, a_{i-1}\}$;

 (iii) $a_i b_i \in M$.

By the marriage condition, our sequence cannot end in a vertex of A: the i vertices $a_0, \ldots, a_{i-1}$ together have at least i neighbours in B, so we can always find a new vertex $b_i \neq b_1, \ldots, b_{i-1}$ that satisfies (ii). Let $b_k \in B$ be the last vertex of the sequence. By (i)–(iii),

$$P := b_k a_{f(k)} b_{f(k)} a_{f^2(k)} b_{f^2(k)} a_{f^3(k)} \ldots a_{f^r(k)}$$

with $f^r(k) = 0$ is an alternating path.

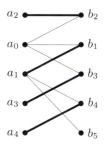

Fig. 2.1.4. Proving the marriage theorem by alternating paths

What is it that prevents us from extending our sequence further? If b_k is matched, say to a, we can indeed extend it by setting $a_k := a$, unless $a = a_i$ with $0 < i < k$, in which case (iii) would imply $b_k = b_i$ with a contradiction. So b_k is unmatched, and hence P is an augmenting path between a_0 and b_k.   □

Third proof. We apply induction on $|A|$. For $|A| = 1$ the assertion is true. Now let $|A| \geqslant 2$, and assume that the marriage condition is sufficient for the existence of a matching of A when $|A|$ is smaller.

If $|N(S)| \geqslant |S| + 1$ for every non-empty set $S \subsetneq A$, we pick an edge $ab \in G$ and consider the graph $G' := G - \{a, b\}$. Then every non-empty set $S \subseteq A \smallsetminus \{a\}$ satisfies

$$|N_{G'}(S)| \geqslant |N_G(S)| - 1 \geqslant |S|\,,$$

so by the induction hypothesis G' contains a matching of $A \smallsetminus \{a\}$. Together with the edge ab, this yields a matching of A in G.

Suppose now that A has a non-empty proper subset A' with $|B'| = |A'|$ for $B' := N(A')$. By the induction hypothesis, $G' := G[A' \cup B']$ contains a matching of A'. But $G - G'$ satisfies the marriage condition too: for any set $S \subseteq A \smallsetminus A'$ with $|N_{G-G'}(S)| < |S|$ we would have $|N_G(S \cup A')| < |S \cup A'|$, contrary to our assumption. Again by induction, $G - G'$ contains a matching of $A \smallsetminus A'$. Putting the two matchings together, we obtain a matching of A in G. $\quad\square$

A', B'
G'

Corollary 2.1.3. *If $|N(S)| \geqslant |S| - d$ for every set $S \subseteq A$ and some fixed $d \in \mathbb{N}$, then G contains a matching of cardinality $|A| - d$.* $\qquad$ [2.2.3]

Proof. We add d new vertices to B, joining each of them to all the vertices in A. By the marriage theorem the new graph contains a matching of A, and at least $|A| - d$ edges in this matching must be edges of G. $\quad\square$

Corollary 2.1.4. *If G is k-regular with $k \geqslant 1$, then G has a 1-factor.*

Proof. If G is k-regular, then clearly $|A| = |B|$; it thus suffices to show by Theorem 2.1.2 that G contains a matching of A. Now every set $S \subseteq A$ is joined to $N(S)$ by a total of $k\,|S|$ edges, and these are among the $k\,|N(S)|$ edges of G incident with $N(S)$. Therefore $k\,|S| \leqslant k\,|N(S)|$, so G does indeed satisfy the marriage condition. $\quad\square$

Despite its seemingly narrow formulation, the marriage theorem counts among the most frequently applied graph theorems, both outside graph theory and within. Often, however, recasting a problem in the setting of bipartite matching requires some clever adaptation. As a simple example, we now use the marriage theorem to derive one of the earliest results of graph theory, a result whose original proof is not all that simple, and certainly not short:

Corollary 2.1.5. (Petersen 1891)
Every regular graph of positive even degree has a 2-factor.

(1.8.1)

Proof. Let G be any $2k$-regular graph ($k \geqslant 1$), without loss of generality connected. By Theorem 1.8.1, G contains an Euler tour $v_0 e_0 \ldots e_{\ell-1} v_\ell$, with $v_\ell = v_0$. We replace every vertex v by a pair (v^-, v^+), and every edge $e_i = v_i v_{i+1}$ by the edge $v_i^+ v_{i+1}^-$ (Fig. 2.1.5). The resulting bipartite graph G' is k-regular, so by Corollary 2.1.4 it has a 1-factor. Collapsing every vertex pair (v^-, v^+) back into a single vertex v, we turn this 1-factor of G' into a 2-factor of G. □

Fig. 2.1.5. Splitting vertices in the proof of Corollary 2.1.5

2.2 Matching in general graphs

$\mathcal{C}_G$
$q(G)$

Given a graph G, let us denote by $\mathcal{C}_G$ the set of its components, and by $q(G)$ the number of its *odd components*, those of odd order. If G has a 1-factor, then clearly

Tutte's
condition

$$q(G - S) \leqslant |S| \qquad \text{for all } S \subseteq V(G),$$

since every odd component of $G - S$ will send a factor edge to S.

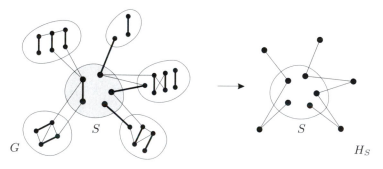

Fig. 2.2.1. Tutte's condition $q(G - S) \leqslant |S|$ for $q = 3$, and the contracted graph H_S from Theorem 2.2.3.

Again, this obvious necessary condition for the existence of a 1-factor is also sufficient:

Theorem 2.2.1. (Tutte 1947)
A graph G has a 1-factor if and only if $q(G - S) \leqslant |S|$ for all $S \subseteq V(G)$.

Proof. Let $G = (V, E)$ be a graph that satisfies the condition of the theorem but has no 1-factor. Adding edges if necessary, we assume that G is edge-maximal without a 1-factor. Then G still satisfies the condition of the theorem: for each odd component D of $G - S$ (after the addition of edges), there is at least one odd component among the components that $V(D)$ induces in the original graph $G - S$, so $G - S$ has at most as many odd components after the addition of edges as before.

Let $K \subseteq V$ be the set of vertices adjacent to every other vertex $\qquad K$
in G, and put $G' := G - K$. We shall prove that all the components of $\qquad G'$
G' are complete. This will give us a 1-factor in G (with a contradiction), as follows. From each of the at most $|K|$ odd components of G we pick one vertex and join it to a vertex in K. Then the remaining vertices in G' can be paired up, so it remains to pair up the vertices of K not yet covered. Since these vertices are pairwise adjacent (by definition of K), it suffices to show that there is an even number of them. Since the edges selected so far cover an even number of vertices, it suffices to check that $|V|$ is even. But this follows from Tutte's condition with $S = \emptyset$.

Suppose then that one of the components of G' is not complete. Then adjacency in G' is not an equivalence relation on $V(G')$, so there are distinct vertices a, b, c in G' such that $ab, bc \in E$ but $ac \notin E$. Since $\qquad a, b, c$
$b \notin K$, there is a vertex $d \in V$ such that $bd \notin E$. By the maximality $\qquad d$
of G, there is a matching M_1 of V in $G + ac$, and a matching M_2 of V $\qquad M_1, M_2$
in $G + bd$.

Let $P = d \ldots v$ be a maximal path in G starting at d with an edge from M_1 and containing alternately edges from M_1 and M_2. If the last edge of P lies in M_1, then $v = b$, since otherwise we could continue P. Let us then set $C := P + bd$. If the last edge of P lies in M_2, then by the maximality of P the M_1-edge at v must be ac, so $v \in \{a, c\}$; then let C be the cycle $dPvbd$. In each case, C is an even cycle with every other edge in M_2, and whose only edge not in E is bd. Replacing in M_2 its edges on C with the edges of $C - M_2$, we obtain a matching of V contained in E, a contradiction. $\qquad\square$

Corollary 2.2.2. (Petersen 1891)
Every bridgeless cubic graph has a 1-factor.

Proof. We show that any bridgeless cubic graph G satisfies Tutte's condition. Let $S \subseteq V(G)$ be given, and consider an odd component C of $G - S$. Since G is cubic, the degrees in C sum to an odd number, but only an even part of this sum arises from edges of C. So G has an odd number of S–C edges, and therefore has at least 3 such edges (since G has no bridge). The total number of edges between S and $G - S$ thus is at least $3q(G - S)$. But it is also at most $3|S|$, because G is cubic. Hence $q(G - S) \leqslant |S|$, as required. $\square$

In order to shed a little more light on the techniques used in matching theory, we now give a second proof of Tutte's theorem. In fact, we shall prove a slightly stronger result, a result that places a structure interesting from the matching point of view on an *arbitrary* graph. If the graph happens to satisfy the condition of Tutte's theorem, this structure will at once yield a 1-factor.

factor-critical   A graph $G = (V, E)$ is called *factor-critical* if $G \neq \emptyset$ and $G - v$ has a 1-factor for every vertex $v \in G$. Then G itself has no 1-factor, *matchable*   because it has odd order. We call a vertex set $S \subseteq V$ *matchable to* $G - S$ if the (bipartite[1]) graph H_S, which arises from G by contracting the components $C \in \mathcal{C}_{G-S}$ to single vertices and deleting all the edges H_S inside S, contains a matching of S. (Formally, H_S is the graph with vertex set $S \cup \mathcal{C}_{G-S}$ and edge set $\{ sC \mid \exists c \in C : sc \in E \}$; see Fig. 2.2.1.)

Theorem 2.2.3. *Every graph $G = (V, E)$ contains a vertex set S with the following two properties:*

   (i) *S is matchable to $G - S$;*

   (ii) *every component of $G - S$ is factor-critical.*

Given any such set S, the graph G contains a 1-factor if and only if $|S| = |\mathcal{C}_{G-S}|$.

For any given G, the assertion of Tutte's theorem follows easily from this result. Indeed, by (i) and (ii) we have $|S| \leqslant |\mathcal{C}_{G-S}| = q(G - S)$ (since factor-critical graphs have odd order); thus Tutte's condition of $q(G - S) \leqslant |S|$ implies $|S| = |\mathcal{C}_{G-S}|$, and the existence of a 1-factor follows from the last statement of Theorem 2.2.3.

(2.1.3)   **Proof of Theorem 2.2.3.** Note first that the last assertion of the theorem follows at once from the assertions (i) and (ii): if G has a 1-factor, we have $q(G - S) \leqslant |S|$ and hence $|S| = |\mathcal{C}_{G-S}|$ as above; conversely if $|S| = |\mathcal{C}_{G-S}|$, then the existence of a 1-factor follows straight from (i) and (ii).

We now prove the existence of a set S satisfying (i) and (ii). We apply induction on $|G|$. For $|G| = 0$ we may take $S = \emptyset$. Now let G be given with $|G| > 0$, and assume the assertion holds for graphs with fewer vertices.

d   Let $d \in \mathbb{N}$ be minimum such that

$$q(G - T) \leqslant |T| + d \qquad \text{for every } T \subseteq V. \qquad (*)$$

---

[1]   except for the—permitted—case that S or $\mathcal{C}_{G-S}$ is empty

Then there exists a set T for which equality holds in $(*)$: this follows from the minimality of d if $d > 0$, and from $q(G - \emptyset) \geq |\emptyset| + 0$ if $d = 0$.
Let S be such a set T of maximum cardinality, and let $\mathcal{C} := \mathcal{C}_{G-S}$. $S, \mathcal{C}$

We first show that every component $C \in \mathcal{C}$ is odd. If $|C|$ is even, pick a vertex $c \in C$, and let $S' := S \cup \{c\}$ and $C' := C - c$. Then C' has odd order, and thus has at least one odd component. Hence, $q(G - S') \geq q(G - S) + 1$. Since $T := S$ satisfies $(*)$ with equality, we obtain

$$q(G - S') \geq q(G - S) + 1 = |S| + d + 1 = |S'| + d \underset{(*)}{\geq} q(G - S')$$

with equality, which contradicts the maximality of S.

Next we prove the assertion (ii), that every $C \in \mathcal{C}$ is factor-critical. Suppose there exist $C \in \mathcal{C}$ and $c \in C$ such that $C' := C - c$ has no 1-factor. By the induction hypothesis (and the fact that, as shown earlier, for fixed G our theorem implies Tutte's theorem) there exists a set $T' \subseteq V(C')$ with

$$q(C' - T') > |T'| \, .$$

Since $|C|$ is odd and hence $|C'|$ is even, the numbers $q(C' - T')$ and $|T'|$ are either both even or both odd, so they cannot differ by exactly 1. We may therefore sharpen the above inequality to

$$q(C' - T') \geq |T'| + 2 \, .$$

For $T := S \cup \{c\} \cup T'$ we thus obtain

$$\begin{aligned}
q(G - T) &= q(G - S) - 1 + q(C' - T') \\
&\geq |S| + d - 1 + |T'| + 2 \\
&= |T| + d \\
&\underset{(*)}{\geq} q(G - T)
\end{aligned}$$

with equality, again contradicting the maximality of S.

It remains to show that S is matchable to $G - S$. If $S = \emptyset$, this is trivial, so we assume that $S \neq \emptyset$. Since $T := S$ satisfies $(*)$ with equality, this implies that $\mathcal{C}$ too is non-empty. We now apply Corollary 2.1.3 to $H := H_S$, but 'backwards', i.e. with $A := \mathcal{C}$. Given $\mathcal{C}' \subseteq \mathcal{C}$, H
set $S' := N_H(\mathcal{C}') \subseteq S$. Since every $C \in \mathcal{C}'$ is an odd component also of $G - S'$, we have

$$|N_H(\mathcal{C}')| = |S'| \underset{(*)}{\geq} q(G - S') - d \geq |\mathcal{C}'| - d \, .$$

By Corollary 2.1.3, then, H contains a matching of cardinality

$$|\mathcal{C}| - d = q(G - S) - d = |S| \, ,$$

which is therefore a matching of S. $\square$

S
$\mathcal{C}$

k_S, k_C

Let us consider once more the set S from Theorem 2.2.3, together with any matching M in G. As before, we write $\mathcal{C} := \mathcal{C}_{G-S}$. Let us denote by k_S the number of edges in M with at least one end in S, and by k_C the number of edges in M with both ends in $G - S$. Since each $C \in \mathcal{C}$ is odd, at least one of its vertices is not incident with an edge of the second type. Therefore every matching M satisfies

$$k_S \leqslant |S| \quad \text{and} \quad k_C \leqslant \tfrac{1}{2}\left(|V| - |S| - |\mathcal{C}|\right). \tag{1}$$

M_0

Moreover, G contains a matching M_0 with equality in both cases: first choose $|S|$ edges between S and $\bigcup \mathcal{C}$ according to (i), and then use (ii) to find a suitable set of $\tfrac{1}{2}\left(|C| - 1\right)$ edges in every component $C \in \mathcal{C}$. This matching M_0 thus has exactly

$$|M_0| = |S| + \tfrac{1}{2}\left(|V| - |S| - |\mathcal{C}|\right) \tag{2}$$

edges.

Now (1) and (2) together imply that *every* matching M of maximum cardinality satisfies both parts of (1) with equality: by $|M| \geqslant |M_0|$ and (2), M has at least $|S| + \tfrac{1}{2}\left(|V| - |S| - |\mathcal{C}|\right)$ edges, which implies by (1) that neither of the inequalities in (1) can be strict. But equality in (1), in turn, implies that M has the structure described above: by $k_S = |S|$, every vertex $s \in S$ is the end of an edge $st \in M$ with $t \in G - S$, and by $k_C = \tfrac{1}{2}\left(|V| - |S| - |\mathcal{C}|\right)$ exactly $\tfrac{1}{2}\left(|C| - 1\right)$ edges of M lie in C, for every $C \in \mathcal{C}$. Finally, since these latter edges miss only one vertex in each C, the ends t of the edges st above lie in different components C for different s.

The seemingly technical Theorem 2.2.3 thus hides a wealth of structural information: it contains the essence of a detailed description of all maximum-cardinality matchings in all graphs.[2]

2.3 Path covers

Let us return for a moment to König's duality theorem for bipartite graphs, Theorem 2.1.1. If we orient every edge of G from A to B, the theorem tells us how many disjoint directed paths we need in order to cover all the vertices of G: every directed path has length 0 or 1, and clearly the number of paths in such a 'path cover' is minimum when it contains as many paths of length 1 as possible—in other words, when it contains a maximum-cardinality matching.

In this section we put the above question more generally: how many paths in a given directed graph will suffice to cover its entire vertex set?

---

[2] A reference to the full statement of this structural result, known as the *Gallai-Edmonds matching theorem*, will be given in the notes at the end of this chapter.

Of course, this could be asked just as well for undirected graphs. As it turns out, however, the result we shall prove is rather more trivial in the undirected case (exercise), and the directed case will also have an interesting corollary.

A *directed path* is a directed graph $P \neq \emptyset$ with distinct vertices $x_0, \ldots, x_k$ and edges $e_0, \ldots, e_{k-1}$ such that e_i is an edge directed from x_i to x_{i+1}, for all $i < k$. We denote the last vertex x_k of P by $\mathrm{ter}(P)$. $\quad$ ter(P)
In this section, *path* will always mean 'directed path'. A *path cover* of a $\quad$ path
directed graph G is a set of disjoint paths in G which together contain $\quad$ path cover
all the vertices of G. Let us denote the maximum cardinality of an
independent set of vertices in G by $\alpha(G)$. $\quad\quad$ $\alpha(G)$

Theorem 2.3.1. (Gallai & Milgram 1960)
Every directed graph G has a path cover by at most $\alpha(G)$ paths.

Proof. Given two sets $\mathcal{P}_1$ and $\mathcal{P}_2$ of disjoint paths in G, we write $\mathcal{P}_1 < \mathcal{P}_2$ $\quad$ $\mathcal{P}_1 < \mathcal{P}_2$
if $\{\, \mathrm{ter}(P) \mid P \in \mathcal{P}_1 \,\} \subseteq \{\, \mathrm{ter}(P) \mid P \in \mathcal{P}_2 \,\}$ and $|\mathcal{P}_1| < |\mathcal{P}_2|$. We shall
prove the following:

> If $\mathcal{P}$ is a $<$-minimal path cover of G, then G contains an
> independent set $\{\, v_P \mid P \in \mathcal{P} \,\}$ of vertices with $v_P \in P$ for $\quad\quad$ $(*)$
> every $P \in \mathcal{P}$.

Clearly, $(*)$ implies the assertion of the theorem.

We prove $(*)$ by induction on $|G|$. Let $\mathcal{P} = \{\, P_1, \ldots, P_m \,\}$ be given $\quad$ $\mathcal{P}, P_i, m$
as in $(*)$, and let $v_i := \mathrm{ter}(P_i)$ for every i. If $\{\, v_i \mid 1 \leqslant i \leqslant m \,\}$ is $\quad$ v_i
independent, there is nothing more to show; we may therefore assume
that G has an edge from v_2 to v_1. Since $P_2 v_2 v_1$ is again a path, the
minimality of $\mathcal{P}$ implies that v_1 is not the only vertex of P_1; let v be $\quad$ v
the vertex preceding v_1 on P_1. Then $\mathcal{P}' := \{\, P_1 v, P_2, \ldots, P_m \,\}$ is a path $\quad$ $\mathcal{P}'$
cover of $G' := G - v_1$ (Fig. 2.3.1). We first show that $\mathcal{P}'$ is $<$-minimal $\quad$ G'
with this property.

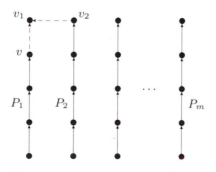

Fig. 2.3.1. The path cover $\mathcal{P}'$ of G'

Suppose that $\mathcal{P}'' < \mathcal{P}'$ is another path cover of G'. If a path $P \in \mathcal{P}''$ ends in v, we may replace P in $\mathcal{P}''$ by Pvv_1 to obtain a smaller path cover of G than $\mathcal{P}$, a contradiction to the minimality of $\mathcal{P}$. If a path $P \in \mathcal{P}''$ ends in v_2 (but none in v), we replace P in $\mathcal{P}''$ by Pv_2v_1, again contradicting the minimality of $\mathcal{P}$. Hence $\{\, \mathrm{ter}(P) \mid P \in \mathcal{P}'' \,\} \subseteq \{\, v_3, \ldots, v_m \,\}$, and in particular $|\mathcal{P}''| \leqslant |\mathcal{P}| - 2$. But now $\mathcal{P}''$ and the trivial path $\{\, v_1 \,\}$ together form a path cover of G that contradicts the minimality of $\mathcal{P}$.

Hence $\mathcal{P}'$ is minimal, as claimed. By the induction hypothesis, $\{\, V(P) \mid P \in \mathcal{P}' \,\}$ has an independent set of representatives. But this is also a set of representatives for $\mathcal{P}$, and $(*)$ is proved. □

As a corollary to Theorem 2.3.1 we now deduce a classic result from the theory of partial orders. A subset of a partially ordered set $(P, \leqslant)$ is *chain*   a *chain* in P if its elements are pairwise comparable; it is an *antichain*
antichain   if they are pairwise incomparable.

Corollary 2.3.2. (Dilworth 1950)
In every finite partially ordered set $(P, \leqslant)$, the minimum number of chains covering P is equal to the maximum cardinality of an antichain in P.

Proof. If A is an antichain in P of maximum cardinality, then clearly P cannot be covered by fewer than $|A|$ chains. The fact that $|A|$ chains will suffice follows from Theorem 2.3.1 applied to the directed graph on P with the edge set $\{\, (x, y) \mid x < y \,\}$. □

Exercises

1.   Let M be a matching in a bipartite graph G. Show that if M is sub-optimal, i.e. contains fewer edges than some other matching in G, then G contains an augmenting path with respect to M. Does this fact generalize to matchings in non-bipartite graphs?

   (Hint. Symmetric difference.)

2.   Describe an algorithm that finds, as efficiently as possible, a matching of maximum cardinality in any bipartite graph.

3.   Find an infinite counterexample to the statement of the marriage theorem.

4.   Let k be an integer. Show that any two partitions of a finite set into k-sets have a common system of distinct representatives.

5. Let A be a finite set with subsets $A_1, \ldots, A_n$, and let $d_1, \ldots, d_n \in \mathbb{N}$. Show that there are disjoint subsets $D_k \subseteq A_k$, with $|D_k| = d_k$ for all $k \leqslant n$, if and only if

$$\left| \bigcup_{i \in I} A_i \right| \geqslant \sum_{i \in I} d_i$$

for all $I \subseteq \{1, \ldots, n\}$.

6.+ Prove *Sperner's lemma*: in an n-set X there are never more than $\binom{n}{\lfloor n/2 \rfloor}$ subsets such that none of these contains another.

(Hint. Construct $\binom{n}{\lfloor n/2 \rfloor}$ chains covering the power set lattice of X.)

7. Describe the set S in Theorem 2.2.3 for the case that G is a forest.

8. Using Theorem 2.2.3, show that a k-connected graph with at least $2k$ vertices contains a matching of size k. Is this best possible?

9. A graph G is called (vertex-) *transitive* if, for any two vertices $v, w \in G$, there is an automorphism of G mapping v to w. Using the observations following the proof of Theorem 2.2.3, show that every transitive connected graph is either factor-critical or contains a 1-factor.

(Hint. Consider the cases of $S = \emptyset$ and $S \neq \emptyset$ separately.)

10. Show that a graph G contains k independent edges if and only if $q(G - S) \leqslant |S| + |G| - 2k$ for all sets $S \subseteq V(G)$.

(Hint. For the 'if' direction, suppose that G has no k independent edges, and and apply Tutte's 1-factor theorem to the graph $G * K^{n-2k}$. Alternatively, use Theorem 2.2.3.)

11.⁻ Find a cubic graph without a 1-factor.

12.⁻ Prove the undirected version of the theorem of Gallai & Milgram (without using the directed version).

13. Derive the marriage theorem from the theorem of Gallai & Milgram.

14.⁻ Show that a partially ordered set of at least $rs + 1$ elements contains either a chain of size $r + 1$ or an antichain of size $s + 1$.

15. Prove the following dual version of Dilworth's theorem: in every finite partially ordered set $(P, \leqslant)$, the minimum number of antichains covering P is equal to the maximum cardinality of a chain in P.

16. Derive König's theorem from Dilworth's theorem.

17.+ Find a partially ordered set that has no infinite antichain but cannot be covered by finitely many chains.

(Hint. $\mathbb{N} \times \mathbb{N}$.)

Notes

There is a very readable and comprehensive monograph about matching in finite graphs: L. Lovász & M.D. Plummer, *Matching Theory*, Annals of Discrete Math. **29**, North Holland 1986. All the references for the results in this chapter can be found there.

As we shall see in Chapter 3, König's Theorem of 1931 is no more than the bipartite case of a more general theorem due to Menger, of 1929. At the time, neither of these results was nearly as well known as Hall's marriage theorem, which was proved even later, in 1935. To this day, Hall's theorem remains one of the most applied graph-theoretic results. Its special case that both partition sets have the same size was proved implicitly already by Frobenius (1917) in a paper on determinants.

Our proof of Tutte's 1-factor theorem is based on a proof by Lovász (1975). Our extension of Tutte's theorem, Theorem 2.2.3 (including the informal discussion following it) is a lean version of a comprehensive structure theorem for matchings, due to Gallai (1964) and Edmonds (1965). See Lovász & Plummer for a detailed statement and discussion of this theorem.

Theorem 2.3.1 is due to T. Gallai & A.N. Milgram, Verallgemeinerung eines graphentheoretischen Satzes von Rédei, *Acta Sci. Math. (Szeged)* **21** (1960), 181–186.

3   Connectivity

Our definition of k-connectedness, given in Chapter 1.4, is somewhat
unintuitive. It does not tell us much about 'connections' in a k-connected
graph: all it says is that we need at least k vertices to *disconnect* it. The
following definition—which, incidentally, implies the one above—might
have been more descriptive: 'a graph is k-*connected* if any two of its
vertices can be joined by k independent paths'.

It is one of the classic results of graph theory that these two definitions are in fact equivalent, are dual aspects of the same property. We
shall study this theorem of Menger (1927) in some depth in Section 3.3.

In Sections 3.1 and 3.2, we investigate the structure of the 2-con-
nected and the 3-connected graphs. For these small values of k it is still
possible to give a simple general description of how these graphs can be
constructed.

In the remaining sections of this chapter we look at other concepts of
connectedness, more recent than the standard one but no less important:
the number of H-paths in a graph for a given subgraph H; the number of
edge-disjoint spanning trees; and the existence of disjoint paths linking
up several given pairs of vertices.

3.1 2-Connected graphs and subgraphs

A maximal connected subgraph without a cutvertex is called a *block*.   *block*
Thus, every block of a graph G is either a maximal 2-connected subgraph,
or a bridge (with its ends), or an isolated vertex. Conversely, every such
subgraph is a block. By their maximality, different blocks of G overlap
in at most one vertex, which is then a cutvertex of G. Hence, every edge
of G lies in a unique block, and G is the union of its blocks.

In a sense, blocks are the 2-connected analogues of components, the
maximal connected subgraphs of a graph. While the structure of G is

determined fully by that of its components, however, it is not captured completely by the structure of its blocks: since the blocks need not be disjoint, the way they intersect defines another structure, giving a coarse picture of G as if viewed from a distance.

The following proposition describes this coarse structure of G as formed by its blocks. Let A denote the set of cutvertices of G, and $\mathcal{B}$ the set of its blocks. We then have a natural bipartite graph on $A \cup \mathcal{B}$ formed by the edges aB with $a \in B$. This *block graph* of G is shown in Figure 3.1.1.

block graph

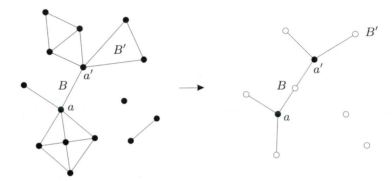

Fig. 3.1.1. A graph and its block graph

Proposition 3.1.1. *The block graph of a connected graph is a tree.*
□

Proposition 3.1.1 reduces the structure of a given graph to that of its blocks. So what can we say about the blocks themselves? The following proposition gives a simple method by which, in principle, a list of all 2-connected graphs could be compiled:

[4.2.5]

Proposition 3.1.2. *A graph is 2-connected if and only if it can be constructed from a cycle by successively adding H-paths to graphs H already constructed (Fig. 3.1.2).*

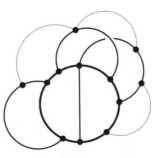

Fig. 3.1.2. The construction of 2-connected graphs

Proof. Clearly, every graph constructed as described is 2-connected. Conversely, let a 2-connected graph G be given. Then G contains a cycle, and hence has a maximal subgraph H constructible as above. Since any edge $xy \in E(G) \smallsetminus E(H)$ with $x, y \in H$ would define an H-path, H is an induced subgraph of G. Thus if $H \neq G$, then by the connectedness of G there is an edge vw with $v \in G - H$ and $w \in H$. As G is 2-connected, $G - w$ contains a v–H path P. Then wvP is an H-path in G, and $H \cup wvP$ is a constructible subgraph of G larger than H. This contradicts the maximality of H. $\qquad\square$

H

3.2 The structure of 3-connected graphs

We start this section with the analogue of Proposition 3.1.2 for 3-connectedness: our first theorem describes how every 3-connected graph can be obtained from a K^4 by a succession of elementary operations preserving 3-connectedness. We then prove a deep result of Tutte about the algebraic structure of the cycle space of 3-connected graphs; this will play an important role again in Chapter 4.5.

Lemma 3.2.1. *If G is 3-connected and $|G| > 4$, then G has an edge e such that G/e is again 3-connected.*

[4.4.3]

Proof. Suppose there is no such edge e. Then, for every edge $xy \in G$, the graph G/xy contains a separating set S of at most 2 vertices. Since $\kappa(G) \geqslant 3$, the contracted vertex v_{xy} of G/xy (see Chapter 1.7) lies in S and $|S| = 2$, i.e. G has a vertex $z \notin \{x, y\}$ such that $\{v_{xy}, z\}$ separates G/xy. Then any two vertices separated by $\{v_{xy}, z\}$ in G/xy are separated in G by $T := \{x, y, z\}$. Since no proper subset of T separates G, every vertex in T has a neighbour in every component C of $G - T$.

xy

z

C

We choose the edge xy, the vertex z, and the component C so that $|C|$ is minimum, and pick a neighbour v of z in C (Fig. 3.2.1). By

v

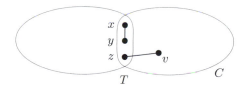

Fig. 3.2.1. Separating vertices in the proof of Lemma 3.2.1

w

assumption, G/zv is again not 3-connected, so again there is a vertex w such that $\{\,z,v,w\,\}$ separates G, and as before every vertex in $\{\,z,v,w\,\}$ has a neighbour in every component of $G - \{\,z,v,w\,\}$.

As x and y are adjacent, $G - \{\,z,v,w\,\}$ has a component D such that $D \cap \{\,x,y\,\} = \emptyset$. Then every neighbour of v in D lies in C (since $v \in C$), so $D \cap C \neq \emptyset$ and hence $D \subsetneq C$ by the choice of D. This contradicts the choice of xy, z and C. $\square$

Theorem 3.2.2. (Tutte 1961)
A graph G is 3-connected if and only if there exists a sequence $G_0, \dots, G_n$ of graphs with the following properties:

 (i) *$G_0 = K^4$ and $G_n = G$;*

 (ii) *G_{i+1} has an edge xy with $d(x), d(y) \geqslant 3$ and $G_i = G_{i+1}/xy$, for every $i < n$.*

Proof. If G is 3-connected, a sequence as in the theorem exists by Lemma 3.2.1. Note that all the graphs in this sequence are 3-connected.

xy

S

C_1, C_2

Conversely, let $G_0, \dots, G_n$ be a sequence of graphs as stated; we show that if $G_i = G_{i+1}/xy$ is 3-connected then so is G_{i+1}, for every $i < n$. Suppose not, let S be a separating set of at most 2 vertices in G_{i+1}, and let C_1, C_2 be two components of $G_{i+1} - S$. As x and y are adjacent, we may assume that $\{\,x,y\,\} \cap V(C_1) = \emptyset$ (Fig. 3.2.2). Then C_2 contains nei-

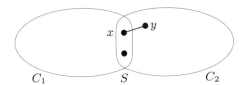

Fig. 3.2.2. The position of $xy \in G_{i+1}$ in the proof of Theorem 3.2.2

ther both vertices x, y nor a vertex $v \notin \{\,x,y\,\}$: otherwise v_{xy} or v would be separated from C_1 in G_i by at most two vertices, a contradiction. But now C_2 contains only one vertex: either x or y. This contradicts our assumption of $d(x), d(y) \geqslant 3$. $\square$

Theorem 3.2.2 is the essential core of a result of Tutte known as his *wheel theorem*.[1] Like Proposition 3.1.2 for 2-connected graphs, it enables us to construct all 3-connected graphs by a simple inductive process depending only on local information: starting with K^4, we pick a vertex v in a graph constructed already, split it into two adjacent vertices v', v'', and join these to the former neighbours of v as we please—provided only that v' and v'' each acquire at least 3 incident edges, and that every former neighbour of v becomes adjacent to at least one of v', v''.

wheel

---

[1] Graphs of the form $C^n * K^1$ are called *wheels*; thus, K^4 is the smallest wheel.

Theorem 3.2.3. (Tutte 1963)

The cycle space of a 3-connected graph is generated by its non-separating   [4.5.2]
induced cycles.

Proof. We apply induction on the order of the graph G considered.   (1.9.1)
In K^4, every cycle is a triangle or (in terms of edges) the symmetric
difference of triangles. As these are both induced and non-separating,
the assertion holds for $|G| = 4$.

For the induction step, let $e = xy$ be an edge of G for which $e = xy$
$G' := G/e$ is again 3-connected; cf. Lemma 3.2.1. Then every edge G'
$e' \in E(G') \smallsetminus E(G)$ is of the form $e' = v_e w$, where at least one of the two
edges xw and yw lies in G. We pick one that does (either xw or yw),
and identify it notationally with the edge e': from now on, we often use
one and the same symbol to refer to the edge e' of G' and to the edge
xw or yw of G identified with e'. In this way, we may regard $E(G')$ as
a subset of $E(G)$, and $\mathcal{E}(G')$ as a subspace of $\mathcal{E}(G)$. (In particular, all
vector operations will take place unambiguously in $\mathcal{E}(G)$.)

Let us consider an induced cycle $C \subseteq G$. If $e \in C$ and $C = C^3$, we
call C a *fundamental triangle*; then $C/e = K^2$. If $e \in C$ but $C \neq C^3$,   *fundamental*
then C/e is a cycle in G'. Finally if $e \notin C$, then at most one of x, y lies   *triangle*
on C (otherwise e would be a chord), so the vertices of C in order also
form a cycle in G' if we replace x or y by v_e as necessary; this cycle,
too, will be denoted by C/e. Thus, as long as C is not a fundamental
triangle, C/e will always denote a unique cycle in G'. Note that, in the C/e
case of $e \notin C$, the edge set of C/e when viewed as a subset of $E(G)$ need
not coincide with $E(C)$: an edge $yw \in E(C)$, for example, gives rise to
the edge $v_e w \in E(C/e)$, but if that edge was identified with xw rather
than with yw, then $E(C/e)$ defines an x–y path in G, not the cycle C
(Fig. 3.2.3).

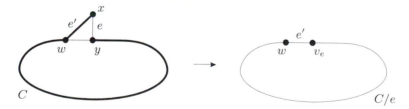

Fig. 3.2.3. One possibility for C/e when $e \notin C$

Let us refer to the non-separating induced cycles in G or G' as *basic*   *basic cycles*
cycles. An element of $\mathcal{C}(G)$ will be called *good* if it is a linear combination   *good*
of basic cycles in G. We thus want to show that every element of $\mathcal{C}(G)$
is good.

We start by proving three auxiliary facts.

$$\text{Every fundamental triangle is a basic cycle in } G. \qquad (1)$$

A fundamental triangle, $wxyw$ say, is clearly induced in G. If it separated G, then $\{v_e, w\}$ would separate G', which contradicts the choice of e. This proves (1).

> If $C \subseteq G$ is an induced cycle but not a fundamental triangle, then $C + C/e + D \in \{\emptyset, \{e\}\}$ for some good $D \in \mathcal{C}(G)$. $\qquad$ (2)

The gist of (2) is that, in terms of 'generatability', C and C/e differ only a little: after the addition of a permissible error term D, by at most the edge e. In which other edges, then, can C and C/e differ? Clearly only in edges incident with v_e in G', and with x or y in G, respectively. Indeed, if $e' = v_e w \in C/e$, and e' is identified with $xw \in G$, then C may contain the edge yw rather than e' (Fig. 3.2.3). In this case, we choose as D the fundamental triangle $wxyw$: then $C + D$ has gained (compared with C) the edge $xw \in D \smallsetminus C$ (which is identified with $e' \in C/e$ and hence desired), and it has lost the edge $yw \in D \cap C$ (which is not in C/e and hence undesired). In general, we can always find a suitable D as the sum of at most two fundamental triangles $wxyw$, one for every edge $v_e w \in E(C/e) \smallsetminus E(C)$. By (1), these fundamental triangles are basic, and so (2) is proved.

> For every basic cycle $C' \subseteq G'$ there exists a basic cycle $C = C(C') \subseteq G$ with $C/e = C'$. $\qquad$ (3)

If $v_e \notin C'$, then (3) is satisfied with $C := C'$. So we assume that $v_e \in C'$.

u, w

P

Let u and w be the two neighbours of v_e on C', and let P be the u–w path in C' avoiding v_e (Fig. 3.2.4). Then $P \subseteq G$.

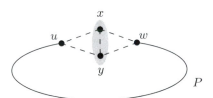

Fig. 3.2.4. The search for a basic cycle C with $C/e = C'$

C_x, C_y

We first assume that $\{ux, uy, wx, wy\} \subseteq E(G)$, and consider (as candidates for C) the cycles $C_x := uPwxu$ and $C_y := uPwyu$. Both are induced cycles in G (because C' is induced in G'), and clearly $C_x/e = C_y/e = C'$. Moreover, neither of these cycles separates two vertices of $G - (V(P) \cup \{x, y\})$ in G, since C' does not separate such vertices in G'. Thus, if C_x (say) is a separating cycle in G, then one of the components of $G - C_x$ consists just of y. Likewise, if C_y separates G then one of the arising components contains only x. However, this cannot happen for both C_x and C_y at once: otherwise $N_G(\{x, y\}) \subseteq V(P)$ and hence $N_G(\{x, y\}) = \{u, w\}$ (since C' has no chord), which contradicts $\kappa(G) \geqslant 3$. Hence, at least one of C_x, C_y is a basic cycle in G.

It remains to consider the case that $\{ux, uy, wx, wy\} \nsubseteq E(G)$, say $ux \notin E(G)$. Then, as above, either $uPwyu$ or $uPwxyu$ is a basic cycle in G, according as wy is an edge of G or not. This completes the proof of (3).

We now come to the main part of our proof, the proof that every $C \in \mathcal{C}(G)$ is good. By Proposition 1.9.1 we may assume that C is an induced cycle in G. By (1) we may further assume that C is not a fundamental triangle; so C/e is a cycle. Our aim is to argue as follows: by (2), C differs from C/e at most in e (disregarding the permissible error term D); by (3), the basic cycles of G' summing to C/e by induction differ at most in e from suitable basic cycles in G; hence, C is the sum of basic cycles—except that the edge e will need some special attention.

By the induction hypothesis, C/e has a representation

$$C/e = C_1' + \ldots + C_k'$$

in $\mathcal{C}(G')$, where every C_i' is a basic cycle in G'. For each i, we obtain from (3) a basic cycle $C(C_i') \subseteq G$ with $C(C_i')/e = C_i'$ (in particular, $C(C_i')$ is not a fundamental triangle), and from (2) some good $D_i \in \mathcal{C}(G)$ such that

$$C(C_i') + C_i' + D_i \in \{\emptyset, \{e\}\}. \tag{4}$$

We let

$$C_i := C(C_i') + D_i\,;$$

then C_i is good, and by (4) it differs from C_i' at most in e. Again by (2), we have

$$C + C/e + D \in \{\emptyset, \{e\}\}$$

for some good $D \in \mathcal{C}(G)$, i.e. $C + D$ differs from C/e at most in e. But then $C + D + C_1 + \ldots + C_k$ differs from $C/e + C_1' + \ldots + C_k' = \emptyset$ at most in e, that is,

$$C + D + C_1 + \ldots + C_k \in \{\emptyset, \{e\}\}\,.$$

Since $C + D + C_1 + \ldots + C_k \in \mathcal{C}(G)$ but $\{e\} \notin \mathcal{C}(G)$, this means that in fact

$$C + D + C_1 + \ldots + C_k = \emptyset\,,$$

so $C = D + C_1 + \ldots + C_k$ is good. $\qquad\square$

3.3 Menger's theorem

The following classic theorem of Menger is one of the corner-stones of graph theory. We give two proofs.[2]

[3.6.2]
[8.1.2]
[12.4.5]

Theorem 3.3.1. (Menger 1927)
Let $G = (V, E)$ be a graph and $A, B \subseteq V$. Then the minimum number of vertices separating A from B in G is equal to the maximum number of disjoint A–B paths in G.

k

First proof. For given G, A, B, let $k = k(G, A, B)$ be the minimum number of vertices separating A from B in G. Then, clearly, G cannot contain more than k disjoint A–B paths; we show by induction on $|G| + \|G\|$ that G does indeed contain k such paths. For all G, A, B with $k \in \{0, 1\}$ this is true. For the induction step let G, A, B with $k \geqslant 2$ be given, and assume that the assertion holds for graphs with fewer vertices or edges.

If there is a vertex $x \in A \cap B$, then $G - x$ contains $k - 1$ disjoint A–B paths by the induction hypothesis. (Why?) Together with the trivial path $\{x\}$, these form the desired paths in G. We shall therefore assume that

$$A \cap B = \emptyset. \tag{1}$$

X

C_A, C_B

G_A, G_B

We first construct the desired paths for the case that A and B are separated by a set $X \subseteq V$ with $|X| = k$ and $X \neq A, B$. Let C_A be the union of all the components of $G - X$ meeting A; note that $C_A \neq \emptyset$, since $|A| \geqslant k = |X|$ but $A \neq X$. The subgraph C_B defined likewise is not empty either, and $C_A \cap C_B = \emptyset$. Let us write $G_A := G[V(C_A) \cup X]$ and $G_B := G[V(C_B) \cup X]$. Since every A–B path in G contains an A–X path in G_A, we cannot separate A from X in G_A by fewer than k vertices. Thus, by the induction hypothesis, G_A contains k disjoint A–X paths (Fig. 3.3.1). In the same way, there are k disjoint X–B paths in G_B. As

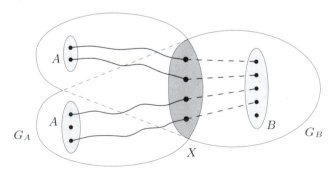

Fig. 3.3.1. Disjoint A–X paths in G_A

---

[2] Another proof will follow as an easy exercise from a theorem on network flows in Chapter 6.

$|X| = k$, we can put these paths together to form k disjoint A–B paths.

For the general case, let P be any A–B path in G. By (1), P has an edge ab with $a \notin B$ and $b \notin A$. Let Y be a set of as few vertices as possible separating A from B in $G - ab$ (Fig. 3.3.2). Then $Y_a := Y \cup \{a\}$ and $Y_b := Y \cup \{b\}$ both separate A from B in G, and by definition of k we have

$$|Y_a|, |Y_b| \geqslant k.$$

If equality holds here, we may assume by the case already treated that $\{Y_a, Y_b\} \subseteq \{A, B\}$, so $\{Y_a, Y_b\} = \{A, B\}$ since $a \notin B$ and $b \notin A$. Thus, $Y = A \cap B$. Since $|Y| \geqslant k - 1 \geqslant 1$, this contradicts (1).

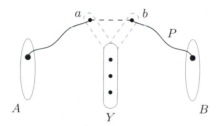

Fig. 3.3.2. Separating A from B in $G - ab$

We therefore have either $|Y_a| > k$ or $|Y_b| > k$, and hence $|Y| \geqslant k$. By the induction hypothesis, then, there are k disjoint A–B paths even in $G - ab \subseteq G$. □

Applied to a bipartite graph, Menger's theorem specializes to the assertion of König's theorem (2.1.1). For our second proof, we now adapt the alternating path proof of König's theorem to the more general set-up of Theorem 3.3.1. Let again G, A, B be given, and let $\mathcal{P}$ be a set of disjoint A–B paths in G. We write

$$V[\mathcal{P}] := \bigcup\{V(P) \mid P \in \mathcal{P}\}$$

$$E[\mathcal{P}] := \bigcup\{E(P) \mid P \in \mathcal{P}\}.$$

A walk $W = x_0 e_0 x_1 e_1 \ldots e_{n-1} x_n$ in G with $e_i \neq e_j$ for $i \neq j$ is said to be *alternating* with respect to $\mathcal{P}$ if the following three conditions are satisfied for all $i < n$ (Fig. 3.3.3):

(i) if $e_i = e \in E[\mathcal{P}]$, then W traverses the edge e backwards, i.e. $x_{i+1} \in P\mathring{x}_i$ for some $P \in \mathcal{P}$;

(ii) if $x_i = x_j$ with $i \neq j$, then $x_i \in V[\mathcal{P}]$;

(iii) if $x_i \in V[\mathcal{P}]$, then $\{e_{i-1}, e_i\} \cap E[\mathcal{P}] \neq \emptyset$.[3]

---

[3] For $i = 0$ we let $\{e_{i-1}, e_i\} := \{e_0\}$.

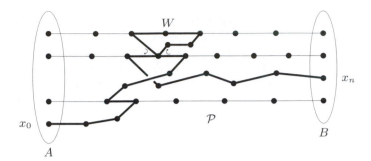

Fig. 3.3.3. An alternating walk from A to B

W, x_i, e_i

Let us consider a walk $W = x_0 e_0 x_1 e_1 \ldots e_{n-1} x_n$ from $A \smallsetminus V[\mathcal{P}]$ to $B \smallsetminus V[\mathcal{P}]$, alternating with respect to $\mathcal{P}$. By (ii), any vertex outside $V[\mathcal{P}]$ occurs at most once on W. Since the edges e_i of W are all distinct, (iii) implies that any vertex in $V[\mathcal{P}]$ occurs at most twice on W. This can happen in two ways: if $x_i = x_j$ with $0 < i < j < n$, say, then

$$either \ e_{i-1}, e_j \in E[\mathcal{P}] \ and \ e_i, e_{j-1} \notin E[\mathcal{P}]$$
$$or \ e_i, e_{j-1} \in E[\mathcal{P}] \ and \ e_{i-1}, e_j \notin E[\mathcal{P}].$$

Lemma 3.3.2. *If such a walk W exists, then G contains $|\mathcal{P}| + 1$ disjoint A–B paths.*

Proof. Let H be the graph on $V[\mathcal{P}] \cup \{x_0, \ldots, x_n\}$ whose edge set is the symmetric difference of $E[\mathcal{P}]$ with $\{e_0, \ldots, e_{n-1}\}$. In H, the ends of the paths in $\mathcal{P}$ and of W have degree 1 (or 0, if the path or W is trivial), and all other vertices have degree 0 or 2. For each of the $|\mathcal{P}| + 1$ vertices

P

$a \in (A \cap V[\mathcal{P}]) \cup \{x_0\}$, therefore, the component of H containing a is a path, $P = v_0 \ldots v_k$ say, which starts in a and ends in A or B. Using conditions (i) and (iii), one easily shows by induction on $i = 0, \ldots, k-1$ that P traverses each of its edges $e = v_i v_{i+1}$ in the forward direction with respect to $\mathcal{P}$ or W. (Formally: if $e \in P'$ with $P' \in \mathcal{P}$, then $v_i \in P' \mathring{v}_{i+1}$; if $e = e_j \in W$, then $v_i = x_j$ and $v_{i+1} = x_{j+1}$.) Hence, P ends in B. As we have $|\mathcal{P}| + 1$ disjoint such paths P, this completes the proof. $\square$

$\mathcal{P}$

Second proof of Menger's theorem. Let $\mathcal{P}$ be a set of as many disjoint A–B paths in G as possible. Unless otherwise stated, all alternating walks considered are alternating with respect to $\mathcal{P}$. We set

A_1, A_2

$$A_1 := A \cap V[\mathcal{P}] \quad \text{and} \quad A_2 := A \smallsetminus A_1,$$

and

B_1, B_2

$$B_1 := B \cap V[\mathcal{P}] \quad \text{and} \quad B_2 := B \smallsetminus B_1.$$

x_P

For every path $P \in \mathcal{P}$, let x_P be the last vertex of P that lies on some alternating walk starting in A_2; if no such vertex exists, let x_P be the first vertex of P. Clearly, the set

$$X := \{\, x_P \mid P \in \mathcal{P} \,\}$$ X

has cardinality $|\mathcal{P}|$; it thus suffices to show that X separates A from B.

Let Q be any A–B path in G; we show that Q meets X. Suppose Q
not. By the maximality of $\mathcal{P}$, the path Q meets $V[\mathcal{P}]$. Since the A–
$V[\mathcal{P}]$ path in Q is trivially an alternating walk, Q also meets the vertex
set $V[\mathcal{P}']$ of

$$\mathcal{P}' := \{\, Px_P \mid P \in \mathcal{P} \,\};$$ $\mathcal{P}'$

let y be the last vertex of Q in $V[\mathcal{P}']$, let P be the path in $\mathcal{P}$ containing y, y, P
and let $x := x_P$. Finally, let W be an alternating walk from A_2 to x, x, W
as in the definition of x_P. By assumption, Q avoids X and hence x, so
$y \in P\mathring{x}$, and $W \cup xPyQ$ is a walk from A_2 to B (Fig. 3.3.4). If this walk
is alternating and ends in B_2, we are home: then G contains $|\mathcal{P}| + 1$
disjoint A–B paths by Lemma 3.3.2, contrary to the maximality of $\mathcal{P}$.

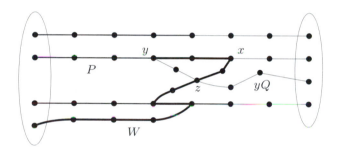

Fig. 3.3.4. Alternating walks in the second proof of Menger's
theorem

How could $W \cup xPyQ$ fail to be an alternating walk? For a start,
W might already use an edge of xPy. But if x' is the first vertex of W
on $xP\mathring{y}$, then $W' := Wx'Py$ is an alternating walk from A_2 to y. (By x', W'
Wx' we mean the initial segment of W ending at the first occurrence of
x' on W; from there onwards, W' follows P back to y.) Even our new
walk $W'yQ$ need not yet be alternating: W' might still meet $\mathring{y}Q$. By
definition of $\mathcal{P}'$ and W, however, and the choice of y on Q, we have

$$V(W') \cap V[\mathcal{P}] \subseteq V[\mathcal{P}'] \quad \text{and} \quad V(\mathring{y}Q) \cap V[\mathcal{P}'] = \emptyset.$$

Thus, W' and $\mathring{y}Q$ can meet only outside $\mathcal{P}$.

If W' does indeed meet $\mathring{y}Q$, let z be the first vertex of W' on $\mathring{y}Q$. As z
z lies outside $V[\mathcal{P}]$, it occurs only once on W' (condition (ii)), and we let
$W'' := W'zQ$. On the other hand if $W' \cap \mathring{y}Q = \emptyset$, we set $W'' := W' \cup yQ$. W''
In both cases, W'' is alternating with respect to $\mathcal{P}'$, because W' is and $\mathring{y}Q$

avoids $V[\mathcal{P}']$. (Note that W'' satisfies condition (iii) at y in the second case, while in the first case (iii) is not applicable to z.) By definition of $\mathcal{P}'$, therefore, W'' avoids $V[\mathcal{P}] \smallsetminus V[\mathcal{P}']$; in particular, $V(\mathring{y}Q) \cap V[\mathcal{P}] = \emptyset$. Thus W'' is also alternating with respect to $\mathcal{P}$, and it ends in B_2. (Note that y cannot be the last vertex of W'', since $y \in P\mathring{x}$ and hence $y \notin B$.) Furthermore, W'' starts in A_2, because W does. We may therefore use W'' with Lemma 3.3.2 to obtain the desired contradiction to the maximality of $\mathcal{P}$.   □

fan   A set of a–B paths is called an a–B *fan* if any two of the paths have only a in common.

[10.1.2]   **Corollary 3.3.3.** *For $B \subseteq V$ and $a \in V \smallsetminus B$, the minimum number of vertices $\neq a$ separating a from B in G is equal to the maximum number of paths forming an a–B fan in G.*

Proof. Apply Theorem 3.3.1 with $A := N(a)$.   □

Corollary 3.3.4. *Let a and b be two distinct vertices of G.*

(i) *If $ab \notin E$, then the minimum number of vertices $\neq a, b$ separating a from b in G is equal to the maximum number of independent a–b paths in G.*

(ii) *The minimum number of edges separating a from b in G is equal to the maximum number of edge-disjoint a–b paths in G.*

Proof. (i) Apply Theorem 3.3.1 with $A := N(a)$ and $B := N(b)$.

(ii) Apply Theorem 3.3.1 to the line graph of G, with $A := E(a)$ and $B := E(b)$.   □

[6.6.1]
[9.4.2]   **Theorem 3.3.5.** (Global Version of Menger's Theorem)

(i) *A graph is k-connected if and only if it contains k independent paths between any two vertices.*

(ii) *A graph is k-edge-connected if and only if it contains k edge-disjoint paths between any two vertices.*

Proof. (i) If a graph G contains k independent paths between any two vertices, then $|G| > k$ and G cannot be separated by fewer than k vertices; thus, G is k-connected.

Conversely, suppose that G is k-connected (and, in particular, has more than k vertices) but contains vertices a, b not linked by k independent paths. By Corollary 3.3.4 (i), a and b are adjacent; let $G' := G - ab$. Then G' contains at most $k - 2$ independent a–b paths. By Corollary 3.3.4 (i), we can separate a and b in G' by a set X of at most $k - 2$ vertices. As $|G| > k$, there is at least one further vertex $v \notin X \cup \{a, b\}$ in G. Now X separates v in G' from either a or b—say, from a. But

a, b

G'

X

v

then $X \cup \{b\}$ is a set of at most $k-1$ vertices separating v from a in G, contradicting the k-connectedness of G.

(ii) follows straight from Corollary 3.3.4 (ii). $\qquad\square$

3.4 Mader's theorem

In analogy to Menger's theorem we may consider the following question: given a graph G with an induced subgraph H, up to how many independent H-paths can we find in G?

In this section, we present without proof a deep theorem of Mader, which solves the above problem in a fashion similar to Menger's theorem. Again, the theorem says that an upper bound on the number of such paths that arises naturally from the size of certain separators is indeed attained by some suitable set of paths.

What could such an upper bound look like? Clearly, if $X \subseteq V(G-H)$ and $F \subseteq E(G-H)$ are such that every H-path in G has a vertex or an edge in $X \cup F$, then G cannot contain more than $|X \cup F|$ independent H-paths. Hence, the least cardinality of such a set $X \cup F$ is a natural upper bound for the maximum number of independent H-paths. (Note that every H-path meets $G - H$, because H is induced in G and edges of H do not count as H-paths.)

In contrast to Menger's theorem, this bound can still be improved. Clearly, we may assume that no edge in F has an end in X: otherwise this edge would not be needed in the separator. Let $Y := V(G-H) \smallsetminus X$, and denote by $\mathcal{C}_F$ the set of components of the graph (Y, F). Since every H-path avoiding X contains an edge from F, it has at least two vertices in ∂C for some $C \in \mathcal{C}_F$, where ∂C denotes the set of vertices in C with a neighbour in $G - X - C$ (Fig. 3.4.1). The number of independent

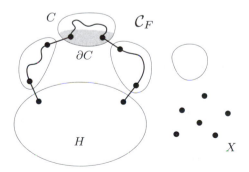

Fig. 3.4.1. An H-path in $G - X$

H-paths in G is therefore bounded above by

$M_G(H)$

$$M_G(H) := \min \left(|X| + \sum_{C \in \mathcal{C}_F} \lfloor \tfrac{1}{2} |\partial C| \rfloor \right),$$

X

where the minimum is taken over all X and F as described above: $X \subseteq V(G - H)$ and $F \subseteq E(G - H - X)$ such that every H-path in G has a vertex or an edge in $X \cup F$.

Now Mader's theorem says that this upper bound is always attained by some set of independent H-paths:

Theorem 3.4.1. (Mader 1978)
Given a graph G with an induced subgraph H, there are always $M_G(H)$ independent H-paths in G.

In order to obtain direct analogues to the vertex and edge version of Menger's theorem, let us consider the two special cases of the above problem where either F or X is required to be empty. Given an induced
$\kappa_G(H)$
subgraph $H \subseteq G$, we denote by $\kappa_G(H)$ the least cardinality of a vertex set $X \subseteq V(G - H)$ that meets every H-path in G. Similarly, we let
$\lambda_G(H)$
$\lambda_G(H)$ denote the least cardinality of an edge set $F \subseteq E(G)$ that meets every H-path in G.

Corollary 3.4.2. *Given a graph G with an induced subgraph H, there are at least $\tfrac{1}{2}\kappa_G(H)$ independent H-paths and at least $\tfrac{1}{2}\lambda_G(H)$ edge-disjoint H-paths in G.*

k

Proof. To prove the first assertion, let k be the maximum number of independent H-paths in G. By Theorem 3.4.1, there are sets $X \subseteq V(G - H)$ and $F \subseteq E(G - H - X)$ with

$$k = |X| + \sum_{C \in \mathcal{C}_F} \lfloor \tfrac{1}{2} |\partial C| \rfloor$$

such that every H-path in G has a vertex in X or an edge in F. For every $C \in \mathcal{C}_F$ with $\partial C \neq \emptyset$, pick a vertex $v \in \partial C$ and let $Y_C := \partial C \smallsetminus \{v\}$; if $\partial C = \emptyset$, let $Y_C := \emptyset$. Then $\lfloor \tfrac{1}{2} |\partial C| \rfloor \geqslant \tfrac{1}{2} |Y_C|$ for all $C \in \mathcal{C}_F$. Moreover,
Y
for $Y := \bigcup_{C \in \mathcal{C}_F} Y_C$ every H-path has a vertex in $X \cup Y$. Hence

$$k \geqslant |X| + \sum_{C \in \mathcal{C}_F} \tfrac{1}{2} |Y_C| \geqslant \tfrac{1}{2} |X \cup Y| \geqslant \tfrac{1}{2}\kappa_G(H)$$

as claimed.

The second assertion follows from the first by considering the line graph of G (Exercise 16). $\square$

It may come as a surprise to see that the bounds in Corollary 3.4.2 are best possible (as general bounds): one can find examples for G and H where G contains no more than $\frac{1}{2}\kappa_G(H)$ independent H-paths or no more than $\frac{1}{2}\lambda_G(H)$ edge-disjoint H-paths (Exercises 17 and 18).

3.5 Edge-disjoint spanning trees

The edge version of Menger's theorem tells us when a graph G contains k edge-disjoint paths between any two vertices. The actual routes of these paths within G may depend a lot on the choice of those two vertices: having found the paths for one pair of endvertices, we are not necessarily better placed to find them for another pair.

In a situation where quick access to a set of k edge-disjoint paths between any two vertices is desirable, it may be a good idea to ask for more than just k-edge-connectedness. For example, if G has k *edge-disjoint spanning trees*, there will be k canonical such paths between any two vertices, one in each tree.

When do such trees exist? If they do, the graph is clearly k-edge-connected. The converse is easily seen to be false; indeed, it is not even clear whether any edge-connectivity, however large, will imply the existence of k edge-disjoint spanning trees. Our first aim in this section will be to study conditions under which k edge-disjoint spanning trees exist.

As before, it is easy to write down some obvious necessary conditions for the existence of k edge-disjoint spanning trees. With respect to any partition of $V(G)$ into r sets, every spanning tree of G has at least $r-1$ *cross-edges*, edges whose ends lie in different partition sets (why?). Thus if G has k edge-disjoint spanning trees, it has at least $k\,(r-1)$ cross-edges.

cross-edges

Once more, this obvious necessary condition is also sufficient:

Theorem 3.5.1. (Tutte 1961; Nash-Williams 1961)
A multigraph contains k edge-disjoint spanning trees if and only if for every partition P of its vertex set it has at least $k\,(|P|-1)$ cross-edges.

Before we prove Theorem 3.5.1, let us note a surprising corollary: to ensure the existence of k edge-disjoint spanning trees, it suffices to raise the edge-connectivity to just $2k$:

Corollary 3.5.2. *Every $2k$-edge-connected multigraph G has k edge-disjoint spanning trees.*

[6.4.4]

Proof. Every set in a vertex partition of G is joined to other partition sets by at least $2k$ edges. Hence, for any partition into r sets, G has at least $\frac{1}{2} \sum_{i=1}^{r} 2k = kr$ cross-edges. The assertion thus follows from Theorem 3.5.1. $\qquad\qquad\square$

$G = (V, E)$

$k, \mathcal{F}$

$E[F], \|F\|$

For the proof of Theorem 3.5.1, let a multigraph $G = (V, E)$ and $k \in \mathbb{N}$ be given. Let $\mathcal{F}$ be the set of all k-tuples $F = (F_1, \ldots, F_k)$ of edge-disjoint spanning forests in G with the maximum total number of edges, i.e. such that $\|F\| := |E[F]|$ with $E[F] := E(F_1) \cup \ldots \cup E(F_k)$ is as large as possible.

edge
replacement

$x F_i y$

If $F = (F_1, \ldots, F_k) \in \mathcal{F}$ and $e \in E \smallsetminus E[F]$, then every $F_i + e$ contains a cycle ($i = 1, \ldots, k$): otherwise we could replace F_i by $F_i + e$ in F and obtain a contradiction to the maximality of $\|F\|$. Let us consider an edge $e' \neq e$ of this cycle, for some fixed i. Putting $F_i' := F_i + e - e'$, and $F_j' := F_j$ for all $j \neq i$, we see that $F' := (F_1', \ldots, F_k')$ is again in $\mathcal{F}$; we say that F' has been obtained from F by the *replacement* of the edge e' with e. Note that the component of F_i containing e' keeps its vertex set when it changes into a component of F_i'. Hence for every path $x \ldots y \subseteq F_i'$ there is a unique path $x F_i y$ in F_i; this will be used later.

F^0

$\mathcal{F}^0$

E^0

We now consider a fixed k-tuple $F^0 = (F_1^0, \ldots, F_k^0) \in \mathcal{F}$. The set of all k-tuples in $\mathcal{F}$ that can be obtained from F^0 by a series of edge replacements will be denoted by $\mathcal{F}^0$. Finally, we let

$$E^0 := \bigcup_{F \in \mathcal{F}^0} (E \smallsetminus E[F])$$

G^0

and $G^0 := (V, E^0)$.

Lemma 3.5.3. *For every $e^0 \in E \smallsetminus E[F^0]$ there exists a set $U \subseteq V$ that is connected in every F_i^0 ($i = 1, \ldots, k$) and contains the ends of e^0.*

C^0

U

i

Proof. As $F^0 \in \mathcal{F}^0$, we have $e^0 \in E^0$; let C^0 be the component of G^0 containing e^0. We shall prove the assertion for $U := V(C^0)$.

Let $i \in \{1, \ldots, k\}$ be given; we have to show that U is connected in F_i^0. To this end, we first prove the following:

> Let $F = (F_1, \ldots, F_k) \in \mathcal{F}^0$, and let $(F_1', \ldots, F_k')$ have been obtained from F by the replacement of an edge of F_i. If $\quad\quad$ (1)
> x, y are the ends of a path in $F_i' \cap C^0$, then also $x F_i y \subseteq C^0$.

Let $e = vw$ be the new edge in $E(F_i') \smallsetminus E[F]$; this is the only edge of F_i' not lying in F_i. We assume that $e \in x F_i' y$: otherwise we would have $x F_i y = x F_i' y$ and nothing to show. It suffices to show that $v F_i w \subseteq C^0$: then $(x F_i' y - e) \cup v F_i w$ is a connected subgraph of $F_i \cap C^0$ that contains x, y, and hence also $x F_i y$. Let e' be any edge of $v F_i w$. Since we could replace e' in $F \in \mathcal{F}^0$ by e and obtain an element of $\mathcal{F}^0$ not containing e', we have $e' \in E^0$. Thus $v F_i w \subseteq G^0$, and hence $v F_i w \subseteq C^0$ since $v, w \in x F_i' y \subseteq C^0$. This proves (1).

In order to prove that $U = V(C^0)$ is connected in F_i^0, we show that $x F_i^0 y \subseteq C^0$ for every edge $xy \in C^0$. As C^0 is connected, the union of all these paths will then be a connected spanning subgraph of $F_i^0[U]$.

So let $e = xy \in C^0$ be given. As $e \in E^0$, there exist an $s \in \mathbb{N}$ and k-tuples $F^r = (F_1^r, \ldots, F_k^r)$ for $r = 1, \ldots, s$ such that each F^r is obtained from F^{r-1} by edge replacement and $e \in E \setminus E[F^s]$. Setting $F := F^s$ in (1), we may think of e as a path of length 1 in $F_i' \cap C^0$. Successive applications of (1) to $F = F^s, \ldots, F^0$ then give $x F_i^0 y \subseteq C^0$ as desired. $\qquad\square$

Proof of Theorem 3.5.1. We prove the backward implication by induction on $|G|$. For $|G| = 2$ the assertion holds. For the induction step, we now suppose that for every partition P of V there are at least $k(|P| - 1)$ cross-edges, and construct k edge-disjoint spanning trees in G. $\hfill (1.5.3)$

Pick a k-tuple $F^0 = (F_1^0, \ldots, F_k^0) \in \mathcal{F}$. If every F_i^0 is a tree, we are done. If not, we have $\hfill F^0$

$$\|F^0\| = \sum_{i=1}^{k} \|F_i^0\| < k(|G| - 1)$$

by Corollary 1.5.3. On the other hand, we have $\|G\| \geqslant k(|G| - 1)$ by assumption: consider the partition of V into single vertices. So there exists an edge $e^0 \in E \setminus E[F^0]$. By Lemma 3.5.3, there exists a set $U \subseteq V$ that is connected in every F_i^0 and contains the ends of e_0; in particular, $|U| \geqslant 2$. Since every partition of the contracted multigraph G/U induces a partition of G with the same cross-edges,[4] G/U has at least $k(|P| - 1)$ cross-edges with respect to any partition P. By the induction hypothesis, therefore, G/U has k edge-disjoint spanning trees $T_1, \ldots, T_k$. Replacing in each T_i the vertex v_U contracted from U by the spanning tree $F_i^0 \cap G[U]$ of $G[U]$, we obtain k edge-disjoint spanning trees in G. $\qquad\square$ $\hfill e^0$ $\hfill U$

Let us say that subgraphs $G_1, \ldots, G_k$ of a graph G *partition* G if their edge sets form a partition of $E(G)$. Our spanning tree problem may then be recast as follows: into how *many* connected spanning subgraphs can we partition a given graph? The excuse for rephrasing our simple tree problem in this more complicated way is that it now has an obvious dual (cf. Theorem 1.5.1): into how *few* acyclic (spanning) subgraphs can we partition a given graph? Or for given k: which graphs can be partitioned into at most k forests? $\hfill$ *graph* $\hfill$ *partition*

An obvious necessary condition now is that every set $U \subseteq V(G)$ induces at most $k(|U| - 1)$ edges, no more than $|U| - 1$ for each forest.

---

[4] see Chapter 1.10 on the contraction of multigraphs

Once more, this condition turns out to be sufficient too. And surprisingly, this can be shown with the help of Lemma 3.5.3, which was designed for the proof of our theorem on edge-disjoint spanning trees:

Theorem 3.5.4. (Nash-Williams 1964)
A multigraph $G = (V, E)$ can be partitioned into at most k forests if and only if $\|G[U]\| \leqslant k(|U| - 1)$ for every non-empty set $U \subseteq V$.

(1.5.3) *Proof.* The forward implication was shown above. Conversely, we show that every k-tuple $F = (F_1, \ldots, F_k) \in \mathcal{F}$ partitions G, i.e. that $E[F] = E$. If not, let $e \in E \smallsetminus E[F]$. By Lemma 3.5.3, there exists a set $U \subseteq V$ that is connected in every F_i and contains the ends of e. Then $G[U]$ contains $|U| - 1$ edges from each F_i, and in addition the edge e. Thus $\|G[U]\| > k(|U| - 1)$, contrary to our assumption. $\qquad\square$

arboricity
The least number of forests forming a partition of a graph G is called the *arboricity* of G. By Theorem 3.5.4, the arboricity is a measure for the maximum local density: a graph has small arboricity if and only it if is 'nowhere dense', i.e. if and only if it has no subgraph H with $\varepsilon(H)$ large.

3.6 Paths between given pairs of vertices

k-linked
A graph with at least $2k$ vertices is said to be k-*linked* if for every $2k$ distinct vertices $s_1, \ldots, s_k, t_1, \ldots, t_k$ it contains k disjoint paths $P_1, \ldots, P_k$ with $P_i = s_i \ldots t_i$ for all i. Thus unlike in Menger's theorem, we are not merely asking for k disjoint paths between two *sets* of vertices: we insist that each of these paths shall link a specified pair of endvertices.

Clearly, every k-linked graph is k-connected. The converse, however, is far from true: being k-linked is generally a much stronger property than k-connectedness. But still, the two properties are related: our aim in this section is to prove the existence of a function $f: \mathbb{N} \to \mathbb{N}$ such that every $f(k)$-connected graph is k-linked.

As a lemma, we need a result that would otherwise belong in Chapter 8:

Theorem 3.6.1. (Mader 1967)
There is a function $h: \mathbb{N} \to \mathbb{N}$ such that every graph with average degree at least $h(r)$ contains K^r as a topological minor, for every $r \in \mathbb{N}$.

r *Proof.* For $r \leqslant 2$, the assertion holds with $h(r) = 1$; we now assume that $r \geqslant 3$. We show by induction on $m = r, \ldots, \binom{r}{2}$ that every connected graph G with average degree $d(G) \geqslant 2^m$ has a topological minor X with r vertices and m edges; for $m = \binom{r}{2}$ this implies the assertion with $h(r) = 2^{\binom{r}{2}}$.

If $m = r$ then, by Propositions 1.2.2 and 1.3.1, G contains a cycle of length at least $\varepsilon(G) + 1 \geqslant 2^{r-1} + 1 \geqslant r + 1$, and the assertion follows with $X = C^r$.

Now let $r < m \leqslant \binom{r}{2}$, and assume the assertion holds for smaller m. Let a connected graph G with $d(G) \geqslant 2^m$ be given; thus, $\varepsilon(G) \geqslant 2^{m-1}$. Consider a maximal set $U \subseteq V(G)$ such that U is connected in G and $\varepsilon(G/U) \geqslant 2^{m-1}$; such a set U exists, because G itself has the form G/U with $|U| = 1$.

$\qquad$ U

Let $H := G[N(U)]$. If H has a vertex v of degree $d_H(v) < 2^{m-1}$, we may add it to U and obtain a contradiction to the maximality of U: when we contract the edge vv_U in G/U, we lose one vertex and $d_H(v) + 1 \leqslant 2^{m-1}$ edges, so ε will still be at least 2^{m-1}. Therefore $d(H) \geqslant \delta(H) \geqslant 2^{m-1}$. By the induction hypothesis, H contains a TY with $|Y| = r$ and $\|Y\| = m - 1$. Let x, y be two branch vertices of this TY that are non-adjacent in Y. Since x and y lie in $N(U)$ and U is connected in G, G contains an x–y path whose inner vertices lie in U. Adding this path to the TY, we obtain the desired TX. $\qquad\square$

$\qquad$ H

How can Theorem 3.6.1 help with our aim to show that high connectivity will make a graph k-linked? Since high connectivity forces the average degree up (even the minimum degree), we may assume by the theorem that our graph contains a subdivision K of a large complete graph. Our plan now is to use Menger's theorem to link the given vertices s_i and t_i disjointly to branch vertices of K, and then to join up the correct pairs of those branch vertices inside K.

Theorem 3.6.2. (Jung 1970; Larman & Mani 1970)
There is a function $f: \mathbb{N} \to \mathbb{N}$ such that every $f(k)$-connected graph is k-linked, for all $k \in \mathbb{N}$.

Proof. We prove the assertion for $f(k) = h(3k) + 2k$, where h is a function as in Theorem 3.6.1. Let G be an $f(k)$-connected graph. Then $d(G) \geqslant \delta(G) \geqslant \kappa(G) \geqslant h(3k)$; choose $K = TK^{3k} \subseteq G$ as in Theorem 3.6.1, and let U denote its set of branch vertices.

$\qquad$ (3.3.1)

$\qquad$ G

$\qquad$ K

$\qquad$ U

For the proof that G is k-linked, let distinct vertices $s_1, \ldots, s_k$ and $t_1, \ldots, t_k$ of G be given. By definition of $f(k)$, we have $\kappa(G) \geqslant 2k$. Hence by Menger's theorem (3.3.1), G contains disjoint paths $P_1, \ldots, P_k$, $Q_1, \ldots, Q_k$, such that each P_i starts in s_i, each Q_i starts in t_i, and all these paths end in U but have no inner vertices in U. Let the set $\mathcal{P}$ of these paths be chosen so that their total number of edges outside $E(K)$ is minimum.

$\qquad$ s_i, t_i

$\qquad$ P_i, Q_i

$\qquad$ $\mathcal{P}$

Let $u_1, \ldots, u_k$ be those k vertices in U that are not an end of a path in $\mathcal{P}$. For each $i = 1, \ldots, k$, let L_i be the U-path in K (i.e., the subdivided edge of the K^{3k}) from u_i to the end of P_i in U, and let v_i be the first vertex of L_i on P_i. Likewise, let M_i be the U-path in K from u_i to the end of Q_i in U, and let w_i be the first vertex of M_i on Q_i

$\qquad$ u_i

$\qquad$ L_i, v_i

$\qquad$ M_i, w_i

(Fig. 3.6.1). We claim that the paths $s_i P_i v_i L_i u_i M_i w_i Q_i t_i$ are disjoint for different i, showing that G is k-linked.

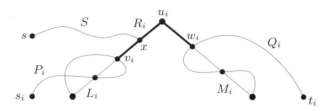

Fig. 3.6.1. Constructing an s_i–t_i path via u_i

R_i

x, S

The only way in which the above paths might fail to be disjoint is that, for some i, an inner vertex x of $R_i := v_i L_i u_i M_i w_i$ lies on one of the paths $S \in \mathcal{P}$; let us choose x and S so that the distance $d_K(x, u_i)$ is minimum. We consider the case that $x \in L_i$; the other case, that $x \in M_i$, is analogous. By the choice of x, no path in $\mathcal{P}$ meets $\mathring{x} L_i u_i$. Thus when S was chosen as a path for $\mathcal{P}$, the path $S x L_i u_i$ was also a candidate. The fact that S was chosen instead implies that xS has all its edges in $E(K)$. But then $S = S x L_i$, and hence $S = P_i$ by definition of L_i. Thus, $x \in P_i$. Since x is an inner vertex of R_i and hence of $L_i v_i$, this contradicts the choice of v_i. $\square$

In our proof of Theorem 3.6.2 we did not try to find any particularly good bound on the connectivity needed to force a graph to be k-linked; the function f we used grows exponentially in k. Not surprisingly, this is far from being best possible. It is still remarkable, though, that f can in fact be chosen linear: as Bollobás & Thomason (1996) have shown, every $22k$-connected graph is k-linked.

Exercises

For the first four exercises, let G be a graph and $a, b \in V(G)$. Suppose that $X \subseteq V(G) \setminus \{a, b\}$ separates a from b in G. We say that X separates a from b *minimally* if no proper subset of X separates a from b in G.

1.⁻ Show that X separates a from b minimally if and only if every vertex in X has a neighbour in the component C_a of $G - X$ containing a, and another in the component C_b of $G - X$ containing b.

2. Let $X' \subseteq V(G) \setminus \{a, b\}$ be another set separating a from b, and define C'_a and C'_b correspondingly. Show that both

$$Y_a := (X \cap C'_a) \cup (X \cap X') \cup (X' \cap C_a)$$

and

$$Y_b := (X \cap C'_b) \cup (X \cap X') \cup (X' \cap C_b)$$

separate a from b (see figure).

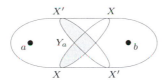

3. Do Y_a and Y_b separate a from b minimally if X and X' do? Are $|Y_a|$ and $|Y_b|$ minimum for vertex sets separating a from b if $|X|$ and $|X'|$ are?

4.+ Suppose that X and X' separate a from b minimally, and that X meets at least two components of $G - X'$. Show that X' meets all the components of $G - X$, and that X meets all the components of $G - X'$.

5.− Prove the elementary properties of blocks mentioned at the beginning of Section 3.1.

6. Show that the block graph of any connected graph is a tree.

7. Show, without using Menger's theorem, that any two vertices of a 2-connected graph lie on a common cycle.

8. For edges $e, e' \in G$ write $e \sim e'$ if either $e = e'$ or e and e' lie on some common cycle in G. Show that $\sim$ is an equivalence relation on $E(G)$ whose equivalence classes are the edge sets of the non-trivial blocks of G.

9. Let G be a 2-connected graph but not a triangle, and let e be an edge of G. Show that either $G - e$ or G/e is again 2-connected.

10. Let G be a 3-connected graph, and let xy be an edge of G. Show that G/xy is 3-connected if and only if $G - \{x, y\}$ is 2-connected.

11. (i) Show that every cubic 3-edge-connected graph is 3-connected.

    (ii) Show that a graph is cubic and 3-connected if and only if it can be constructed from a K^4 by successive applications of the following operation: subdivide two edges by inserting a new vertex on each of them, and join the two new subdividing vertices by an edge.

12.− Show that Menger's theorem is equivalent to the following statement. For every graph G and vertex sets $A, B \subseteq V(G)$, there exist a set $\mathcal{P}$ of disjoint A–B paths in G and a set $X \subseteq V(G)$ separating A from B in G such that X has the form $X = \{ x_P \mid P \in \mathcal{P} \}$ with $x_P \in P$ for all $P \in \mathcal{P}$.

13. Work out the details of the proof of Corollary 3.3.4 (ii).

14. Let $k \geqslant 2$. Show that every k-connected graph of order at least $2k$ contains a cycle of length at least $2k$.

15. Let $k \geqslant 2$. Show that in a k-connected graph any k vertices lie on a common cycle.

16. Derive the edge part of Corollary 3.4.2 from the vertex part.

     (Hint. Consider the H-paths in the graph obtained from the disjoint
     union of H and the line graph $L(G)$ by adding all the edges he such
     that h is a vertex of H and $e \in E(G) \smallsetminus E(H)$ is an edge at h.)

17.⁻ To the disjoint union of the graph $H = \overline{K^{2m+1}}$ with k copies of K^{2m+1}
     add edges joining H bijectively to each of the K^{2m+1}. Show that the
     resulting graph G contains at most $km = \frac{1}{2}\kappa_G(H)$ independent H-
     paths.

18. Find a bipartite graph G, with partition classes A and B say, such that
     for $H := G[A]$ there are at most $\frac{1}{2}\lambda_G(H)$ edge-disjoint H-paths in G.

19.⁺ Derive Tutte's 1-factor theorem (2.2.1) from Mader's theorem.

     (Hint. Extend the given graph G to a graph G' by adding, for each
     vertex $v \in G$, a new vertex v' and joining v' to v. Choose $H \subseteq G'$ so that
     the 1-factors in G correspond to the large enough sets of independent
     H-paths in G'.)

Notes

Although connectivity theorems are doubtless among the most natural, and
also the most applicable, results in graph theory, there is still no comprehensive
monograph on this subject. Some areas are covered in B. Bollobás, *Extremal
Graph Theory*, Academic Press 1978, in R. Halin, *Graphentheorie*, Wissenschaftliche Buchgesellschaft 1980, and in A. Frank's chapter of the *Handbook of
Combinatorics* (R.L. Graham, M. Grötschel & L. Lovász, eds.), North-Holland
1995. A survey specifically of techniques and results on minimally k-connected
graphs (see below) is given by W. Mader, On vertices of degree n in minimally
n-connected graphs and digraphs, in (D. Miklos, V.T. Sós & T. Szőnyi, eds.)
Paul Erdős is 80, Vol. 2, Proc. Colloq. Math. Soc. János Bolyai, Budapest 1996.

     Our proof of Tutte's Theorem 3.2.3 is due to C. Thomassen, Planarity and
duality of finite and infinite graphs, *J. Combin. Theory B* **29** (1980), 244–271.
This paper also contains Lemma 3.2.1 and its short proof from first principles.
(The lemma's assertion, of course, follows from Tutte's wheel theorem—its
significance lies in its independent proof, which has shortened the proofs of
both of Tutte's theorems considerably.)

     An approach to the study of connectivity not touched upon in this chap-
ter is the investigation of *minimal* k-connected graphs, those that lose their
k-connectedness as soon as we delete an edge. Like all k-connected graphs,
these have minimum degree at least k, and by a fundamental result of Halin
(1969), their minimum degree is exactly k. The existence of a vertex of small
degree can be particularly useful in induction proofs about k-connected graphs.
Halin's theorem was the starting point for a series of more and more sophisticated studies of minimal k-connected graphs; see the books of Bollobás and
Halin cited above, and in particular Mader's survey.

     Our first proof of Menger's theorem is due to J.S. Pym, A proof of Men-
ger's theorem, *Monatshefte Math.* **73** (1969), 81–88; the second to T. Grünwald
(later Gallai), Ein neuer Beweis eines Mengerschen Satzes, *J. London Math.*

Soc. **13** (1938), 188–192. The global version of Menger's theorem (Theorem 3.3.5) was first stated and proved by Whitney (1932).

Mader's Theorem 3.4.1 is taken from W. Mader, Über die Maximalzahl kreuzungsfreier H-Wege, *Arch. Math.* **31** (1978), 387–402. The theorem may be viewed as a common generalization of Menger's theorem and Tutte's 1-factor theorem (Exercise 19). Theorem 3.5.1 was proved independently by Nash-Williams and by Tutte; both papers are contained in *J. London Math. Soc.* **36** (1961). Theorem 3.5.4 is due to C. St. J. A. Nash-Williams, Decompositions of finite graphs into forests, *J. London Math. Soc.* **39** (1964), 12. Our proofs follow an account by Mader (personal communication). Both results can be elegantly expressed and proved in the setting of matroids; see § 18 in B. Bollobás, *Combinatorics*, Cambridge University Press 1986.

In Chapter 8.1 we shall prove that, in order to force a topological K^r minor in a graph G, we do not need an average degree of G as high as $h(r) = 2^{\binom{r}{2}}$ (as used in our proof of Theorem 3.6.1): the average degree required can be bounded above by a function quadratic in r (Theorem 8.1.1). The improvement of Theorem 3.6.2 mentioned in the text is due to B. Bollobás & A. G. Thomason, Highly linked graphs, *Combinatorica* **16** (1996), 313–320. N. Robertson & P. D. Seymour, Graph Minors XIII: The disjoint paths problem, *J. Combin. Theory B* **63** (1995), 65-110, showed that, for every fixed k, there is an $O(n^3)$ algorithm that decides whether a given graph of order n is k-linked. If k is taken as part of the input, the problem becomes NP-hard.

4   Planar Graphs

When we draw a graph on a piece of paper, we naturally try to do this as transparently as possible. One obvious way to limit the mess created by all the lines is to avoid intersections. For example, we may ask if we can draw the graph in such a way that no two edges meet in a point other than a common end.

Graphs drawn in this way are called *plane graphs*; abstract graphs that *can* be drawn in this way are called *planar*. In this chapter we study both plane and planar graphs—as well as the relationship between the two: the question of how an abstract graph might be drawn in fundamentally different ways. After collecting together in Section 4.1 the few basic topological facts that will enable us later to prove all results rigorously without too much technical ado, we begin in Section 4.2 by studying the structural properties of plane graphs. In Section 4.3, we investigate how two drawings of the same graph can differ. The main result of that section is that 3-connected planar graphs have essentially only one drawing, in some very strong and natural topological sense. The next two sections are devoted to the proofs of all the classical planarity criteria, conditions telling us when an abstract graph is planar. We complete the chapter with a section on *plane duality*, a notion with fascinating links to algebraic, colouring, and flow properties of graphs (Chapters 1.9 and 6.5).

The traditional notion of a graph drawing is that its vertices are represented by points in the Euclidean plane, its edges are represented by curves between these points, and different curves meet only in common endpoints. To avoid unnecessary topological complication, however, we shall only consider curves that are piecewise linear; it is not difficult to show that any drawing can be straightened out in this way, so the two notions come to the same thing.

4.1 Topological prerequisites

In this section we briefly review some basic topological definitions and facts needed later. All these facts have (by now) easy and well-known proofs; see the notes for sources. Since those proofs contain no graph theory, we do not repeat them here: indeed our aim is to collect precisely those topological facts that we need but do *not* want to prove. Later, all proofs will follow strictly from the definitions and facts stated here (and be guided by but not rely on geometric intuition), so the material presented now will help to keep elementary topological arguments in those proofs to a minimum.

A *straight line segment* in the Euclidean plane is a subset of $\mathbb{R}^2$ that has the form $\{\, p + \lambda(q - p) \mid 0 \leqslant \lambda \leqslant 1 \,\}$ for distinct points $p, q \in \mathbb{R}^2$.

polygon   A *polygon* is a subset of $\mathbb{R}^2$ which is the union of finitely many straight line segments and is homeomorphic to the unit circle. Here, as later, any subset of a topological space is assumed to carry the subspace topology. A *polygonal arc* is a subset of $\mathbb{R}^2$ which is the union of finitely many straight line segments and is homeomorphic to the closed unit interval $[0, 1]$. The images of 0 and of 1 under such a homeomorphism are the *endpoints* of this polygonal arc, which *links* them and runs *between* them.

arc   Instead of 'polygonal arc' we shall simply say *arc*. If P is an arc between x and y, we denote the point set $P \smallsetminus \{\, x, y \,\}$, the *interior* of P, by $\mathring{P}$.

Let $O \subseteq \mathbb{R}^2$ be an open set. Being linked by an arc in O defines an equivalence relation on O. The corresponding equivalence classes are

region   again open; they are the *regions* of O. A closed set $X \subseteq \mathbb{R}^2$ is said to
separate   *separate* O if $O \smallsetminus X$ has more than one region. The *frontier* of a set
frontier $X \subseteq \mathbb{R}^2$ is the set Y of all points $y \in \mathbb{R}^2$ such that every neighbourhood of y meets both X and $\mathbb{R}^2 \smallsetminus X$.

The frontier of a region O of $\mathbb{R}^2 \smallsetminus X$, where X is a finite union of points and arcs, has two important properties. The first is accessibility: if $x \in X$ lies on the frontier of O, then x can be linked to some point in O by a straight line segment whose interior lies wholly inside O. As a consequence, any two points on the frontier of O can be linked by an arc whose interior lies in O (why?). The second notable property of the frontier of O is that it separates O from the rest of $\mathbb{R}^2$. Indeed, if $\varphi \colon [0, 1] \to P \subseteq \mathbb{R}^2$ is continuous, with $\varphi(0) \in O$ and $\varphi(1) \notin O$, then P meets the frontier of O at least in the point $\varphi(y)$ for $y := \inf \{\, x \mid \varphi(x) \notin O \,\}$, the *first point*

[4.2.1]
[4.2.4]
[4.2.5]
[4.2.10]
[4.3.1]
[4.5.1]
[4.6.1]
[5.1.2]

of P in $\mathbb{R}^2 \smallsetminus O$.

Theorem 4.1.1. (Jordan Curve Theorem for Polygons)
For every polygon $P \subseteq \mathbb{R}^2$, the set $\mathbb{R}^2 \smallsetminus P$ has exactly two regions, of which exactly one is bounded. Each of the two regions has the entire polygon P as its frontier.

With the help of Theorem 4.1.1, it is not difficult to prove the following lemma.

[4.2.5]
[4.2.6]
[4.2.10]

Lemma 4.1.2. *Let P_1, P_2, P_3 be three arcs, between the same two end-point but otherwise disjoint.*

(i) $\mathbb{R}^2 \smallsetminus (P_1 \cup P_2 \cup P_3)$ *has exactly three regions, with frontiers $P_1 \cup P_2$, $P_2 \cup P_3$ and $P_1 \cup P_3$.*

(ii) *If P is an arc between a point in $\mathring{P}_1$ and a point in $\mathring{P}_3$ whose interior lies in the region of $\mathbb{R}^2 \smallsetminus (P_1 \cup P_3)$ that contains $\mathring{P}_2$, then $\mathring{P} \cap \mathring{P}_2 \neq \emptyset$.*

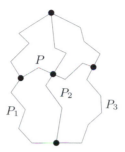

Fig. 4.1.1. The arcs in Lemma 4.1.2 (ii)

Our next lemma complements the Jordan curve theorem by saying that an arc does *not* separate the plane. For easier application later, we phrase this a little more generally:

[4.2.1]
[4.2.3]

Lemma 4.1.3. *Let $X_1, X_2 \subseteq \mathbb{R}^2$ be disjoint sets, each the union of finitely many points and arcs, and let P be an arc between a point in X_1 and one in X_2 whose interior $\mathring{P}$ lies in a region O of $\mathbb{R}^2 \smallsetminus (X_1 \cup X_2)$. Then $O \smallsetminus \mathring{P}$ is a region of $\mathbb{R}^2 \smallsetminus (X_1 \cup P \cup X_2)$.*

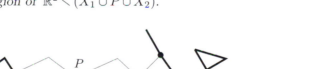

Fig. 4.1.2. P does not separate the region O of $\mathbb{R}^2 \smallsetminus (X_1 \cup X_2)$

It remains to introduce a few terms and facts that will be used only once, when we consider notions of equivalence for graph drawings in Chapter 4.3.

As usual, we denote by S^n the n-dimensional sphere, the set of points in $\mathbb{R}^{n+1}$ at distance 1 from the origin. The 2-sphere minus its 'north pole' $(0,0,1)$ is homeomorphic to the plane; let us choose a fixed such homeomorphism $\pi \colon S^2 \smallsetminus \{ (0,0,1) \} \to \mathbb{R}^2$ (for example, stereograph-ic projection). If $P \subseteq \mathbb{R}^2$ is a polygon and O is the bounded region of

S^n

π

$\mathbb{R}^2 \smallsetminus P$, let us call $C := \pi^{-1}(P)$ a *circle on* S^2, and the sets $\pi^{-1}(O)$ and $S^2 \smallsetminus \pi^{-1}(P \cup O)$ the *regions* of C.

Our last tool is the theorem of Jordan and Schoenflies, again adapted slightly for our purposes:

[4.3.1]

Theorem 4.1.4. *Let* $\varphi \colon C_1 \to C_2$ *be a homeomorphism between two circles on* S^2, *let* O_1 *be a region of* C_1, *and let* O_2 *be a region of* C_2. *Then* φ *can be extended to a homeomorphism* $C_1 \cup O_1 \to C_2 \cup O_2$.

4.2 Plane graphs

plane graph

A *plane graph* is a pair (V, E) of finite sets with the following properties (the elements of V are again called *vertices*, those of E *edges*):

(i) $V \subseteq \mathbb{R}^2$;

(ii) every edge is an arc between two vertices;

(iii) different edges have different sets of endpoints;

(iv) the interior of an edge contains no vertex and no point of any other edge.

A plane graph (V, E) defines a graph G on V in a natural way. As long as no confusion can arise, we shall use the name G of this abstract graph also for the plane graph (V, E), or for the point set $V \cup \bigcup E$; similar notational conventions will be used for abstract versus plane edges, for subgraphs, and so on.[1]

For every plane graph G, the set $\mathbb{R}^2 \smallsetminus G$ is open; its regions are the *faces* of G. Since G is bounded—i.e., lies inside some sufficiently large disc D—exactly one of its faces is unbounded: the face[2] that contains $\mathbb{R}^2 \smallsetminus D$. This face is the *outer face* of G; the other faces are its *inner faces*. We denote the set of faces of G by $F(G)$.

faces

F(G)

In order to lay the foundations for the (easy but) rigorous introduction to plane graphs that this section aims to provide, let us descend once now into the realm of truly elementary topology of the plane, and prove what seems entirely obvious:[3] that the frontier of a face of a plane graph G is always a subgraph of G—not, say, half an edge. The following lemma states this formally, together with two similarly 'obvious' properties of plane graphs:

---

[1] However, we shall continue to use $\smallsetminus$ for differences of point sets and $-$ for graph differences—which may help a little to keep the two apart.

[2] Why is there such a face? Why is it unique?

[3] Note that even the best intuition can only ever be 'accurate', i.e., coincide with what the technical definitions imply, inasmuch as those definitions do indeed formalize what is intuitively intended. Given the complexity of definitions in elementary topology, this can hardly be taken for granted.

Lemma 4.2.1. *Let G be a plane graph and e an edge of G.*

[4.5.1]
[4.5.2]

  (i) *If X is the frontier of a face of G, then either $e \subseteq X$ or $X \cap \mathring{e} = \emptyset$.*

  (ii) *If e lies on a cycle $C \subseteq G$, then e lies on the frontier of exactly two faces of G, and these are contained in distinct faces of C.*

 (iii) *If e lies on no cycle, then e lies on the frontier of exactly one face of G.*

Proof. We prove all three assertions together. Let us start by considering one point $x_0 \in \mathring{e}$. We show that x_0 lies on the frontier of either exactly two faces or exactly one, according as e lies on a cycle in G or not. We then show that every other point in $\mathring{e}$ lies on the frontier of exactly the same faces as x_0. Then the endpoints of e will also lie on the frontier of these faces—simply because every neighbourhood of an endpoint of e is also the neighbourhood of an inner point of e.

(4.1.1)
(4.1.3)

G is the union of finitely many straight line segments; we may assume that any two of these intersect in at most one point. Around every point $x \in \mathring{e}$ we can find an open disc D_x, with centre x, which meets only those (one or two) straight line segments that contain x.

D_x

Let us pick an inner point x_0 from a straight line segment $S \subseteq e$. Then $D_{x_0} \cap G = D_{x_0} \cap S$, so $D_{x_0} \setminus G$ is the union of two open half-discs. Since these half-discs do not meet G, they each lie in a face of G. Let us denote these faces by f_1 and f_2; they are the only faces of G with x_0 on their frontier, and they may coincide (Fig. 4.2.1).

x_0, S

f_1, f_2

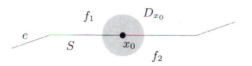

Fig. 4.2.1. Faces f_1, f_2 of G in the proof of Lemma 4.2.1

If e lies on a cycle $C \subseteq G$, then D_{x_0} meets both faces of C (Theorem 4.1.1). The faces f_1, f_2 of G are therefore contained in distinct faces of C—since $C \subseteq G$, every face of G is a subset of a face of C—and in particular $f_1 \neq f_2$. If e does not lie on any cycle, then e is a bridge and thus links two disjoint point sets X_1, X_2 with $X_1 \cup X_2 = G \setminus \mathring{e}$. Clearly, $f_1 \cup \mathring{e} \cup f_2$ is the subset of a face f of $G - e$. (Why?) By Lemma 4.1.3, $f \setminus \mathring{e}$ is a face of G. But $f \setminus \mathring{e}$ contains f_1 and f_2 by definition of f, so $f_1 = f \setminus \mathring{e} = f_2$ since f_1, f_2 and f are all faces of G.

Now consider any other point $x_1 \in \mathring{e}$. Let P be the arc from x_0 to x_1 contained in e. Since P is compact, finitely many of the discs D_x with $x \in P$ cover P. Let us enumerate these discs as $D_0, \ldots, D_n$ in the natural order of their centres along P; adding D_{x_0} or D_{x_1} as necessary, we may assume that $D_0 = D_{x_0}$ and $D_n = D_{x_1}$. By induction on n, one easily proves that every point $y \in D_n \setminus e$ can be linked by an arc inside

x_1

P

$D_0, \ldots, D_n$

y

z

$(D_0 \cup \ldots \cup D_n) \smallsetminus e$ to a point $z \in D_0 \smallsetminus e$ (Fig. 4.2.2); then y and z are
equivalent in $\mathbb{R}^2 \smallsetminus G$. Hence, every point of $D_n \smallsetminus e$ lies in f_1 or in f_2, so
x_1 cannot lie on the frontier of any other face of G. Since both half-discs
of $D_0 \smallsetminus e$ can be linked to $D_n \smallsetminus e$ in this way (swap the roles of D_0
and D_n), we find that x_1 lies on the frontier of both f_1 and f_2. $\quad\square$

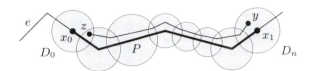

Fig. 4.2.2. An arc from y to D_0, close to P

Corollary 4.2.2. *The frontier of a face is always the point set of a
subgraph.* $\quad\square$

boundary
$G[f]$

The subgraph of G whose point set is the frontier of a face f is said to
bound f and is called its *boundary*; we denote it by $G[f]$. A face is said
to be *incident* with the vertices and edges of its boundary. Note that
every face of a plane graph is also a face of its own boundary—a fact
used frequently in the proofs to come.

[4.6.1]

Proposition 4.2.3. *A plane forest has exactly one face.*

(4.1.3)

Proof. Apply induction on the number of edges and use Lemma 4.1.3. $\quad\square$

With just one exception, different faces of a plane graph have different boundaries:

[4.3.1]

Lemma 4.2.4. *If a plane graph has different faces with the same boundary, then the graph is a cycle.*

(4.1.1)

Proof. Let G be a plane graph, and let $H \subseteq G$ be the boundary of
distinct faces f_1, f_2 of G. Since f_1 and f_2 are also faces of H, Proposition
4.2.3 implies that H contains a cycle C. By Lemma 4.2.1 (ii), f_1 and f_2
are contained in different faces of C. Since f_1 and f_2 both have all of H
as boundary, this implies that $H = C$: any further vertex or edge of H
would lie in one of the faces of C and hence not on the boundary of the
other. Thus, f_1 and f_2 are distinct faces of C. As C has only two faces,
it follows that $f_1 \cup C \cup f_2 = \mathbb{R}^2$ and hence $G = C$. $\quad\square$

[4.3.1]
[4.4.3]
[4.5.1]
[4.5.2]

Proposition 4.2.5. *In a 2-connected plane graph, every face is bounded
by a cycle.*

Proof. Let f be a face in a 2-connected plane graph G. We show by induction on $|G|$ that $G[f]$ is a cycle. If G is itself a cycle, this holds by Theorem 4.1.1; we therefore assume that G is not a cycle.

By Proposition 3.1.2, there exist a 2-connected plane graph $H \subseteq G$ and a plane H-path P such that $G = H \cup P$. The interior of P lies in a face f' of H, which by the induction hypothesis is bounded by a cycle C.

If f is also a face of H, we are home by the induction hypothesis. If not, then the frontier of f meets $P \smallsetminus H$, so $f \subseteq f'$. Therefore f is a face of $C \cup P$, and is hence bounded by a cycle (Lemma 4.1.2 (i)).   □

(3.1.2)
(4.1.1)
(4.1.2)

H
P
f', C

A plane graph G is called *maximally plane*, or just *maximal*, if we cannot add a new edge to form a plane graph $G' \supsetneq G$ with $V(G') = V(G)$. We call G a *plane triangulation* if every face of G (including the outer face) is bounded by a triangle.

*maximal
plane graph*

*plane
triangulation*

Proposition 4.2.6. *A plane graph of order at least 3 is maximally plane if and only if it is a plane triangulation.*

[4.4.1]
[5.4.2]

Proof. Let G be a plane graph of order at least 3. It is easy to see that if every face of G is bounded by a triangle, then G is maximally plane. Indeed, any additional edge e would have its interior inside a face of G and its ends on the boundary of that face. Hence these ends are already adjacent in G, so $G \cup e$ cannot satisfy condition (iii) in the definition of a plane graph.

(4.1.2)

Conversely, assume that G is maximally plane and let $f \in F(G)$ be a face; let us write $H := G[f]$. Since G is maximal as a plane graph, $G[H]$ is complete: any two vertices of H that are not already adjacent in G could be linked by an arc through f, extending G to a larger plane graph. Thus $G[H] = K^n$ for some n—but we do not know yet which edges of $G[H]$ lie in H.

f
H

n

Let us show first that H contains a cycle. If not, then $G \smallsetminus H \neq \emptyset$: by $G \supseteq K^n$ if $n \geqslant 3$, or else by $|G| \geqslant 3$. On the other hand we have $f \cup H = \mathbb{R}^2$ by Proposition 4.2.3 and hence $G = H$, a contradiction.

Since H contains a cycle, it suffices to show that $n \leqslant 3$: then $H = K^3$ as claimed. Suppose $n \geqslant 4$, and let $C = v_1 v_2 v_3 v_4 v_1$ be a cycle in $G[H]$ ($= K^n$). By $C \subseteq G$, our face f is contained in a face f_C of C; let f'_C be the other face of C. Since the vertices v_1 and v_3 lie on the boundary of f, they can be linked by an arc whose interior lies in f_C and avoids G.

C, v_i

f_C, f'_C

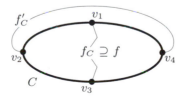

Fig. 4.2.3. The edge $v_2 v_4$ of G runs through the face f'_C

Hence by Lemma 4.1.2 (ii), the plane edge v_2v_4 of $G[H]$ runs through f'_C rather than f_C (Fig. 4.2.3). Analogously, since $v_2, v_4 \in G[f]$, the edge v_1v_3 runs through f'_C. But the edges v_1v_3 and v_2v_4 are disjoint, so this contradicts Lemma 4.1.2 (ii). $\square$

The following classic result of Euler (1752)—here stated in its simplest form, for the plane—marks one of the common origins of graph theory and topology. The theorem relates the number of vertices, edges and faces in a plane graph: taken with the correct signs, these numbers always add up to 2. The general form of Euler's theorem asserts the same for graphs suitably embedded in other (closed) surfaces, too: the sum obtained is always a fixed number depending only on the surface, not on the graph, and this number differs for distinct surfaces. Hence, any two surfaces can be distinguished by a simple arithmetic invariant of the graphs embedded in them![4]

Let us then prove Euler's theorem in its simplest form:

Theorem 4.2.7. (Euler's Formula)

[4.5.1] *Let G be a connected plane graph with n vertices, m edges, and ℓ faces. Then*

$$n - m + \ell = 2 .$$

(1.5.1)
(1.5.3) *Proof.* We fix n and apply induction on m. For $m \leqslant n-1$, G is a tree and $m = n-1$ (why?), so the assertion follows from Proposition 4.2.3.

e, G' Now let $m \geqslant n$. Then G has an edge e lying on a cycle; let $G' :=$ $G - e$. By Lemma 4.2.1 (ii), e lies on the boundary of exactly two faces

f_1, f_2
$f_{1,2}$ f_1, f_2 of G; we put $f_{1,2} := f_1 \cup \mathring{e} \cup f_2$. We shall prove that

$$F(G) \smallsetminus \{ f_1, f_2 \} = F(G') \smallsetminus \{ f_{1,2} \}, \qquad (*)$$

without assuming that $f_{1,2} \in F(G')$. However, since $\mathring{e}$ must lie in some face of G' and this will not be a face of G, by $(*)$ it can only be $f_{1,2}$. Thus again by $(*)$, G' has one face less than G. As G' also has one edge less than G, the assertion then follows from the induction hypothesis for G'.

For our proof of $(*)$ we first consider any $f \in F(G) \smallsetminus \{ f_1, f_2 \}$. By Lemma 4.2.1 (i), we have $G[f] \subseteq G \smallsetminus \mathring{e} = G'$. So f is also a face of G' (but obviously not equal to $f_{1,2}$) and hence lies in $F(G') \smallsetminus \{ f_{1,2} \}$.

f' Conversely, let a face $f' \neq f_{1,2}$ of G' be given. Since e lies on the boundary of both f_1 and f_2, we can link any two points of $f_{1,2}$ by an

$f'_{1,2}$ arc in $\mathbb{R}^2 \smallsetminus G'$, so $f_{1,2}$ lies inside a face $f'_{1,2}$ of G'. Our assumption

---

[4] This fundamental connection between graphs and surfaces lies at the heart of the proof of the famous Robertson-Seymour *minor theorem*; see the last few pages of this book.

of $f' \neq f_{1,2}$ therefore implies $f' \not\subseteq f_{1,2}$ (as otherwise $f' \subseteq f_{1,2} \subseteq f'_{1,2}$ and hence $f' = f_{1,2} = f'_{1,2}$); let x be a point in $f' \setminus f_{1,2}$. Then x lies in some face $f \neq f_1, f_2$ of G. As shown above, f is also a face of G'. Hence $x \in f \cap f'$ implies $f = f'$, and we have $f' \in F(G) \setminus \{f_1, f_2\}$ as desired.   □

    x
    f

Corollary 4.2.8. *A plane graph with $n \geqslant 3$ vertices has at most $3n - 6$ edges. Every plane triangulation with n vertices has $3n - 6$ edges.*

[4.4.1]
[5.1.2]

Proof. By Proposition 4.2.6 it suffices to prove the second assertion. In a plane triangulation G, every face boundary contains exactly three edges, and every edge lies on the boundary of exactly two faces (Lemma 4.2.1). The bipartite graph on $E(G) \cup F(G)$ with edge set $\{ef \mid e \subseteq G[f]\}$ thus has exactly $2|E(G)| = 3|F(G)|$ edges. According to this identity we may replace ℓ with $2m/3$ in Euler's formula, and obtain $m = 3n - 6$.   □

    Euler's formula can be useful for showing that certain graphs cannot occur as plane graphs. The graph K^5, for example, has $10 > 3 \cdot 5 - 6$ edges, more than allowed by Corollary 4.2.8. Similarly, $K_{3,3}$ cannot be a plane graph. For since $K_{3,3}$ is 2-connected but contains no triangle, every face of a plane $K_{3,3}$ would be bounded by a cycle of length $\geqslant 4$ (Proposition 4.2.5). As in the proof of Corollary 4.2.8 this implies $2m \geqslant 4\ell$, which yields $m \leqslant 2n - 4$ when substituted in Euler's formula. But $K_{3,3}$ has $9 > 2 \cdot 6 - 4$ edges.

    Clearly, along with K^5 and $K_{3,3}$ themselves, their subdivisions cannot occur as plane graphs either:

Corollary 4.2.9. *A plane graph contains neither K^5 nor $K_{3,3}$ as a topological minor.*   □

[4.4.5]
[4.4.6]

Surprisingly, it turns out that this simple property of plane graphs identifies them among all other graphs: as Section 4.4 will show, an arbitrary graph can be drawn in the plane if and only if it has no (topological) K^5 or $K_{3,3}$ minor.

    As we have seen, every face boundary in a 2-connected plane graph is a cycle. In a 3-connected graph, these cycles can be identified combinatorially:

Proposition 4.2.10. *The face boundaries in a 3-connected plane graph are precisely its non-separating induced cycles.*

[4.3.2]
[4.5.2]

Proof. Let G be a 3-connected plane graph, and let $C \subseteq G$. If C is a non-separating induced cycle, then by the Jordan curve theorem its two faces cannot both contain points of $G \setminus C$. Therefore it bounds a face of G.

(4.1.1)
(4.1.2)

C, f

Conversely, suppose that C bounds a face f. By Proposition 4.2.5, C is a cycle. If x, x', y, y' are four vertices along C in this order (but not necessarily adjacent on C), then any two C-paths $x \ldots y$ and $x' \ldots y'$ in G run through the other face of C (not f), and hence meet by Lemma 4.1.2 (ii); we shall call such paths *crossing C-paths*.

If C has a chord xy, then the components of $C - \{x, y\}$ are linked by a C-path $x' \ldots y'$ in G, because G is 3-connected. By the observation above, this path has to meet the edge xy, which is impossible. Hence C has no chords, i.e. C is induced in G.

P, x, y

D

z

Suppose now that C separates G. Let $P = x \ldots y$ be a path in C containing all the neighbours of some component D of $G - C$; we choose D and P so that P has minimum length. Then x and y are among those neighbours, and since G is 3-connected, P contains a further neighbour z of D (Fig. 4.2.4).

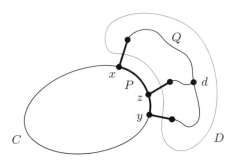

Fig. 4.2.4. The proof of Proposition 4.2.10

As D is not the only component of $G - C$, clearly $G - (D \cup P) \neq \emptyset$. Now $\{x, y\}$ does not separate G, but C has no chord, so some component

D'

$D' \neq D$ of $G - C$ has a neighbour z' in $\mathring{P}$. In fact, P contains all the neighbours v of D': otherwise G would contain crossing C-paths $v \ldots z'$ through D' and $x \ldots y$ through D, contradicting $D \cap D' = \emptyset$. But now the minimality of P implies that x and y are neighbours not only of D but also of D'. Thus if $z \neq z'$, we again have two crossing C-paths through D and D' (with ends x, y, z, z'). Therefore $z = z'$, i.e. all three vertices x, y, z are neighbours of both D and D'.

Q

Let $Q = x \ldots y$ be a C-path through D and $z \ldots d$ a C–$\mathring{Q}$ path; together, these paths form a d–$\{x, y, z\}$ fan with vertices in D and $\{x, y, z\}$. Similarly, there exists a d'–$\{x, y, z\}$ fan for some $d' \in D'$, with vertices in D' and $\{x, y, z\}$. Let us extend G to a larger plane graph G' by adding a vertex $\delta \in f$ and joining it to x, y and z (how exactly?). This δ–$\{x, y, z\}$ fan combines with the other two fans to form a $TK_{3,3}$ in the plane, contradicting Corollary 4.2.9. $\qquad\square$

4.3 Drawings

An embedding in the plane, or *planar embedding*, of an (abstract) graph *planar*
G is an isomorphism between G and a plane graph $\tilde{G}$. The latter will *embedding*
be called a *drawing* of G. We shall not always distinguish notationally *drawing*
between the vertices and edges of G and of $\tilde{G}$.

In this section we investigate how two planar embeddings of a graph
can differ. For this to make sense, we first have to agree when two embeddings are to be considered the same: for example, if we compose one
embedding with a simple rotation of the plane, the resulting embedding
will hardly count as a genuinely different way of drawing that graph.

To prepare the ground, let us first consider three possible notions
of equivalence for plane graphs (refining abstract isomorphism), and see
how they are related. Let $G = (V, E)$ and $G' = (V', E')$ be two plane $G; V, E, F$
graphs, with face sets $F(G) =: F$ and $F(G') =: F'$. Assume that G and $G'; V', E', F'$
G' are isomorphic as abstract graphs, and let $\sigma \colon V \to V'$ be an isomorphism. Setting $xy \mapsto \sigma(x)\sigma(y)$, we may extend σ in a natural way to a σ
bijection $V \cup E \to V' \cup E'$ which maps V to V' and E to E', and which
preserves incidence (and non-incidence) between vertices and edges.

Our first notion of equivalence between plane graphs is perhaps
the most natural one. Intuitively, we would like to call our isomorphism σ 'topological' if it is induced by a homeomorphism from the
plane $\mathbb{R}^2$ to itself. To avoid having to grant the outer faces of G and
G' a special status, however, we take a detour via the homeomorphism
$\pi \colon S^2 \smallsetminus \{(0,0,1)\} \to \mathbb{R}^2$ chosen in Section 4.1: we call σ a *topological* π
isomorphism between the plane graphs G and G' if there exists a homeo-
morphism $\varphi \colon S^2 \to S^2$ such that $\psi := \pi \circ \varphi \circ \pi^{-1}$ induces σ on $V \cup E$. *topological*
(More formally: we ask that ψ agree with σ on V, and that it map every *isomorphism*
plane edge $e \in G$ onto the plane edge $\sigma(e) \in G'$. Unless φ fixes the point
$(0,0,1)$, the map ψ will be undefined at $\pi(\varphi^{-1}(0,0,1))$.)

 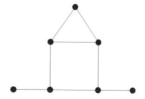

Fig. 4.3.1. Two drawings of a graph that are not topologically
isomorphic—why not?

It can be shown that, up to topological isomorphism, inner and
outer faces are indeed no longer different: if we choose as φ a rotation
of S^2 mapping the π^{-1}-image of a point of some inner face of G to the
north pole $(0,0,1)$ of S^2, then ψ maps the rest of this face to the outer

face of $\psi(G)$. (To ensure that the edges of $\psi(G)$ are again piecewise linear, however, one may have to adjust φ a little.)

If σ is a topological isomorphism as above, then—except possibly for a pair of missing points where ψ or ψ^{-1} is undefined—ψ maps the faces of G onto those of G' (proof?). In this way, σ extends naturally to a bijection $\sigma\colon V \cup E \cup F \to V' \cup E' \cup F'$ which preserves incidence of vertices, edges and faces.

Let us single out this last property of a topological isomorphism as the defining property for our second notion of equivalence for plane graphs: let us call our given isomorphism σ between the abstract graphs *combinatorial* G and G' a *combinatorial isomorphism* of the plane graphs G and G' *isomorphism* if it can be extended to a bijection $\sigma\colon V \cup E \cup F \to V' \cup E' \cup F'$ that preserves incidence not only of vertices with edges but also of vertices and edges with faces. (Formally: we require that a vertex or edge $x \in G$ shall lie on the boundary of a face $f \in F$ if and only if $\sigma(x)$ lies on the boundary of the face $\sigma(f)$.)

Fig. 4.3.2. Two drawings of a graph that are combinatorially isomorphic but not topologically—why not?

If σ is a combinatorial isomorphism of the plane graphs G and G', it maps the face boundaries of G to those of G'. Let us raise this property *graph-* to our third definition of equivalence for plane graphs: we call our isomor- *theoretical* phism σ of the abstract graphs G and G' a *graph-theoretical isomorphism* *isomorphism* of the plane graphs G and G' if

$$\big\{\, \sigma(G[f]) : f \in F \,\big\} = \big\{\, G'[f'] : f' \in F' \,\big\}.$$

Thus, we no longer keep track of *which* face is bounded by a given subgraph: the only information we keep is whether a subgraph bounds some face or not, and we require that σ map the subgraphs that do onto each other. At first glance, this third notion of equivalence may appear a little less natural than the previous two. However, it has the practical advantage of being formally weaker and hence easier to verify, and moreover, it will turn out to be equivalent to the other two notions in most cases.

As we have seen, every topological isomorphism between two plane graphs is also combinatorial, and every combinatorial isomorphism is also graph-theoretical. The following theorem shows that, for most graphs, the converse is true as well:

Theorem 4.3.1.

(i) *Every graph-theoretical isomorphism between two plane graphs is combinatorial. Its extension to a face bijection is unique if and only if the graph is not a cycle.*

(ii) *Every combinatorial isomorphism between two 2-connected plane graphs is topological.*

(4.1.1)
(4.1.4)
(4.2.4)
(4.2.5)

Proof. Let $G = (V, E)$ and $G' = (V', E')$ be two plane graphs, put $F(G) =: F$ and $F(G') =: F'$, and let $\sigma: V \cup E \to V' \cup E'$ be an isomorphism between the underlying abstract graphs.

(i) If G is a cycle, the assertion follows from the Jordan curve theorem. We now assume that G is not a cycle. Let $\mathcal{H}$ and $\mathcal{H}'$ be the sets of all face boundaries in G and G', respectively. If σ is a graph-theoretical isomorphism, then the map $H \mapsto \sigma(H)$ is a bijection between $\mathcal{H}$ and $\mathcal{H}'$. By Lemma 4.2.4, the map $f \mapsto G[f]$ is a bijection between F and $\mathcal{H}$, and likewise for F' and $\mathcal{H}'$. The composition of these three bijections is a bijection between F and F', which we choose as $\sigma: F \to F'$. By construction, this extension of σ to $V \cup E \cup F$ preserves incidences (and is unique with this property), so σ is indeed a combinatorial isomorphism.

(ii) Let us assume that G is 2-connected, and that σ is a combinatorial isomorphism. We have to construct a homeomorphism $\varphi: S^2 \to S^2$ which, for every vertex or plane edge $x \in G$, maps $\pi^{-1}(x)$ to $\pi^{-1}(\sigma(x))$. Since σ is a combinatorial isomorphism, $\tilde{\sigma}: \pi^{-1} \circ \sigma \circ \pi$ is an incidence preserving bijection from the vertices, edges and faces[5] of $\tilde{G} := \pi^{-1}(G)$ to the vertices, edges and faces of $\tilde{G}' := \pi^{-1}(G')$.

σ

$\tilde{\sigma}$

$\tilde{G}, \tilde{G}'$

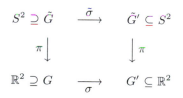

Fig. 4.3.3. Defining $\tilde{\sigma}$ via σ

We construct φ in three steps. Let us first define φ on the vertex set of $\tilde{G}$, setting $\varphi(x) := \tilde{\sigma}(x)$ for all $x \in V(\tilde{G})$. This is trivially a homeomorphism between $V(\tilde{G})$ and $V(\tilde{G}')$.

As the second step, we now extend φ to a homeomorphism between $\tilde{G}$ and $\tilde{G}'$ that induces $\tilde{\sigma}$ on $V(\tilde{G}) \cup E(\tilde{G})$. We may do this edge by edge, as follows. Every edge xy of $\tilde{G}$ is homeomorphic to the edge

---

[5] By the 'vertices, edges and faces' of $\tilde{G}$ and $\tilde{G}'$ we mean the images under π^{-1} of the vertices, edges and faces of G and G' (plus $(0,0,1)$ in the case of the outer face). Their sets will be denoted by $V(\tilde{G})$, $E(\tilde{G})$, $F(\tilde{G})$ and $V(\tilde{G}')$, $E(\tilde{G}')$, $F(\tilde{G}')$, and incidence is defined as inherited from G and G'.

$\tilde{\sigma}(xy) = \varphi(x)\varphi(y)$ of $\tilde{G}'$, by a homeomorphism mapping x to $\varphi(x)$ and y to $\varphi(y)$. Then the union of all these homeomorphisms, one for every edge of $\tilde{G}$, is indeed a homeomorphism between $\tilde{G}$ and $\tilde{G}'$—our desired extension of φ to $\tilde{G}$: all we have to check is continuity at the vertices (where the edge homeomorphisms overlap), and this follows at once from our assumption that the two graphs and their individual edges all carry the subspace topology in $\mathbb{R}^3$.

In the third step we now extend our homeomorphism $\varphi \colon \tilde{G} \to \tilde{G}'$ to all of S^2. This can be done analogously to the second step, face by face. By Proposition 4.2.5, all face boundaries in $\tilde{G}$ and $\tilde{G}'$ are cycles. Now if f is a face of $\tilde{G}$ and C its boundary, then $\tilde{\sigma}(C) := \bigcup \{ \tilde{\sigma}(e) \mid e \in E(C) \}$ bounds the face $\tilde{\sigma}(f)$ of $\tilde{G}'$. By Theorem 4.1.4, we may therefore extend the homeomorphism $\varphi \colon C \to \tilde{\sigma}(C)$ defined so far to a homeomorphism from $C \cup f$ to $\tilde{\sigma}(C) \cup \tilde{\sigma}(f)$. We finally take the union of all these homeomorphisms, one for every face f of $\tilde{G}$, as our desired homeomorphism $\varphi \colon S^2 \to S^2$; as before, continuity is easily checked. □

So far, we have considered ways of comparing plane graphs. We now come to our actual goal, the definition of equivalence for planar embeddings. Let us call two planar embeddings σ_1, σ_2 of a graph G *topologically* (respectively, *combinatorially*) *equivalent* if $\sigma_2 \circ \sigma_1^{-1}$ is a topological (respectively, combinatorial) isomorphism between $\sigma_1(G)$ and $\sigma_2(G)$. If G is 2-connected, the two definitions coincide by Theorem 4.3.1, and we simply speak of *equivalent* embeddings. Clearly, this is indeed an equivalence relation on the set of planar embeddings of any given graph.

equivalent embeddings

Note that two drawings of G resulting from inequivalent embeddings may well be topologically isomorphic (exercise): for the equivalence of two embeddings we ask not only that some (topological or combinatorial) isomorphism exist between the their images, but that the canonical isomorphism $\sigma_2 \circ \sigma_1^{-1}$ be a topological or combinatorial one.

Even in this strong sense, 3-connected graphs have only one embedding up to equivalence:

Theorem 4.3.2. (Whitney 1932)
Any two planar embeddings of a 3-connected graph are equivalent.

(4.2.10)

Proof. Let G be a 3-connected graph with planar embeddings $\sigma_1 \colon G \to G_1$ and $\sigma_2 \colon G \to G_2$. By Theorem 4.3.1 it suffices to show that $\sigma_2 \circ \sigma_1^{-1}$ is a graph-theoretical isomorphism, i.e. that $\sigma_1(C)$ bounds a face of G_1 if and only if $\sigma_2(C)$ bounds a face of G_2, for every subgraph $C \subseteq G$. This follows at once from Proposition 4.2.10. □

4.4 Planar graphs: Kuratowski's theorem

A graph is called *planar* if it can be embedded in the plane: if it is   *planar*
isomorphic to a plane graph. A planar graph is *maximal*, or *maximally
planar*, if it is planar but cannot be extended to a larger planar graph
by adding an edge (but no vertex).

    Drawings of maximal planar graphs are clearly maximally plane.
The converse, however, is not obvious: when we start to draw a planar
graph, could it happen that we get stuck half-way with a proper subgraph
that is already maximally plane? Our first proposition says that this
can never happen, that is, a plane graph is never maximally plane just
because it is badly drawn:

Proposition 4.4.1.

  (i) *Every maximal plane graph is maximally planar.*

  (ii) *A planar graph with $n \geqslant 3$ vertices is maximally planar if and
only if it has $3n - 6$ edges.*

Proof. Apply Proposition 4.2.6 and Corollary 4.2.8.   □   (4.2.6)
                                                                     (4.2.8)

    Which graphs are planar? As we saw in Corollary 4.2.9, no planar
graph contains K^5 or $K_{3,3}$ as a topological minor. Our aim in this section
is to prove the surprising converse, a classic theorem of Kuratowski: any
graph without a topological K^5 or $K_{3,3}$ minor is planar.

    Before we prove Kuratowski's theorem, let us note that it suffices
to consider ordinary minors rather than topological ones:

Proposition 4.4.2. *A graph contains K^5 or $K_{3,3}$ as a minor if and only
if it contains K^5 or $K_{3,3}$ as a topological minor.*

Proof. By Proposition 1.7.2 it suffices to show that every graph G (1.7.2)
with a K^5 minor contains either K^5 as a topological minor or $K_{3,3}$ as
a minor. So suppose that $G \succcurlyeq K^5$, and let $K \subseteq G$ be minimal such
that $K = MK^5$. Then every branch set of K induces a tree in K, and
between any two branch sets K has exactly one edge. If we take the
tree induced by a branch set V_x and add to it the four edges joining it
to other branch sets, we obtain another tree, T_x say. By the minimality
of K, T_x has exactly 4 leaves, the 4 neighbours of V_x in other branch
sets (Fig. 4.4.1).

    If each of the five trees T_x is a $TK_{1,4}$ then K is a TK^5, and we are
done. If one of the T_x is not a $TK_{1,4}$ then it has exactly two vertices
of degree 3. Contracting V_x onto these two vertices, and every other
branch set to a single vertex, we obtain a graph on 6 vertices containing
a $K_{3,3}$. Thus, $G \succcurlyeq K_{3,3}$ as desired.   □

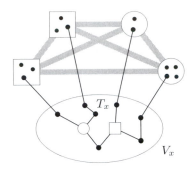

Fig. 4.4.1. Every MK^5 contains a TK^5 or $MK_{3,3}$

We first prove Kuratowski's theorem for 3-connected graphs. This is the heart of the proof: the general case will then follow easily.

Lemma 4.4.3. *Every 3-connected graph G without a K^5 or $K_{3,3}$ minor is planar.*

(3.2.1)
(4.2.5)

xy

$\tilde{G}$

f, C

X, Y

$\tilde{G}'$

Proof. We apply induction on $|G|$. For $|G| = 4$ we have $G = K^4$, and the assertion holds. Now let $|G| > 4$, and assume the assertion is true for smaller graphs. By Lemma 3.2.1, G has an edge xy such that G/xy is again 3-connected. Since the minor relation is transitive, G/xy has no K^5 or $K_{3,3}$ minor either. Thus, by the induction hypothesis, G/xy has a drawing $\tilde{G}$ in the plane. Let f be the face of $\tilde{G} - v_{xy}$ containing the point v_{xy}, and let C be the boundary of f. Let $X := N_G(x) \smallsetminus \{y\}$ and $Y := N_G(y) \smallsetminus \{x\}$; then $X \cup Y \subseteq V(C)$, because $v_{xy} \in f$. Clearly,

$$\tilde{G}' := \tilde{G} - \{v_{xy}v \mid v \in Y \smallsetminus X\}$$

may be viewed as a drawing of $G - y$, in which the vertex x is represented by the point v_{xy} (Fig. 4.4.2). Our aim is to add y to this drawing to obtain a drawing of G.

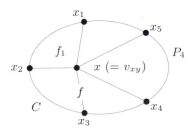

Fig. 4.4.2. $\tilde{G}'$ as a drawing of $G - y$: the vertex x is represented by the point v_{xy}

Since $\tilde{G}$ is 3-connected, $\tilde{G} - v_{xy}$ is 2-connected, so C is a cycle (Proposition 4.2.5). Let $x_1, \ldots, x_k$ be an enumeration along this cycle of the vertices in X, and let $P_i = x_i \ldots x_{i+1}$ be the X-paths on C between them ($i = 1, \ldots, k$; with $x_{k+1} := x_1$). For each i, the set $C \smallsetminus P_i$ is contained in one of the two faces of the cycle $C_i := x x_i P_i x_{i+1} x$; we denote the *other* face of C_i by f_i. Since f_i contains points of f (close to x) but no points of C, we have $f_i \subseteq f$. Moreover, the plane edges $x x_j$ with $j \notin \{i, i+1\}$ meet C_i only in x and end outside f_i in $C \smallsetminus P_i$, so f_i meets none of those edges. Hence $f_i \subseteq \mathbb{R}^2 \smallsetminus \tilde{G}'$, that is, f_i is contained in a face of $\tilde{G}'$. (In fact, f_i is a face of $\tilde{G}'$, but we do not need this.)

In order to turn $\tilde{G}'$ into a drawing of G, let us try to find an i such that $Y \subseteq V(P_i)$; we may then embed y into f_i and link it up to its neighbours by arcs inside f_i. Suppose there is no such i: how then can the vertices of Y be distributed around C? If y had a neighbour in some $\mathring{P}_i$, it would also have one in $C - P_i$, so G would contain a $TK_{3,3}$ (with branch vertices x, y, x_i, x_{i+1} and those two neighbours of y). Hence $Y \subseteq X$. Now if $|Y| = |Y \cap X| \geqslant 3$, we have a TK^5 in G. So $|Y| \leqslant 2$; in fact, $|Y| = 2$, because $d(y) \geqslant \kappa(G) \geqslant 3$. Since the two vertices of Y lie on no common P_i, we can once more find a $TK_{3,3}$ in G, a contradiction. $\qquad\square$

Compared with other proofs of Kuratowski's theorem, the above proof has the attractive feature that it can easily be adapted to produce a drawing in which every inner face is convex (exercise); in particular, every edge can be drawn straight. Note that 3-connectedness is essential here: a 2-connected planar graph need not have a drawing with all inner faces convex (example?), although it always has a straight-line drawing (Exercise 9).

It is not difficult, in principle, to reduce the general Kuratowski theorem to the 3-connected case by manipulating and combining partial drawings assumed to exist by induction. For example, if $\kappa(G) = 2$ and $G = G_1 \cup G_2$ with $V(G_1 \cap G_2) = \{x, y\}$, and if G has no TK^5 or $TK_{3,3}$ subgraph, then neither $G_1 + xy$ nor $G_2 + xy$ has such a subgraph, and we may try to combine drawings of these graphs to one of $G + xy$. (If xy is already an edge of G, the same can be done with G_1 and G_2.) For $\kappa(G) \leqslant 1$, things become even simpler. However, the geometric operations involved require some cumbersome shifting and scaling, even if all the plane edges occurring are assumed to be straight.

The following more combinatorial route is just as easy, and may be a welcome alternative.

Lemma 4.4.4. *Let $\mathcal{X}$ be a set of 3-connected graphs. Let G be a graph with $\kappa(G) \leqslant 2$, and let G_1, G_2 be proper induced subgraphs of G such that $G = G_1 \cup G_2$ and $|G_1 \cap G_2| = \kappa(G)$. If G is edge-maximal without a topological minor in $\mathcal{X}$, then so are G_1 and G_2, and $G_1 \cap G_2 = K^2$.*

[8.3.1]

S

X

P

Proof. Note first that every vertex $v \in S := V(G_1 \cap G_2)$ has a neighbour in every component of $G_i - S$, $i = 1, 2$: otherwise $S \smallsetminus \{v\}$ would separate G, contradicting $|S| = \kappa(G)$. By the maximality of G, every edge e added to G lies in a $TX \subseteq G + e$ with $X \in \mathcal{X}$. For all the choices of e considered below, the 3-connectedness of X will imply that the branch vertices of this TX all lie in the same G_i, say in G_1. (The position of e will always be symmetrical with respect to G_1 and G_2, so this assumption entails no loss of generality.) Then the TX meets G_2 at most in a path P corresponding to an edge of X.

If $S = \emptyset$, we obtain an immediate contradiction by choosing e with one end in G_1 and the other in G_2. If $S = \{v\}$ is a singleton, let e join a neighbour v_1 of v in $G_1 - S$ to a neighbour v_2 of v in $G_2 - S$ (Fig. 4.4.3). Then P contains both v and the edge $v_1 v_2$; replacing vPv_1 with the edge vv_1, we obtain a TX in $G_1 \subseteq G$, a contradiction.

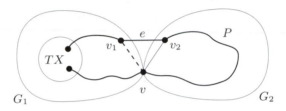

Fig. 4.4.3. If $G + e$ contains a TX, then so does G_1 or G_2

x, y

So $|S| = 2$, say $S = \{x, y\}$. If $xy \notin G$, we let $e := xy$, and in the arising TX replace e by an x–y path through G_2; this gives a TX in G, a contradiction. Hence $xy \in G$, and $G[S] = K^2$ as claimed.

It remains to show that G_1 and G_2 are edge-maximal without a topological minor in $\mathcal{X}$. So let e' be an additional edge for G_1, say. Replacing xPy with the edge xy if necessary, we obtain a TX either in $G_1 + e'$ (which shows the edge-maximality of G_1, as desired) or in G_2 (which contradicts $G_2 \subseteq G$). $\qquad\square$

Lemma 4.4.5. *If $|G| \geqslant 4$ and G is edge-maximal with $TK^5, TK_{3,3} \not\subseteq G$, then G is 3-connected.*

(4.2.9)

G_1, G_2

x, y

f_i

z_i

K

Proof. We apply induction on $|G|$. For $|G| = 4$, we have $G = K^4$ and the assertion holds. Now let $|G| > 4$, and let G be edge-maximal without a TK^5 or $TK_{3,3}$. Suppose $\kappa(G) \leqslant 2$, and choose G_1 and G_2 as in Lemma 4.4.4. For $\mathcal{X} := \{K^5, K_{3,3}\}$, the lemma says that $G_1 \cap G_2$ is a K^2, with vertices x, y say. By Lemmas 4.4.4, 4.4.3 and the induction hypothesis, G_1 and G_2 are planar. For each $i = 1, 2$ separately, choose a drawing of G_i, a face f_i with the edge xy on its boundary, and a vertex $z_i \neq x, y$ on the boundary of f_i. Let K be a TK^5 or $TK_{3,3}$ in the abstract graph $G + z_1 z_2$ (Fig. 4.4.4).

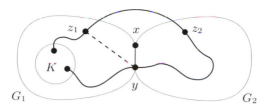

Fig. 4.4.4. A TK^5 or $TK_{3,3}$ in $G + z_1 z_2$

If all the branch vertices of K lie in the same G_i, then either $G_i + xz_i$ or $G_i + yz_i$ (or G_i itself, if z_i is already adjacent to x or y, respectively) contains a TK^5 or $TK_{3,3}$; this contradicts Corollary 4.2.9, since these graphs are planar by the choice of z_i. Since $G + z_1 z_2$ does not contain four independent paths between $(G_1 - G_2)$ and $(G_2 - G_1)$, these subgraphs cannot both contain a branch vertex of a TK^5, and cannot both contain two branch vertices of a $TK_{3,3}$. Hence K is a $TK_{3,3}$ with only one branch vertex v in, say, $G_2 - G_1$. But then also the graph $G_1 + v + \{vx, vy, vz_1\}$, which is planar by the choice of z_1, contains a $TK_{3,3}$. This contradicts Corollary 4.2.9.   □

Theorem 4.4.6. (Kuratowski 1930; Wagner 1937)
The following assertions are equivalent for graphs G: [4.5.1]

   (i) *G is planar;*

   (ii) *G contains neither K^5 nor $K_{3,3}$ as a minor;*

   (iii) *G contains neither K^5 nor $K_{3,3}$ as a topological minor.*

Proof. Combine Corollary 4.2.9 and Proposition 4.4.2 with Lemmas   (4.2.9)
4.4.3 and 4.4.5.   □

Corollary 4.4.7. *Every maximal planar graph with at least four vertices is 3-connected.*

Proof. Apply Lemma 4.4.5 and Theorem 4.4.6.   □

4.5 Algebraic planarity criteria

In this section we show that planarity can be characterized in purely algebraic terms, by a certain abstract property of its cycle space. Theorems relating such seemingly distant graph properties are rare, and their significance extends beyond their immediate applicability. In a sense, they indicate that both ways of viewing a graph—in our case, the topological and the algebraic way—are not just formal curiosities: if both are natural enough that, quite unexpectedly, each can be expressed in terms of the other, the indications are that they have the power to reveal some genuine insights into the structure of graphs and are worth pursuing.

simple

Let $G = (V, E)$ be a graph. We call a subset $\mathcal{F}$ of its edge space $\mathcal{E}(G)$ *simple* if every edge of G lies in at most two sets of $\mathcal{F}$. For example, the cut space $\mathcal{C}^*(G)$ has a simple basis: according to Proposition 1.9.3 it is generated by the cuts $E(v)$ formed by all the edges at a given vertex v, and an edge $xy \in G$ lies in $E(v)$ only for $v = x$ and for $v = y$.

Theorem 4.5.1. (MacLane 1937)

[4.6.3]

A graph is planar if and only if its cycle space has a simple basis.

(1.9.2)
(1.9.6)
(4.1.1)
(4.2.1)
(4.2.5)
(4.4.6)

Proof. The assertion being trivial for graphs of order at most 2, we consider a graph G of order at least 3. If $\kappa(G) \leqslant 1$, then G is the union of two proper induced subgraphs G_1, G_2 with $|G_1 \cap G_2| \leqslant 1$. Then $\mathcal{C}(G)$ is the direct sum of $\mathcal{C}(G_1)$ and $\mathcal{C}(G_2)$, and hence has a simple basis if and only if both $\mathcal{C}(G_1)$ and $\mathcal{C}(G_2)$ do (proof?). Moreover, G is planar if and only if both G_1 and G_2 are: this follows at once from Kuratowski's theorem, but also from easy geometrical considerations. The assertion for G thus follows inductively from those for G_1 and G_2. For the rest of the proof, we now assume that G is 2-connected.

We first assume that G is planar and choose a drawing. By Lemma 4.2.5, the face boundaries of G are cycles, so they are elements of $\mathcal{C}(G)$. We shall show that the face boundaries generate all the cycles in G; then $\mathcal{C}(G)$ has a simple basis by Lemma 4.2.1. Let $C \subseteq G$ be any cycle, and let f be its inner face. By Lemma 4.2.1, every edge e with $\mathring{e} \subseteq f$ lies on exactly two face boundaries $G[f']$ with $f' \subseteq f$, and every edge of C lies on exactly one such face boundary. Hence the sum in $\mathcal{C}(G)$ of all those face boundaries is exactly C.

Conversely, let $\{C_1, \ldots, C_k\}$ be a simple basis of $\mathcal{C}(G)$. Then, for every edge $e \in G$, also $\mathcal{C}(G - e)$ has a simple basis. Indeed, if e lies in just one of the sets C_i, say in C_1, then $\{C_2, \ldots, C_k\}$ is a simple basis of $\mathcal{C}(G - e)$; if e lies in two of the C_i, say in C_1 and C_2, then $\{C_1 + C_2, C_3, \ldots, C_k\}$ is such a basis. (Note that the two bases are indeed subsets of $\mathcal{C}(G - e)$ by Proposition 1.9.2.) Thus every subgraph of G has a cycle space with a simple basis. For our proof that G is planar, it thus suffices to show that the cycle spaces of K^5 and $K_{3,3}$ (and hence those of their subdivisions) do *not* have a simple basis: then G cannot contain a TK^5 or $TK_{3,3}$, and so is planar by Kuratowski's theorem.

Let us consider K^5 first. By Theorem 1.9.6, $\dim \mathcal{C}(K^5) = 6$; let $\mathcal{B} = \{C_1, \ldots, C_6\}$ be a simple basis, and put $C_0 := C_1 + \ldots + C_6$. As $\mathcal{B}$ is linearly independent, none of the sets $C_0, \ldots, C_6$ is empty, and so each of them contains at least three edges (cf. Proposition 1.9.2). The simplicity of $\mathcal{B}$ therefore implies

$$18 = 6 \cdot 3 \leqslant |C_1| + \ldots + |C_6|$$
$$\leqslant 2 \, \|K^5\| - |C_0|$$
$$\leqslant 2 \cdot 10 - 3 = 17,$$

a contradiction; for the middle inequality note that every edge in C_0 lies in just one of the sets $C_1, \ldots, C_6$.

For $K_{3,3}$, Theorem 1.9.6 gives $\dim \mathcal{C}(K_{3,3}) = 4$; let $\mathcal{B} = \{C_1, \ldots, C_4\}$ be a simple basis, and put $C_0 := C_1 + \ldots + C_4$. Since $K_{3,3}$ has girth 4, each C_i contains at least four edges, so

$$
\begin{aligned}
16 = 4 \cdot 4 &\leqslant |C_1| + \ldots + |C_4| \\
&\leqslant 2 \, \|K_{3,3}\| - |C_0| \\
&\leqslant 2 \cdot 9 - 4 = 14 \,,
\end{aligned}
$$

a contradiction. $\square$

It is one of the hidden beauties of planarity theory that two such abstract and seemingly unintuitive results about generating sets in cycle spaces as MacLane's theorem and Tutte's theorem 3.2.3 conspire to produce a very tangible planarity criterion for 3-connected graphs:

Theorem 4.5.2. (Tutte 1963)
A 3-connected graph is planar if and only if every edge lies on at most (equivalently: exactly) two non-separating induced cycles.

Proof. The forward implication follows from Propositions 4.2.10 and 4.2.1 (and Proposition 4.2.5 for the 'exactly two' version); the backward implication follows from Theorems 3.2.3 and 4.5.1. $\square$

(3.2.3)
(4.2.1)
(4.2.5)
(4.2.10)

4.6 Plane duality

In this section we shall use MacLane's theorem to uncover another connection between planarity and algebraic structure: a connection between the duality of plane graphs, defined below, and the duality of the cycle and cut space hinted at in Chapters 1.9 and 3.5.

A *plane multigraph* is a pair $G = (V, E)$ of finite sets (of *vertices* and *edges*, respectively) satisfying the following conditions:

plane multigraph

(i) $V \subseteq \mathbb{R}^2$;

(ii) every edge is either an arc between two vertices or a polygon containing exactly one vertex (its *endpoint*);

(iii) apart from its own endpoint(s), an edge contains no vertex and no point of any other edge.

We shall use terms defined for plane graphs freely for plane multigraphs. Note that, as in abstract multigraphs, both loops and double edges count as cycles.

Let us consider the plane multigraph G shown in Figure 4.6.1. Let us place a new vertex inside each face of G and link these new vertices up to form another plane multigraph G^*, as follows: for every edge e of G we link the two new vertices in the faces incident with e by an edge e^* crossing e; if e is incident with only one face, we attach a loop e^* to the new vertex in that face, again crossing the edge e. The plane multigraph G^* formed in this way is then dual to G in the following sense: if we apply the same procedure as above to G^*, we obtain a plane multigraph very similar to G; in fact, G itself may be reobtained from G^* in this way.

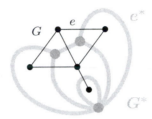

Fig. 4.6.1. A plane graph and its dual

To make this idea more precise, let $G = (V, E)$ and (V^*, E^*) be any two plane multigraphs, and put $F(G) =: F$ and $F((V^*, E^*)) =: F^*$. We call (V^*, E^*) a *plane dual* of G, and write $(V^*, E^*) =: G^*$, if there are bijections

$$F \to V^* \qquad\qquad E \to E^* \qquad V \to F^*$$
$$f \mapsto v^*(f) \qquad\quad e \mapsto e^* \qquad v \mapsto f^*(v)$$

satisfying the following conditions:

(i) $v^*(f) \in f$ for all $f \in F$;

(ii) $|e^* \cap G| = |\mathring{e}^* \cap \mathring{e}| = |e \cap G^*| = 1$ for all $e \in E$;

(iii) $v \in f^*(v)$ for all $v \in V$.

The existence of such bijections implies that both G and G^* are connected (exercise). Conversely, every connected plane multigraph G has a plane dual G^*: if we pick from each face f of G a point $v^*(f)$ as a vertex for G^*, we can always link these vertices up by independent arcs as required by condition (ii), and there is always a bijection $V \to F^*$ satisfying (iii) (exercise).

If G_1^* and G_2^* are two plane duals of G, then clearly $G_1^* \simeq G_2^*$; in fact, one can show that the natural bijection $v_1^*(f) \mapsto v_2^*(f)$ is a topological isomorphism between G_1^* and G_2^*. In this sense, we may speak of *the* plane dual G^* of G.

Finally, G is in turn a plane dual of G^*. Indeed, this is witnessed by the inverse maps of the bijections from the definition of G^*: setting $v^*(f^*(v)) := v$ and $f^*(v^*(f)) := f$ for $f^*(v) \in F^*$ and $v^*(f) \in V^*$, we see that conditions (i) and (iii) for G^* transform into (iii) and (i) for G, while condition (ii) is symmetrical in G and G^*. Thus, the term 'dual' is also formally justified.

Plane duality is fascinating not least because it establishes a connection between two natural but very different kinds of edge sets in a multigraph, between cycles and cuts:

Proposition 4.6.1. *For any connected plane multigraph G, an edge set* $E \subseteq E(G)$ *is the edge set of a cycle in G if and only if $E^* := \{\, e^* \mid e \in E \,\}$ is a minimal cut in G^*.*

[6.5.2]

Proof. By conditions (i) and (ii) in the definition of G^*, two vertices $v^*(f_1)$ and $v^*(f_2)$ of G^* lie in the same component of $G^* - E^*$ if and only if f_1 and f_2 lie in the same region of $\mathbb{R}^2 \setminus \bigcup E$: every $v^*(f_1)$–$v^*(f_2)$ path in $G^* - E^*$ is an arc between f_1 and f_2 in $\mathbb{R}^2 \setminus \bigcup E$, and conversely every such arc P (with $P \cap V(G) = \emptyset$) defines a walk in $G^* - E^*$ between $v^*(f_1)$ and $v^*(f_2)$.

(4.1.1)
(4.2.3)

Now if $C \subseteq G$ is a cycle and $E = E(C)$ then, by the Jordan curve theorem and the above correspondence, $G^* - E^*$ has exactly two components, so E^* is a minimal cut in G^*.

Conversely, if $E \subseteq E(G)$ is such that E^* is a cut in G^*, then, by Proposition 4.2.3 and the above correspondence, E contains the edges of a cycle $C \subseteq G$. If E^* is minimal as a cut, then E cannot contain any further edges (by the implication shown before), so $E = E(C)$. □

Proposition 4.6.1 suggests the following generalization of plane duality to a notion of duality for abstract multigraphs. Let us call a multigraph G^* an *abstract dual* of a multigraph G if $E(G^*) = E(G)$ and the minimal cuts in G^* are precisely the edge sets of cycles in G. Note that any abstract dual of a multigraph is connected.

abstract dual

Proposition 4.6.2. *If G^* is an abstract dual of G, then the cut space of G^* is the cycle space of G, i.e.*

$$\mathcal{C}^*(G^*) = \mathcal{C}(G).$$

Proof. By Lemma 1.9.4,[6] $\mathcal{C}^*(G^*)$ is the subspace of $\mathcal{E}(G^*) = \mathcal{E}(G)$ generated by the minimal cuts in G^*. By assumption, these are precisely the edge sets of the cycles in G, and these generate $\mathcal{C}(G)$ in $\mathcal{E}(G)$. □

(1.9.4)

---

[6] Although the lemma was stated for graphs only, its proof remains the same for multigraphs.

We finally come to one of the highlights of classical planarity theory: the planar graphs are characterized by the fact that they have an abstract dual. Although less obviously intuitive, this duality is just as fundamental a property as planarity itself; indeed the following theorem may well be seen as a topological characterization of the graphs that have a dual:

Theorem 4.6.3. (Whitney 1933)
A graph is planar if and only if it has an abstract dual.

(1.9.3)
(4.5.1)

Proof. Let G be a graph. If G is plane, then every component C of G has a plane dual C^*. Let us consider these C^* as abstract multigraphs, pick a vertex in each of them, and identify these vertices. In the connected multigraph G^* obtained, the set of minimal cuts is the union of the sets of minimal cuts in the multigraphs C^*. By Proposition 4.6.1, these cuts are precisely the edge sets of the cycles in G, so G^* is an abstract dual of G.

Conversely, suppose that G has an abstract dual G^*. By Theorem 4.5.1 and Proposition 4.6.2 it suffices to show that $\mathcal{C}^*(G^*)$ has a simple basis, which it has by Proposition 1.9.3.   □

Exercises

1.   Show that every graph can be embedded in $\mathbb{R}^3$ with all edges straight.

2.⁻ Show directly by Lemma 4.1.2 that $K_{3,3}$ is not planar.
     (Hint. Figure 1.6.2.)

3.   Find an Euler formula for disconnected graphs.

4.   Show that every planar graph is a union of three forests.
     (Hint. Theorem 3.5.4.)

5.⁻ Show that the two plane graphs in Fig. 4.3.1 are not combinatorially (and hence not topologically) isomorphic.

6.   Show that the two graphs in Fig. 4.3.2 are combinatorially but not topologically isomorphic.

7.⁻ Show that our definition of equivalence for planar embeddings does indeed define an equivalence relation.

8.   Find a 2-connected planar graph whose drawings are all topologically isomorphic but whose planar embeddings are not all equivalent.

9.⁺ Show that every plane graph is combinatorially isomorphic to a plane graph whose edges are all straight.
     (Hint. Given a plane triangulation, construct inductively a graph-theoretically isomorphic plane graph whose edges are straight. Which additional property of the inner faces could help with the induction?)

Do not use Kuratowski's theorem in the following two exercises.

10.   Show that any minor of a planar graph is planar.

11.   (i) Show that the planar graphs can in principle be characterized as in Kuratowski's theorem, i.e., that there exists a set $\mathcal{X}$ of graphs such that a graph G is planar if and only if G has no topological minor in $\mathcal{X}$.

(ii) More generally, which graph properties can be characterized in this way?

12.$^-$ Does every planar graph have a drawing with all inner faces convex?

13.   Modify the proof of Lemma 4.4.3 so that all inner faces become convex.

14.   Does every minimal non-planar graph G (i.e., every non-planar graph G whose proper subgraphs are all planar) contain an edge e such that $G - e$ is maximally planar? Does the answer change if we define 'minimal' with respect to topological minors rather than subgraphs?

15.   Show that adding a new edge to a maximal planar graph of order at least 6 always produces both a TK^5 and a $TK_{3,3}$ subgraph.

16.   A graph is called *outerplanar* if it has a drawing in which every vertex lies on the boundary of the outer face. Show that a graph is outerplanar if and only if it contains neither K^4 nor $K_{2,3}$ as a minor.

17.   Let $G = G_1 \cup G_2$, where $|G_1 \cap G_2| \leqslant 1$. Show that $\mathcal{C}(G)$ has a simple basis if both $\mathcal{C}(G_1)$ and $\mathcal{C}(G_2)$ have one.

18.$^+$ Find a simple cycle space basis among the face boundaries of a 2-connected plane graph.

19.   Show that a 2-connected plane graph is bipartite if and only if every face is bounded by an even cycle.

20.$^-$ Let G be a connected plane multigraph, and let G^* be its plane dual. Prove the following two statements for every edge $e \in G$:

(i) If e lies on the boundary of two distinct faces f_1, f_2 of G, then $e^* = v^*(f_1)\, v^*(f_2)$.

(ii) If e lies on the boundary of exactly one face f of G, then e^* is a loop at $v^*(f)$.

21.$^-$ What does the plane dual of a plane tree look like?

22.$^-$ Show that the plane dual of a plane multigraph is connected.

23.$^+$ Show that a plane multigraph has a plane dual if and only if it is connected.

24.   Let G, G^* be mutually dual plane multigraphs, and let $e \in E(G)$. Prove the following statements (with a suitable definition of G/e):

(i) If e is not a bridge, then G^*/e^* is a plane dual of $G - e$.

(ii) If e is not a loop, then $G^* - e^*$ is a plane dual of G/e.

25. Show that any two plane duals of a plane multigraph are combinatorially isomorphic.

26. Let G, G^* be mutually dual plane graphs. Prove the following statements:

  (i) If G is 2-connected, then G^* is 2-connected.

  (ii) If G is 3-connected, then G^* is 3-connected.

  (iii) If G is 4-connected, then G^* need not be 4-connected.

27. Let G, G^* be mutually dual plane graphs. Let $B_1, \ldots, B_n$ be the blocks of G. Show that $B_1^*, \ldots, B_n^*$ are the blocks of G^*.

28. Find a direct proof for planar graphs of Tutte's theorem on the cycle space of 3-connected graphs (Theorem 3.2.3).

  (Hint. Proposition 1.9.3.)

29. Show that if G^* is an abstract dual of a multigraph G, then G is an abstract dual G^*.

30. Show that a connected graph $G = (V, E)$ is planar if and only if there exists a connected multigraph $G' = (V', E)$ (i.e. with the same edge set) such that the following holds for every set $F \subseteq E$: the graph (V, F) is a tree if and only if $(V', E \smallsetminus F)$ is a tree.

Notes

There is an excellent monograph on the embedding of graphs in surfaces, including the plane: B. Mohar & C. Thomassen, *Graphs on Surfaces*, Johns Hopkins University Press 1997. Proofs of the results cited in Section 4.1, as well as all references for this chapter, can be found there. A good account of the Jordan curve theorem, both polygonal and general, is given also in J. Stillwell, *Classical topology and combinatorial group theory*, Springer 1980.

  The short proof of Corollary 4.2.8 uses a trick that deserves special mention: the so-called *double counting* of pairs, illustrated in the text by a bipartite graph whose edges can be counted alternatively by summing its degrees on the left or on the right. Double counting is a technique widely used in combinatorics, and there will be more examples later in the book.

  The material of Section 4.3 is not normally standard for an introductory graph theory course, and the rest of the chapter can be read independently of this section. However, the results of Section 4.3 are by no means unimportant. In a way, they have fallen victim to their own success: the shift from a topological to a combinatorial setting for planarity problems which they achieve has made the topological techniques developed there dispensable for most of planarity theory.

  In its original version, Kuratowski's theorem was stated only for topological minors; the version for general minors was added by Wagner in 1937. Our proof of the 3-connected case (Lemma 4.4.3) can easily be strengthened to make all the inner faces convex (exercise); see C. Thomassen, Planarity and duality of finite and infinite graphs, *J. Combin. Theory B* **29** (1980), 244–271.

The existence of such 'convex' drawings for 3-connected planar graphs follows already from the theorem of Steinitz (1922) that these graphs are precisely the 1-skeletons of 3-dimensional convex polyhedra. Compare also W.T. Tutte, How to draw a graph, *Proc. London Math. Soc.* **13** (1963), 743–767.

As one readily observes, adding an edge to a maximal planar graph (of order at least 6) produces not only a topological K^5 or $K_{3,3}$, but both. In Chapter 8.3 we shall see that, more generally, every graph with n vertices and more than $3n - 6$ edges contains a TK^5 and, with one easily described class of exceptions, also a $TK_{3,3}$.

The simple cycle space basis constructed in the proof of MacLane's theorem, which consists of the inner face boundaries, is canonical in the following sense: for every simple basis $\mathcal{B}$ of the cycle space of a 2-connected planar graph there exists a drawing of that graph in which $\mathcal{B}$ is precisely the set of inner face boundaries. (This is proved in Mohar & Thomassen, who also mention some further planarity criteria.) Our proof of the backward direction of MacLane's theorem is based on Kuratowski's theorem. A more direct approach, in which a planar embedding is actually constructed from a simple basis, is adopted in K. Wagner, *Graphentheorie*, BI Hochschultaschenbücher 1972.

The proper setting for duality phenomena between cuts and cycles in abstract graphs (and beyond) is the theory of *matroids*; see J.G. Oxley, *Matroid Theory*, Oxford University Press 1992.

5   Colouring

How many colours do we need to colour the countries of a map in such a way that adjacent countries are coloured differently? How many days have to be scheduled for committee meetings of a parliament if every committee intends to meet for one day and some members of parliament serve on several committees? How can we find a school timetable of minimum total length, based on the information of how often each teacher has to teach each class?

A *vertex colouring* of a graph $G = (V, E)$ is a map $c: V \to S$ such that $c(v) \neq c(w)$ whenever v and w are adjacent. The elements of the set S are called the available *colours*. All that interests us about S is its size: typically, we shall be asking for the smallest integer k such that G has a *k-colouring*, a vertex colouring $c: V \to \{1, \ldots, k\}$. This k is the *(vertex-) chromatic number* of G; it is denoted by $\chi(G)$. A graph G with $\chi(G) = k$ is called *k-chromatic*; if $\chi(G) \leqslant k$, we call G *k-colourable*.

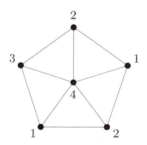

Fig. 5.0.1. A vertex colouring $V \to \{1, \ldots, 4\}$

Note that a k-colouring is nothing but a vertex partition into k independent sets, now called *colour classes*; the non-trivial 2-colourable graphs, for example, are precisely the bipartite graphs. Historically, the colouring terminology comes from the map colouring problem stated above, which leads to the problem of determining the maximum chromatic number of planar graphs. The committee scheduling problem, too, can be phrased as a vertex colouring problem—how?

edge colouring

An *edge colouring* of $G = (V, E)$ is a map $c \colon E \to S$ with $c(e) \neq c(f)$ for any adjacent edges e, f. The smallest integer k for which a k-*edge-colouring* exists, i.e. an edge colouring $c \colon E \to \{1, \ldots, k\}$, is the *edge-chromatic number*, or *chromatic index*, of G; it is denoted by $\chi'(G)$. The third of our introductory questions can be modelled as an edge colouring problem in a bipartite multigraph (how?).

chromatic index
$\chi'(G)$

Clearly, every edge colouring of G is a vertex colouring of its line graph $L(G)$, and vice versa; in particular, $\chi'(G) = \chi(L(G))$. The problem of finding good edge colourings may thus be viewed as a restriction of the more general vertex colouring problem to this special class of graphs. As we shall see, this relationship between the two types of colouring problem is reflected by a marked difference in our knowledge about their solutions: while there are only very rough estimates for χ, its sister χ' always takes one of two values, either Δ or $\Delta + 1$.

5.1 Colouring maps and planar graphs

If any result in graph theory has a claim to be known to the world outside, it is the following *four colour theorem* (which implies that every map can be coloured with at most four colours):

Theorem 5.1.1. (Four Colour Theorem)
Every planar graph is 4-colourable.

Some remarks about the proof of the four colour theorem and its history can be found in the notes at the end of this chapter. Here, we prove the following weakening:

Proposition 5.1.2. (Five Colour Theorem)
Every planar graph is 5-colourable.

(4.1.1)
(4.2.8)

Proof. Let G be a plane graph with $n \geqslant 6$ vertices and m edges. We assume inductively that every plane graph with fewer than n vertices can be 5-coloured. By Corollary 4.2.8,

n, m

$$d(G) = 2m/n \leqslant 2\,(3n - 6)/n < 6\,;$$

v
H
c

let $v \in G$ be a vertex of degree at most 5. By the induction hypothesis, the graph $H := G - v$ has a vertex colouring $c \colon V(H) \to \{1, \ldots, 5\}$. If c uses at most 4 colours for the neighbours of v, we can extend it to a 5-colouring of G. Let us assume, therefore, that v has exactly 5 neighbours, and that these have distinct colours.

Let D be an open disc around v, so small that it meets only those five straight edge segments of G that contain v. Let us enumerate these segments according to their cyclic position in D as $s_1, \ldots, s_5$, and let vv_i be the edge containing s_i ($i = 1, \ldots, 5$; Fig. 5.1.1). Without loss of generality we may assume that $c(v_i) = i$ for each i.

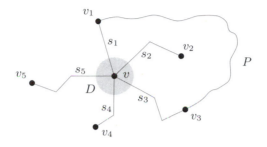

Fig. 5.1.1. The proof of the five colour theorem

Let us show first that every v_1–v_3 path $P \subseteq H$ separates v_2 from v_4 in H. Clearly, this is the case if and only if the cycle $C := vv_1 P v_3 v$ separates v_2 from v_4 in G. We prove this by showing that v_2 and v_4 lie in different faces of C.

Consider the two regions of $D \smallsetminus (s_1 \cup s_3)$. One of these regions meets s_2, the other s_4. Since $C \cap D \subseteq s_1 \cup s_3$, the two regions are each contained within a face of C. Moreover, these faces are distinct: otherwise, D would meet only one face of C, contrary to the fact that v lies on the boundary of both faces (Theorem 4.1.1). Thus $D \cap s_2$ and $D \cap s_4$ lie in distinct faces of C. As C meets the edges $vv_2 \supseteq s_2$ and $vv_4 \supseteq s_4$ only in v, the same holds for v_2 and v_4.

Given $i, j \in \{1, \ldots, 5\}$, let $H_{i,j}$ be the subgraph of H induced by the vertices coloured i or j. We may assume that the component C_1 of $H_{1,3}$ containing v_1 also contains v_3. Indeed, if we interchange the colours 1 and 3 at all the vertices of C_1, we obtain another 5-colouring of H; if $v_3 \notin C_1$, then v_1 and v_3 are both coloured 3 in this new colouring, and we may assign colour 1 to v. Thus, $H_{1,3}$ contains a v_1–v_3 path P. As shown above, P separates v_2 from v_4 in H. Since $P \cap H_{2,4} = \emptyset$, this means that v_2 and v_4 lie in different components of $H_{2,4}$. In the component containing v_2, we now interchange the colours 2 and 4, thus recolouring v_2 with colour 4. Now v no longer has a neighbour coloured 2, and we may give it this colour. $\square$

As a backdrop to the two famous theorems above, let us cite another well-known result:

Theorem 5.1.3. (Grötzsch 1959)
Every planar graph not containing a triangle is 3-colourable.

5.2 Colouring vertices

How do we determine the chromatic number of a given graph? How can we *find* a vertex-colouring with as few colours as possible? How does the chromatic number relate to other graph invariants, such as average degree, connectivity or girth?

Straight from the definition of the chromatic number we may derive the following upper bound:

Proposition 5.2.1. *Every graph G with m edges satisfies*

$$\chi(G) \leqslant \tfrac{1}{2} + \sqrt{2m + \tfrac{1}{4}}.$$

Proof. Let c be a vertex colouring of G with $k = \chi(G)$ colours. Then G has at least one edge between any two colour classes: if not, we could have used the same colour for both classes. Thus, $m \geqslant \tfrac{1}{2}k(k-1)$. Solving this inequality for k, we obtain the assertion claimed. □

greedy
algorithm One obvious way to colour a graph G with not too many colours is the following *greedy algorithm*: starting from a fixed vertex enumeration $v_1, \ldots, v_n$ of G, we consider the vertices in turn and colour each v_i with the first available colour—e.g., with the smallest positive integer not already used to colour any neighbour of v_i among $v_1, \ldots, v_{i-1}$. In this way, we never use more than $\Delta(G) + 1$ colours, even for unfavourable choices of the enumeration $v_1, \ldots, v_n$. If G is complete or an odd cycle, then this is even best possible.

In general, though, this upper bound of $\Delta + 1$ is rather generous, even for greedy colourings. Indeed, when we come to colour the vertex v_i in the above algorithm, we only need a supply of $d_{G[v_1,\ldots,v_i]}(v_i) + 1$ rather than $d_G(v_i) + 1$ colours to proceed; recall that, at this stage, the algorithm ignores any neighbours v_j of v_i with $j > i$. Hence in most graphs, there will be scope for an improvement of the $\Delta + 1$ bound by choosing a particularly suitable vertex ordering to start with: one that picks vertices of large degree early (when most neighbours are ignored) and vertices of small degree last. Locally, the number $d_{G[v_1,\ldots,v_i]}(v_i) + 1$ of colours required will be smallest if v_i has minimum degree in $G[v_1, \ldots, v_i]$. But this is easily achieved: we just choose v_n first, with $d(v_n) = \delta(G)$, then choose as v_{n-1} a vertex of minimum degree in $G - v_n$, and so on.

We thus have the following improved bound:

Proposition 5.2.2. *Every graph G satisfies*

$$\chi(G) \leqslant 1 + \max\{\,\delta(H) \mid H \subseteq G\,\}.$$

□

[9.2.1]
[9.2.3]
[11.2.3]

Corollary 5.2.3. *Every graph G has a subgraph of minimum degree at least $\chi(G) - 1$.* $\qquad\square$

As one easily checks (exercise), the bound of Proposition 5.2.2 is always at least as good as that of Proposition 5.2.1.

The number $1 + \max_{H \subseteq G} \delta(H)$ featuring in Proposition 5.2.2 is sometimes called the *colouring number* of G. As pointed out after Theorem 3.5.4, the colouring number of a graph is closely related to its arboricity.

colouring number

As we have seen, every graph G satisfies $\chi(G) \leqslant \Delta(G) + 1$, with equality for complete graphs and odd cycles. In all other cases, this general bound can be improved a little:

Theorem 5.2.4. (Brooks 1941)
Let G be a connected graph. If G is neither complete nor an odd cycle, then

$$\chi(G) \leqslant \Delta(G).$$

Proof. We apply induction on $|G|$. If $\Delta(G) \leqslant 2$, then G is a path or a cycle, and the assertion is trivial. We therefore assume that $\Delta := \Delta(G) \geqslant 3$, and that the assertion holds for graphs of smaller order. Suppose that $\chi(G) > \Delta$.

Δ

Let $v \in G$ be a vertex of degree Δ. Then $H := G - v$ satisfies $\chi(H) \leqslant \Delta$. Indeed, by induction, for every component H' of H we have $\chi(H') \leqslant \Delta(H') \leqslant \Delta$, unless H' is complete or an odd cycle—in which case $\chi(H') = \Delta(H') + 1 \leqslant \Delta$, since every vertex of H' has maximum degree in H' and one such vertex is also adjacent to v in G.

v, H

Since H can be Δ-coloured but G cannot, we have the following:

Every Δ-colouring of H uses all the colours $1, \ldots, \Delta$ on the neighbours of v. $\qquad(1)$

Given any Δ-colouring of H, let us denote the neighbour of v coloured i by v_i, $i = 1, \ldots, \Delta$. For all $i \neq j$, let $H_{i,j}$ denote the subgraph of H spanned by all the vertices coloured i or j.

$v_1, \ldots, v_\Delta$
$H_{i,j}$

For all $i \neq j$, the vertices v_i and v_j lie in a common component $C_{i,j}$ of $H_{i,j}$. $\qquad(2)$

$C_{i,j}$

Otherwise we could interchange the colours i and j in one of those components; then v_i and v_j would be coloured the same, contrary to (1).

$$C_{i,j} \text{ is always a } v_i\text{--}v_j \text{ path.} \qquad(3)$$

Indeed, let P be a v_i–v_j path in $C_{i,j}$. As $d_H(v_i) \leqslant \Delta - 1$, the neighbours of v_i have pairwise different colours: otherwise we could recolour v_i, contrary to (1). Hence the neighbour of v_i on P is its only neighbour in $C_{i,j}$, and similarly for v_j. Thus if $C_{i,j} \neq P$, then P has an inner vertex with three identically coloured neighbours in H; let u be the first such vertex on P (Fig. 5.2.1). Since at most $\Delta - 2$ colours are used on the neighbours of u, we may recolour u. But this makes $P\mathring{u}$ into a component of $H_{i,j}$, contradicting (2).

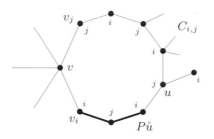

Fig. 5.2.1. The proof of (3) in Brooks's theorem

For distinct i, j, k, the paths $C_{i,j}$ and $C_{i,k}$ meet only in v_i. $\qquad (4)$

For if $u \neq v_i \in C_{i,j} \cap C_{i,k}$, then u has two neighbours coloured j and two coloured k, so we may recolour u. In the new colouring, v_i and v_j lie in different components of $H_{i,j}$, contrary to (2).

The proof of the theorem now follows easily. If the neighbours of v are pairwise adjacent, then each has Δ neighbours in $N(v) \cup \{v\}$ already, so $G = G[N(v) \cup \{v\}] = K^{\Delta+1}$. As G is complete, there is nothing to show. We may thus assume that $v_1 v_2 \notin G$, where $v_1, \dots, v_\Delta$ derive their names from some fixed Δ-colouring c of H. Let $u \neq v_2$ be the neighbour of v_1 on the path $C_{1,2}$; then $c(u) = 2$. Interchanging the colours 1 and 3 in $C_{1,3}$, we obtain a new colouring c' of H; let v_i', $H_{i,j}'$, $C_{i,j}'$ etc. be defined with respect to c' in the obvious way. As a neighbour of $v_1 = v_3'$, our vertex u now lies in $C_{2,3}'$, since $c'(u) = c(u) = 2$. By (4) for c, however, the path $\mathring{v}_1 C_{1,2}$ retained its original colouring, so $u \in \mathring{v}_1 C_{1,2} \subseteq C_{1,2}'$. Hence $u \in C_{2,3}' \cap C_{1,2}'$, contradicting (4) for c'. $\qquad \square$

As we have seen, a graph G of large chromatic number must have large maximum degree: at least $\chi(G) - 1$. What else can we say about the structure of graphs with large chromatic number?

One obvious possible cause for $\chi(G) \geqslant k$ is the presence of a K^k subgraph. This is a local property of G, compatible with arbitrary values of global invariants such as ε and κ. Hence, the assumption of $\chi(G) \geqslant k$ does not tell us anything about those invariants for G itself. It does, however, imply the existence of a subgraph where those invariants are large: by Corollary 5.2.3, G has a subgraph H with $\delta(H) \geqslant k - 1$, and hence by Theorem 1.4.2 a subgraph H' with $\kappa(H') \geqslant \frac{1}{4}(k-1)$.

So are those somewhat denser subgraphs the 'cause' for the large value of χ? Do they, in turn, necessarily contain a graph of high chromatic number—maybe even one from some small collection of *canonical* such graphs, such as K^k? Interestingly, this is not so: those subgraphs of large but 'constant' average degree—bounded below only by a function of k, not of $|G|$—are not nearly dense enough to contain (necessarily) any particular graph of high chromatic number, let alone K^k.[1]

Yet even if the above local structures do not appear to help, it might still be the case that, somehow, a high chromatic number forces the existence of certain canonical highly chromatic subgraphs. That this is in fact not the case will be our main result in Chapter 11: according to a classic result of Erdős, proved by probabilistic methods, *there are graphs of arbitrarily large chromatic number and yet arbitrarily large girth* (Theorem 11.2.2). Thus given any graph H that is not a forest, for every $k \in \mathbb{N}$ there are graphs G with $\chi(G) \geqslant k$ but $H \nsubseteq G$.[2]

Thus, contrary to our initial guess that a large chromatic number might always be caused by some dense local substructure, it can in fact occur as a purely global phenomenon: after all, locally (around each vertex) a graph of large girth looks just like a tree, and is in particular 2-colourable there!

So far, we asked what a high chromatic number implies: it forces the invariants δ, d, Δ and κ up in some subgraph, but it does not imply the existence of any concrete subgraph of large chromatic number. Let us now consider the converse question: from what assumptions could we deduce that the chromatic number of a given graph is large?

Short of a concrete subgraph known to be highly chromatic (such as K^k), there is little or nothing in sight: no values of the invariants studied so far imply that the graph considered has a large chromatic number. (Recall the example of $K_{n,n}$.) So what exactly can cause high chromaticity as a global phenomenon largely remains a mystery!

Nevertheless, there exists a simple—though not always short—procedure to construct all the graphs of chromatic number $\geqslant k$. For each $k \in \mathbb{N}$, let us define the class of *k-constructible* graphs recursively as follows:

k-con-structible

(i) K^k is k-constructible.

(ii) If G is k-constructible and $x, y \in V(G)$ are non-adjacent, then also $(G + xy)/xy$ is k-constructible.

---

[1] This is obvious from the examples of $K_{n,n}$, which are 2-chromatic but whose connectivity and average degree n exceeds any constant bound. Which (non-constant) average degree exactly will force the existence of a given subgraph will be the topic of Chapter 7.

[2] By Corollaries 5.2.3 and 1.5.4, of course, every graph of sufficiently high chromatic number will contain any given forest.

(iii) If G_1, G_2 are k-constructible and there are vertices x, y_1, y_2 such that $G_1 \cap G_2 = \{\, x \,\}$, $xy_1 \in E(G_1)$ and $xy_2 \in E(G_2)$, then also $(G_1 \cup G_2) - xy_1 - xy_2 + y_1y_2$ is k-constructible (Fig. 5.2.2).

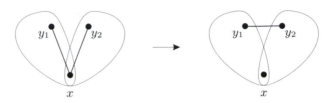

Fig. 5.2.2. The *Hajós construction* (iii)

One easily checks inductively that all k-constructible graphs—and hence their supergraphs—are at least k-chromatic. Indeed, if $(G + xy)/xy$ as in (ii) has a colouring with fewer than k colours, then this defines such a colouring also for G, a contradiction. Similarly, in any colouring of the graph constructed in (iii), the vertices y_1 and y_2 do not both have the same colour as x, so this colouring induces a colouring of either G_1 or G_2 and hence uses at least k colours.

It is remarkable, though, that the converse holds too:

Theorem 5.2.5. (Hajós 1961)
Let G be a graph and $k \in \mathbb{N}$. Then $\chi(G) \geqslant k$ if and only if G has a k-constructible subgraph.

Proof. Let G be a graph with $\chi(G) \geqslant k$; we show that G has a k-constructible subgraph. Suppose not; then $k \geqslant 3$. Adding some edges if necessary, let us make G edge-maximal with the property that none of its subgraphs is k-constructible. Now G is not a complete r-partite graph for any r: for then $\chi(G) \geqslant k$ would imply $r \geqslant k$, and G would contain the k-constructible graph K^k.

Since G is not a complete multipartite graph, non-adjacency is not an equivalence relation on $V(G)$. So there are vertices y_1, x, y_2 such that $y_1x, xy_2 \notin E(G)$ but $y_1y_2 \in E(G)$. Since G is edge-maximal without a k-constructible subgraph, each edge xy_i lies in some k-constructible subgraph H_i of $G + xy_i$ ($i = 1, 2$).

Let H_2' be an isomorphic copy of H_2 that contains x and $H_2 - H_1$ but is otherwise disjoint from G, together with an isomorphism $v \mapsto v'$ from H_2 to H_2' that fixes $H_2 \cap H_2'$ pointwise. Then $H_1 \cap H_2' = \{\, x \,\}$, so

$$H := (H_1 \cup H_2') - xy_1 - xy_2' + y_1y_2'$$

is k-constructible by (iii). One vertex at a time, let us identify in H each vertex $v' \in H_2' - G$ with its partner v; since vv' is never an edge of H,

y_1x, xy_2

H_1, H_2

H_2'

v' etc.

each of these identifications amounts to a construction step of type (ii). Eventually, we obtain the graph

$$(H_1 \cup H_2) - xy_1 - xy_2 + y_1y_2 \subseteq G \,;$$

this is the desired k-constructible subgraph of G. $\square$

5.3 Colouring edges

Clearly, every graph G satisfies $\chi'(G) \geqslant \Delta(G)$. For bipartite graphs, we have equality here:

Proposition 5.3.1. (König 1916)
Every bipartite graph G satisfies $\chi'(G) = \Delta(G)$.

Proof. We apply induction on $\|G\|$. For $\|G\| = 0$ the assertion holds. Now assume that $\|G\| \geqslant 1$, and that the assertion holds for graphs with fewer edges. Let $\Delta := \Delta(G)$, pick an edge $xy \in G$, and choose a Δ-edge-colouring of $G - xy$ by the induction hypothesis. Let us refer to the edges coloured α as α-*edges*, etc.

(1.6.1)

Δ, xy

α-edge

In $G - xy$, each of x and y is incident with at most $\Delta - 1$ edges. Hence there are $\alpha, \beta \in \{1, \ldots, \Delta\}$ such that x is not incident with an α-edge and y is not incident with a β-edge. If $\alpha = \beta$, we can colour the edge xy with this colour and are done; so we may assume that $\alpha \neq \beta$, and that x is incident with a β-edge.

α, β

Let us extend this edge to a maximal walk W whose edges are coloured β and α alternately. Since no such walk contains a vertex twice (why not?), W exists and is a path. Moreover, W does not contain y: if it did, it would end in y on an α-edge (by the choice of β) and thus have even length, so $W + xy$ would be an odd cycle in G (cf. Proposition 1.6.1). We now recolour all the edges on W, swapping α with β. By the choice of α and the maximality of W, adjacent edges of $G - xy$ are still coloured differently. We have thus found a Δ-edge-colouring of $G - xy$ in which neither x nor y is incident with a β-edge. Colouring xy with β, we extend this colouring to a Δ-edge-colouring of G. $\square$

Theorem 5.3.2. (Vizing 1964)
Every graph G satisfies

$$\Delta(G) \leqslant \chi'(G) \leqslant \Delta(G) + 1 \,.$$

Proof. We prove the second inequality by induction on $\|G\|$. For $\|G\| = 0$ it is trivial. For the induction step let $G = (V, E)$ with $\Delta := \Delta(G) > 0$ be

V, E

Δ

given, and assume that the assertion holds for graphs with fewer edges.
Instead of '$(\Delta + 1)$-edge-colouring' let us just say 'colouring'. An edge
coloured α will again be called an α-*edge*.

For every edge $e \in G$ there exists a colouring of $G - e$, by the
induction hypothesis. In such a colouring, the edges at a given vertex
v use at most $d(v) \leqslant \Delta$ colours, so some colour $\beta \in \{1, \ldots, \Delta + 1\}$ is
missing at v. For any other colour α, there is a unique maximal walk
(possibly trivial) starting at v, whose edges are coloured alternately α
and β. This walk is a path; we call it the α/β - *path from* v.

Suppose that G has no colouring. Then the following holds:

> Given $xy \in E$, and any colouring of $G - xy$ in which the
> colour α is missing at x and the colour β is missing at y,   (1)
> the α/β - path from y ends in x.

Otherwise we could interchange the colours α and β along this path and
colour xy with α, obtaining a colouring of G (contradiction).

Let $xy_0 \in G$ be an edge. By induction, $G_0 := G - xy_0$ has a
colouring c_0. Let α be a colour missing at x in this colouring. Further,
let $y_0, y_1, \ldots, y_k$ be a maximal sequence of distinct neighbours of x in G,
such that $c_0(xy_i)$ is missing at y_{i-1} for each $i = 1, \ldots, k$. For each of the
graphs $G_i := G - xy_i$ we define a colouring c_i, setting

$$c_i(e) := \begin{cases} c_0(xy_{j+1}) & \text{for } e = xy_j \text{ with } j \in \{0, \ldots, i-1\} \\ c_0(e) & \text{otherwise;} \end{cases}$$

note that in each of these colourings the same colours are missing at x
as in c_0.

Now let β be a colour missing at y_k in c_0. Clearly, β is still missing
at y_k in c_k. If β were also missing at x, we could colour xy_k with β and
thus extend c_k to a colouring of G. Hence, x is incident with a β-edge
(in every colouring). In fact, by the maximality of k,

$$c_0(xy_i) = \beta \qquad \text{for some } i \in \{1, \ldots, k-1\}. \qquad (2)$$

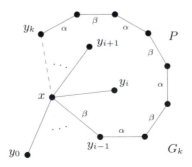

Fig. 5.3.1. The α/β - path P in G_k

Let P be the α/β-path from y_k in G_k (with respect to c_k; Fig. 5.3.1). $\quad P$
By (1), P ends in x, and it does so on a β-edge, since α is missing at x.
As $\beta = c_0(xy_i) = c_k(xy_{i-1})$, this is the edge xy_{i-1}. In c_0, however, and
hence also in c_{i-1}, β is missing at y_{i-1} (by (2) and the choice of y_i); let
P' be the α/β-path from y_{i-1} in G_{i-1} (with respect to c_{i-1}). Since P' $\quad P'$
is uniquely determined, it starts with $y_{i-1}Py_k$; note that the edges of
$P\mathring{x}$ are coloured the same in c_{i-1} as in c_k. But in c_0, and hence in c_{i-1},
there is no β-edge at y_k (by the choice of β). Therefore P' ends in y_k,
contradicting (1). $\hfill\square$

Vizing's theorem divides the finite graphs into two classes according
to their chromatic index; graphs satisfying $\chi' = \Delta$ are called (imagina-
tively) *class 1*, those with $\chi' = \Delta + 1$ are *class 2*.

5.4 List colouring

In this section, we take a look at a relatively recent generalization of the
concepts of colouring studied so far. This generalization may seem a little
far-fetched at first glance, but it turns out to supply a fundamental link
between the classical (vertex and edge) chromatic numbers of a graph
and its other invariants.

Suppose we are given a graph $G = (V, E)$, and for each vertex of
G a list of colours permitted at that particular vertex: when can we
colour G (in the usual sense) so that each vertex receives a colour from
its list? More formally, let $(S_v)_{v \in V}$ be a family of sets. We call a vertex
colouring c of G with $c(v) \in S_v$ for all $v \in V$ a colouring *from the
lists* S_v. The graph G is called *k-list-colourable*, or *k-choosable*, if, for $\quad$ *k-choosable*
every family $(S_v)_{v \in V}$ with $|S_v| = k$ for all v, there is a vertex colouring
of G from the lists S_v. The least integer k for which G is k-choosable is
the *list-chromatic number*, or *choice number* $\mathrm{ch}(G)$ of G. $\quad$ *choice number* $\mathrm{ch}(G)$

List-colourings of edges are defined analogously. The least integer
k such that G has an edge colouring from any family of lists of size k
is the *list-chromatic index* $\mathrm{ch}'(G)$ of G; formally, we just set $\mathrm{ch}'(G) :=$ $\quad \mathrm{ch}'(G)$
$\mathrm{ch}(L(G))$, where $L(G)$ is the line graph of G.

In principle, showing that a given graph is k-choosable is more difficult than proving it to be k-colourable: the latter is just the special case
of the former where all lists are equal to $\{1, \ldots, k\}$. Thus,

$$\mathrm{ch}(G) \geqslant \chi(G) \quad \text{and} \quad \mathrm{ch}'(G) \geqslant \chi'(G)$$

for all graphs G.

In spite of these inequalities, many of the known upper bounds
for the chromatic number have turned out to be valid for the choice
number, too. Examples for this phenomenon include Proposition 5.2.2

(with the same proof), and Brooks's theorem. On the other hand, it is easy to construct graphs for which the two invariants are wide apart (Exercise 24). Taken together, these two facts indicate a little how far those general upper bounds on the chromatic number may be from the truth.

The following theorem shows that, in terms of its relationship to other graph invariants, the choice number differs fundamentally from the chromatic number. As mentioned before, there are 2-chromatic graphs of arbitrarily large minimum degree, e.g. the graphs $K_{n,n}$. The choice number, however, will be forced up by large values of invariants like δ, ε or κ:

Theorem 5.4.1. (Alon 1993)
There exists a function $f: \mathbb{N} \to \mathbb{N}$ such that, given any integer k, all graphs G with average degree $d(G) \geqslant f(k)$ satisfy $\mathrm{ch}(G) \geqslant k$.

The proof of Theorem 5.4.1 uses probabilistic methods as introduced in Chapter 11.

Empirically, the choice number's different character is highlighted by another phenomenon: even in cases where known bounds for the chromatic number could be transferred to the choice number, their proofs have tended to be rather different.

One of the simplest and most impressive examples for this is the list version of the five colour theorem: every planar graph is 5-choosable. This had been conjectured for almost 20 years, before Thomassen found a very simple induction proof. This proof does not use the five colour theorem—which thus gets reproved in a very different way.

Theorem 5.4.2. (Thomassen 1994)
Every planar graph is 5-choosable.

(4.2.6)

Proof. We shall prove the following assertion for all plane graphs G with at least 3 vertices:

> Suppose that every inner face of G is bounded by a triangle and its outer face by a cycle $C = v_1 \ldots v_k v_1$. Suppose further that v_1 has already been coloured with the colour 1, and v_2 has been coloured 2. Suppose finally that with every other vertex of C a list of at least 3 colours is associated, and with every vertex of $G - C$ a list of at least 5 colours. Then the colouring of v_1 and v_2 can be extended to a colouring of G from the given lists. $\qquad(*)$

Let us check first that $(*)$ implies the assertion of the theorem. Let any plane graph be given, together with a list of 5 colours for each vertex. Add edges to this graph until it is a maximal plane graph G.

By Proposition 4.2.6, G is a plane triangulation; let $v_1v_2v_3v_1$ be the boundary of its outer face. We now colour v_1 and v_2 (differently) from their lists, and extend this colouring by $(*)$ to a colouring of G from the lists given.

Let us now prove $(*)$, by induction on $|G|$. If $|G| = 3$, then $G = C$ and the assertion is trivial. Now let $|G| \geqslant 4$, and assume $(*)$ for smaller graphs. If C has a chord vw, then vw lies on two unique cycles $C_1, C_2 \subseteq C + vw$ with $v_1v_2 \in C_1$ and $v_1v_2 \notin C_2$. For $i = 1, 2$, let G_i denote the subgraph of G induced by the vertices lying on C_i or in its inner face (Fig. 5.4.1). Applying the induction hypothesis first to G_1 and then—with the colours now assigned to v and w—to G_2 yields the desired colouring of G.

vw

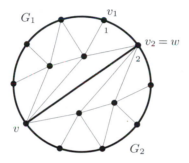

Fig. 5.4.1. The induction step with a chord vw; here the case of $w = v_2$

If C has no chord, let $v_1, u_1, \ldots, u_m, v_{k-1}$ be the neighbours of v_k in their natural cyclic order order around v_k;[3] by definition of C, all those neighbours u_i lie in the inner face of C (Fig. 5.4.2). As the inner faces of C are bounded by triangles, $P := v_1u_1 \ldots u_mv_{k-1}$ is a path in G, and $C' := P \cup (C - v_k)$ a cycle.

$u_1, \ldots, u_m$

C'

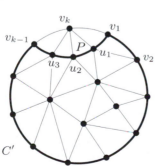

Fig. 5.4.2. The induction step without a chord

---

[3] as in the first proof of the five colour theorem

We now choose two different colours $j, \ell \neq 1$ from the list of v_k and delete these colours from the lists of all the vertices u_i. Then every list of a vertex on C' still has at least 3 colours, so by induction we may colour C' and its interior, i.e. the graph $G - v_k$. At least one of the two colours j, ℓ is not used for v_{k-1}, and we may assign that colour to v_k.   □

As is often the case with induction proofs, the trick of the proof above lies in the delicately balanced strengthening of the assertion proved. Note that the proof uses neither traditional colouring arguments (such as swapping colours along a path) nor the Euler formula implicit in the standard proof of the five colour theorem. This suggests that maybe in other unsolved colouring problems too it might be of advantage to aim straight for their list version, i.e. to prove an assertion of the form $\mathrm{ch}(G) \leqslant k$ instead of the formally weaker $\chi(G) \leqslant k$. Unfortunately, this approach fails for the four colour theorem: planar graphs are *not* in general 4-choosable.

As mentioned before, the chromatic number of a graph and its choice number may differ a lot. Surprisingly, however, no such examples are known for edge colourings. Indeed it has been conjectured that none exist:

List colouring conjecture. *Every graph G satisfies $\mathrm{ch}'(G) = \chi'(G)$.*

We shall prove the list colouring conjecture for bipartite graphs. As a tool we shall use orientations of graphs, defined in Chapter 1.10. If D is a directed graph and $v \in V(D)$, we denote by $N^+(v)$ the set, and by $d^+(v)$ the number, of vertices w such that D contains an edge directed from v to w.

To see how orientations come into play in the context of colouring, let us recall the greedy algorithm from Section 5.2. In order to apply the algorithm to a graph G, we first have to choose a vertex enumeration $v_1, \ldots, v_n$ of G. The enumeration chosen defines an orientation of G: just orient every edge $v_i v_j$ 'backwards', from v_i to v_j if $i > j$. Then, for each vertex v_i to be coloured, the algorithm considers only those edges at v_i that are directed away from v_i: if $d^+(v) < k$ for all vertices v, it will use at most k colours. Moreover, the first colour class U found by the algorithm has the following property: it is an independent set of vertices to which every other vertex sends an edge. The second colour class has the same property in $G - U$, and so on.

The following lemma generalizes this to orientations D of G that do not necessarily come from a vertex enumeration, but may contain some directed cycles. Let us call an independent set $U \subseteq V(D)$ a *kernel* of D if, for every vertex $v \in D - U$, there is an edge in D directed from v to a vertex in U. Note that kernels of non-empty directed graphs are themselves non-empty.

$N^+(v)$
$d^+(v)$

kernel

Lemma 5.4.3. *Let H be a graph and $(S_v)_{v \in V(H)}$ a family of lists. If H has an orientation D with $d^+(v) < |S_v|$ for every v, and such that every induced subgraph of D has a kernel, then H can be coloured from the lists S_v.*

Proof. We apply induction on $|H|$. For $|H| = 0$ we take the empty colouring. For the induction step, let $|H| > 0$. Let α be a colour occurring in one of the lists S_v, and let D be an orientation of H as stated. The vertices v with $\alpha \in S_v$ span a non-empty subgraph D' in D; by assumption, D' has a kernel $U \neq \emptyset$.

   Let us colour the vertices in U with α, and remove α from the lists of all the other vertices of D'. Since each of those vertices sends an edge to U, the modified lists S_v' for $v \in D - U$ again satisfy the condition $d^+(v) < |S_v'|$ in $D - U$. Since $D - U$ is an orientation of $H - U$, we can thus colour $H - U$ from those lists by the induction hypothesis. As none of these lists contains α, this extends our colouring $U \to \{\alpha\}$ to the desired list colouring of H.   □

Theorem 5.4.4. (Galvin 1995)
Every bipartite graph G satisfies $\mathrm{ch}'(G) = \chi'(G)$.

Proof. Let $G =: (X \cup Y, E)$, where $\{X, Y\}$ is a vertex bipartition of G. Let us say that two edges of G *meet in X* if they share an end in X, and correspondingly for Y. Let $\chi'(G) =: k$, and let c be a k-edge-colouring of G.

   Clearly, $\mathrm{ch}'(G) \geqslant k$; we prove that $\mathrm{ch}'(G) \leqslant k$. Our plan is to use Lemma 5.4.3 to show that the line graph H of G is k-choosable. To apply the lemma, it suffices to find an orientation D of H with $d^+(v) < k$ for every vertex v, and such that every induced subgraph of D has a kernel. To define D, consider adjacent $e, e' \in E$, say with $c(e) < c(e')$. If e and e' meet in X, we orient the edge $ee' \in H$ from e' towards e; if e and e' meet in Y, we orient it from e to e' (Fig 5.4.3).

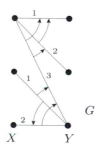

Fig. 5.4.3. Orienting the line graph of G

   Let us compute $d^+(e)$ for given $e \in E = V(D)$. If $c(e) = i$, say, then every $e' \in N^+(e)$ meeting e in X has its colour in $\{1, \dots, i-1\}$, and every $e' \in N^+(e)$ meeting e in Y has its colour in $\{i+1,\ldots,k\}$. As any two neighbours e' of e meeting e either both in X or both in Y are themselves adjacent and hence coloured differently, this implies $d^+(e) < k$ as desired.

It remains to show that every induced subgraph D' of D has a kernel. We show this by induction on $|D'|$. For $D' = \emptyset$, the empty set is a kernel; so let $|D'| \geqslant 1$. Let $E' := V(D') \subseteq E$. For every $x \in X$ at which E' has an edge, let $e_x \in E'$ be the edge at x with minimum c-value, and let U denote the set of all those edges e_x. Then every edge $e' \in E' \smallsetminus U$ meets some $e \in U$ in X, and the edge $ee' \in D'$ is directed from e' to e. If U is independent, it is thus a kernel of D' and we are home; let us assume, therefore, that U is not independent.

Let $e, e' \in U$ be adjacent, and assume that $c(e) < c(e')$. By definition of U, e and e' meet in Y, so the edge $ee' \in D$ is directed from e to e'. By the induction hypothesis, $D' - e$ has a kernel U'. If $e' \in U'$, then U' is also a kernel of D', and we are done. If not, there exists an $e'' \in U'$ such that D' has an edge directed from e' to e''. If e' and e'' met in X, then $c(e'') < c(e')$ by definition of D, contradicting $e' \in U$. Hence e' and e'' meet in Y, and $c(e') < c(e'')$. Since e and e' meet in Y, too, also e and e'' meet in Y, and $c(e) < c(e') < c(e'')$. So the edge ee'' is directed from e towards e'', so again U' is also a kernel of D'. $\qquad\square$

By Proposition 5.3.1, we now know the exact list-chromatic index of bipartite graphs:

Corollary 5.4.5. *Every bipartite graph G satisfies* $\mathrm{ch}'(G) = \Delta(G)$.
$\qquad\square$

5.5 Perfect graphs

As discussed in Section 5.2, a high chromatic number may occur as a purely global phenomenon: even when a graph has large girth, and thus locally looks like a tree, its chromatic number may be arbitrarily high. Since such 'global dependence' is obviously difficult to deal with, one may become interested in graphs where this phenomenon does not occur, i.e. whose chromatic number is high only when there is a local reason for it.

Before we make this precise, let us note two definitions for a graph G. The greatest integer r such that $K^r \subseteq G$ is the *clique number* $\omega(G)$ of G, and the greatest integer r such that $\overline{K^r} \subseteq G$ (induced) is the *independence number* $\alpha(G)$ of G. Clearly, $\alpha(G) = \omega(\overline{G})$ and $\omega(G) = \alpha(\overline{G})$.

A graph is called *perfect* if every induced subgraph $H \subseteq G$ has chromatic number $\chi(H) = \omega(H)$, i.e. if the trivial lower bound of $\omega(H)$ colours always suffices to colour the vertices of H. Thus, while proving an assertion of the form $\chi(G) > k$ may in general be difficult, even in principle, for a given graph G, it can always be done for a perfect graph simply by exhibiting some K^{k+1} subgraph as a 'certificate' for non-colourability with k colours.

At first glance, the structure of the class of perfect graphs appears somewhat contrived: although it is closed under induced subgraphs (if only by explicit definition), it is not closed under taking general subgraphs or supergraphs, let alone minors (examples?). However, perfection is an important notion in graph theory: the fact that several fundamental classes of graphs are perfect (as if by fluke) may serve as a superficial indication of this.[4]

What graphs, then, are perfect? Bipartite graphs are, for instance. Less trivially, the complements of bipartite graphs are perfect, too— a fact equivalent to König's duality theorem 2.1.1 (Exercise 33). The so-called *comparability graphs* are perfect, and so are the *interval graphs* (see the exercises); both these turn up in numerous applications.

In order to study at least one such example in some detail, we prove here that the chordal graphs are perfect: a graph is *chordal* (or *triangulated*) if each of its cycles of length at least 4 has a chord, i.e. if it contains no induced cycles other than triangles.

chordal

To show that chordal graphs are perfect, we shall first characterize their structure. If G is a graph with induced subgraphs G_1, G_2 and S, such that $G = G_1 \cup G_2$ and $S = G_1 \cap G_2$, we say that G arises from G_1 and G_2 by *pasting* these graphs together along S.

pasting

Proposition 5.5.1. *A graph is chordal if and only if it can be constructed recursively by pasting along complete subgraphs, starting from complete graphs.*

[12.3.8]

Proof. If G is obtained from two chordal graphs G_1, G_2 by pasting them together along a complete subgraph, then G is clearly again chordal: any induced cycle in G lies in either G_1 or G_2, and is hence a triangle by assumption. Since complete graphs are chordal, this proves that all graphs constructible as stated are chordal.

Conversely, let G be a chordal graph. We show by induction on $|G|$ that G can be constructed as described. This is trivial if G is complete. We therefore assume that G is not complete, in particular $|G| > 1$, and that all smaller chordal graphs are constructible as stated. Let $a, b \in G$ be two non-adjacent vertices, and let $X \subseteq V(G) \setminus \{a, b\}$ a minimal set of vertices separating a from b. Let C denote the component of $G - X$ containing a, and put $G_1 := G[V(C) \cup X]$ and $G_2 := G - C$.

a, b
X
C
G_1, G_2

---

[4] The class of perfect graphs has duality properties with deep connections to optimization and complexity theory, which are far from understood. Theorem 5.5.5 shows the tip of an iceberg here; for more, the reader is referred to Lovász's survey cited in the notes.

S

Then G arises from G_1 and G_2 by pasting these graphs together along
$S := G[X]$.

s,t

Since G_1 and G_2 are both chordal (being induced subgraphs of G)
and hence constructible by induction, it suffices to show that S is complete. Suppose, then, that $s,t \in S$ are non-adjacent. By the minimality
of $X = V(S)$ as an a–b separator, both s and t have a neighbour in C.
Hence, there is an X-path from s to t in G_1; we let P_1 be a shortest such
path. Analogously, G_2 contains a shortest X-path P_2 from s to t. But
then $P_1 \cup P_2$ is a chordless cycle of length $\geqslant 4$ (Fig. 5.5.1), contradicting
our assumption that G is chordal. □

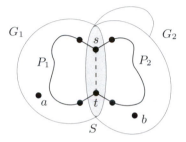

Fig. 5.5.1. If G_1 and G_2 are chordal, then so is G

Proposition 5.5.2. *Every chordal graph is perfect.*

Proof. Since complete graphs are perfect, it suffices by Proposition
5.5.1 to show that any graph G obtained from perfect graphs G_1, G_2 by
pasting them together along a complete subgraph S is again perfect. So
let $H \subseteq G$ be an induced subgraph; we show that $\chi(H) \leqslant \omega(H)$.

Let $H_i := H \cap G_i$ for $i = 1, 2$, and let $T := H \cap S$. Then T is
again complete, and H arises from H_1 and H_2 by pasting along T. As
an induced subgraph of G_i, each H_i can be coloured with $\omega(H_i)$ colours.
Since T is complete and hence coloured injectively, two such colourings,
one of H_1 and one of H_2, may be combined into a colouring of H with
$\max \{ \omega(H_1), \omega(H_2) \} \leqslant \omega(H)$ colours—if necessary by permuting the
colours in one of the H_i. □

We now come to the main result in the theory of perfect graphs, the
perfect graph theorem:

perfect
graph
theorem

Theorem 5.5.3. (Lovász 1972)
A graph is perfect if and only if its complement is perfect.

We shall give two proofs of Theorem 5.5.3. The first of these is Lovász's
original proof, which is still unsurpassed in its clarity and the amount
of 'feel' for the problem it conveys. Our second proof, due to Gasparian
(1996), is in fact a very short and elegant linear algebra proof of another
theorem of Lovász's (Theorem 5.5.5), which easily implies Theorem 5.5.3.

Let us prepare our first proof of the perfect graph theorem by a lemma. Let G be a graph and $x \in G$ a vertex, and let G' be obtained from G by adding a vertex x' and joining it to x and all the neighbours of x. We say that G' is obtained from G by *expanding* the vertex x to an edge xx' (Fig. 5.5.2).

expanding a vertex

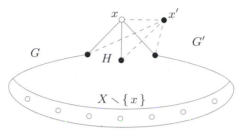

Fig. 5.5.2. Expanding the vertex x in the proof of Lemma 5.5.4

Lemma 5.5.4. *Any graph obtained from a perfect graph by expanding a vertex is again perfect.*

Proof. We use induction on the order of the perfect graph considered. Expanding the vertex of K^1 yields K^2, which is perfect. For the induction step, let G be a non-trivial perfect graph, and let G' be obtained from G by expanding a vertex $x \in G$ to an edge xx'. For our proof that G' is perfect it suffices to show $\chi(G') \leqslant \omega(G')$: every proper induced subgraph H of G' is either isomorphic to an induced subgraph of G or obtained from a proper induced subgraph of G by expanding x; in either case, H is perfect by assumption and the induction hypothesis, and can hence be coloured with $\omega(H)$ colours.

x, x'

Let $\omega(G) =: \omega$; then $\omega(G') \in \{\omega, \omega+1\}$. If $\omega(G') = \omega+1$, then

ω

$$\chi(G') \leqslant \chi(G)+1 = \omega+1 = \omega(G')$$

and we are done. So let us assume that $\omega(G') = \omega$. Then x lies in no $K^\omega \subseteq G$: together with x', this would yield a $K^{\omega+1}$ in G'. Let us colour G with ω colours. Since every $K^\omega \subseteq G$ meets the colour class X of x but not x itself, the graph $H := G - (X \setminus \{x\})$ has clique number $\omega(H) < \omega$ (Fig. 5.5.2). Since G is perfect, we may thus colour H with $\omega - 1$ colours. Now X is independent, so the set $(X \setminus \{x\}) \cup \{x'\} = V(G' - H)$ is also independent. We can therefore extend our $(\omega - 1)$-colouring of H to an ω-colouring of G', showing that $\chi(G') \leqslant \omega = \omega(G')$ as desired. $\square$

X

H

Proof of Theorem 5.5.3. Applying induction on $|G|$, we show that the complement $\overline{G}$ of any perfect graph $G = (V, E)$ is again perfect. For $|G| = 1$ this is trivial, so let $|G| \geqslant 2$ for the induction step. Let $\mathcal{K}$ denote the set of all vertex sets of complete subgraphs of G. Put $\alpha(G) =: \alpha$, and let $\mathcal{A}$ be the set of all independent vertex sets A in G with $|A| = \alpha$.

$G = (V, E)$

$\mathcal{K}$

α

$\mathcal{A}$

Every proper induced subgraph of $\overline{G}$ is the complement of a proper induced subgraph of G, and is hence perfect by induction. For the perfection of $\overline{G}$ it thus suffices to prove $\chi(\overline{G}) \leqslant \omega(\overline{G})\ (=\alpha)$. To this end, we shall find a set $K \in \mathcal{K}$ such that $K \cap A \neq \emptyset$ for all $A \in \mathcal{A}$; then

$$\omega(\overline{G} - K) = \alpha(G - K) < \alpha = \omega(\overline{G}),$$

so by the induction hypothesis

$$\chi(\overline{G}) \leqslant \chi(\overline{G} - K) + 1 = \omega(\overline{G} - K) + 1 \leqslant \omega(\overline{G})$$

as desired.

Suppose there is no such K; thus, for every $K \in \mathcal{K}$ there exists a set $A_K \in \mathcal{A}$ with $K \cap A_K = \emptyset$. Let us replace in G every vertex x by a complete graph G_x of order

$$k(x) := \big| \{\, K \in \mathcal{K} \mid x \in A_K \,\} \big|,$$

joining all the vertices of G_x to all the vertices of G_y whenever x and y are adjacent in G. The graph G' thus obtained has vertex set $\bigcup_{x \in V} V(G_x)$, and two vertices $v \in G_x$ and $w \in G_y$ are adjacent in G' if and only if either $x = y$, or $x \neq y$ and $xy \in E$. Moreover, G' can be obtained by repeated vertex expansion from the graph $G\,[\,\{\, x \in V \mid k(x) > 0 \,\}\,]$. Being an induced subgraph of G, this latter graph is perfect by assumption, so G' is perfect by Lemma 5.5.4. In particular,

$$\chi(G') \leqslant \omega(G').\tag{1}$$

In order to obtain a contradiction with (1), we now compute in turn the actual values of $\omega(G')$ and $\chi(G')$. By construction of G', every maximal complete subgraph of G' has the form $G'\,[\,\bigcup_{x \in X} G_x\,]$ for some $X \in \mathcal{K}$. So there exists a set $X \in \mathcal{K}$ such that

$$\omega(G') = \sum_{x \in X} k(x)$$

$$= \big| \{\, (x, K) : x \in X,\ K \in \mathcal{K},\ x \in A_K \,\} \big|$$

$$= \sum_{K \in \mathcal{K}} |X \cap A_K|$$

$$\leqslant |\mathcal{K}| - 1;\tag{2}$$

the last inequality follows from the fact that $|X \cap A_K| \leqslant 1$ for all K (since A_K is independent but $G\,[\,X\,]$ is complete), and $|X \cap A_X| = 0$ (by the choice of A_X). On the other hand, we have

$$
\begin{aligned}
|G'| &= \sum_{x \in V} k(x) \\
&= \left| \{ (x, K) : x \in V, \ K \in \mathcal{K}, \ x \in A_K \} \right| \\
&= \sum_{K \in \mathcal{K}} |A_K| \\
&= |\mathcal{K}| \cdot \alpha .
\end{aligned}
$$

As $\alpha(G') \leqslant \alpha$ by construction of G', this implies

$$
\chi(G') \geqslant \frac{|G'|}{\alpha(G')} \geqslant \frac{|G'|}{\alpha} = |\mathcal{K}| . \tag{3}
$$

Putting (2) and (3) together we obtain

$$
\chi(G') \geqslant |\mathcal{K}| > |\mathcal{K}| - 1 \geqslant \omega(G') ,
$$

a contradiction to (1). $\square$

Since the following characterization of perfection is symmetrical in G and $\overline{G}$, it clearly implies the perfect graph theorem:

Theorem 5.5.5. (Lovász 1972)
A graph G is perfect if and only if

$$
|H| \leqslant \alpha(H) \cdot \omega(H) \tag{$*$}
$$

for all induced subgraphs $H \subseteq G$.

Proof. Let us write $V(G) =: V =: \{ v_1, \ldots, v_n \}$, and put $\alpha := \alpha(G)$ and $\omega := \omega(G)$. The necessity of $(*)$ is immediate: if G is perfect, then every induced subgraph H of G can be partitioned into at most $\omega(H)$ colour classes each containing at most $\alpha(H)$ vertices, and $(*)$ follows.

    To prove sufficiency, we apply induction on $n = |G|$. Assume that every induced subgraph H of G satisfies $(*)$, and suppose that G is not perfect. By the induction hypothesis, every proper induced subgraph of G is perfect. Hence, every non-empty independent set $U \subseteq V$ satisfies

$$
\chi(G - U) = \omega(G - U) = \omega . \tag{1}
$$

Indeed, while the first equality is immediate from the perfection of $G - U$, the second is easy: '$\leqslant$' is obvious, while $\chi(G - U) < \omega$ would imply $\chi(G) \leqslant \omega$, so G would be perfect contrary to our assumption.

V, v_i, n
α, ω

Let us apply (1) to a singleton $U = \{u\}$ and consider an ω-colouring of $G - u$. Let K be the vertex set of any K^ω in G. Clearly,

$$\text{if } u \notin K \text{ then } K \text{ meets every colour class of } G - u; \qquad (2)$$

$$\text{if } u \in K \text{ then } K \text{ meets all but exactly one colour class of } G - u. \qquad (3)$$

A_0 Let $A_0 = \{u_1, \ldots, u_\alpha\}$ be an independent set in G of size α. Let $A_1, \ldots, A_\omega$ be the colour classes of an ω-colouring of $G - u_1$, let $A_{\omega+1}, \ldots, A_{2\omega}$ be the colour classes of an ω-colouring of $G - u_2$, and A_i so on; altogether, this gives us $\alpha\omega + 1$ independent sets $A_0, A_1, \ldots, A_{\alpha\omega}$ in G. For each $i = 0, \ldots, \alpha\omega$, there exists by (1) a $K^\omega \subseteq G - A_i$; we K_i denote its vertex set by K_i.

Note that if K is the vertex set of any K^ω in G, then

$$K \cap A_i = \emptyset \text{ for exactly one } i \in \{0, \ldots, \alpha\omega + 1\}. \qquad (4)$$

Indeed, if $K \cap A_0 = \emptyset$ then $K \cap A_i \neq \emptyset$ for all $i \neq 0$, by definition of A_i and (2). Similarly if $K \cap A_0 \neq \emptyset$, then $|K \cap A_0| = 1$, so $K \cap A_i = \emptyset$ for exactly one $i \neq 0$: apply (3) to the unique vertex $u \in K \cap A_0$, and (2) to all the other vertices $u \in A_0$.

J Let J be the real $(\alpha\omega + 1) \times (\alpha\omega + 1)$ matrix with zero entries in A the main diagonal and all other entries 1. Let A be the real $(\alpha\omega + 1) \times n$ matrix whose rows are the incidence vectors of the subsets $A_i \subseteq V$: if $a_{i1}, \ldots, a_{in}$ denote the entries of the ith row of A, then $a_{ij} = 1$ if $v_j \in A_i$, B and $a_{ij} = 0$ otherwise. Similarly, let B denote the real $n \times (\alpha\omega + 1)$ matrix whose columns are the incidence vectors of the subsets $K_i \subseteq V$. Now while $|K_i \cap A_i| = 0$ for all i by the choice of K_i, we have $K_i \cap A_j \neq \emptyset$ and hence $|K_i \cap A_j| = 1$ whenever $i \neq j$, by (4). Thus,

$$AB = J.$$

Since J is non-singular, this implies that A has rank $\alpha\omega + 1$. In particular, $n \geqslant \alpha\omega + 1$, which contradicts $(*)$ for $H := G$. $\square$

By definition, every induced subgraph of a perfect graph is again perfect. The property of perfection can therefore be characterized by forbidden induced subgraphs: there exists a set $\mathcal{H}$ of imperfect graphs such that any graph is perfect if and only if it has no induced subgraph isomorphic to an element of $\mathcal{H}$. (For example, we may choose as $\mathcal{H}$ the set of all imperfect graphs with vertices in $\mathbb{N}$.)

Naturally, it would be desirable to keep $\mathcal{H}$ as small as possible. In fact, one of the best known conjectures in graph theory says that $\mathcal{H}$ need only contain two types of graph: the odd cycles of length $\geqslant 5$ and their complements. (Neither of these are perfect—why?) Or, rephrased slightly:

Perfect Graph Conjecture. (Berge 1966)
A graph G is perfect if and only if neither G nor $\overline{G}$ contains an odd cycle of length at least 5 as an induced subgraph.

Clearly, this conjecture implies the perfect graph theorem. In fact, that theorem had also been conjectured by Berge: until its proof, it was known as the 'weak' version of the perfect graph conjecture, the above conjecture being the 'strong' version.

Graphs G such that neither G nor $\overline{G}$ contains an induced odd cycle of length at least 5 have been called *Berge graphs*. Thus all perfect graphs are Berge graphs, and the perfect graph conjecture claims that all Berge graphs are perfect. This has been approximately verified by Prömel & Steger (1992), who proved that the proportion of perfect graphs to Berge graphs on n vertices tends to 1 as $n \to \infty$.

Exercises

1.⁻ Show that the four colour theorem does indeed solve the map colouring problem stated in the first sentence of the chapter. Conversely, does the 4-colourability of every map imply the four colour theorem?

2.⁻ Show that, for the map colouring problem above, it suffices to consider maps such that no point lies on the boundary of more than three countries. How does this affect the proof of the four colour theorem?

3. Try to turn the proof of the five colour theorem into one of the four colour theorem, as follows. Defining v and H as before, assume inductively that H has a 4-colouring; then proceed as before. Where does the proof fail?

4. Calculate the chromatic number of a graph in terms of the chromatic numbers of its blocks.

5.⁻ Show that every graph G has a vertex ordering for which the greedy algorithm uses only $\chi(G)$ colours.

6. For every $n > 1$, find a bipartite graph on $2n$ vertices, ordered in such a way that the greedy algorithm uses n rather than 2 colours.

7. Consider the following approach to vertex colouring. First, find a maximal independent set of vertices and colour these with colour 1; then find a maximal independent set of vertices in the remaining graph and colour these 2, and so on. Compare this algorithm with the greedy algorithm: which is better?

8. Show that the bound of Proposition 5.2.2 is always at least as sharp as that of Proposition 5.2.1.

9. Find a function f such that every graph of arboricity at least $f(k)$ has colouring number at least k, and a function g such that every graph of colouring number at least $g(k)$ has arboricity at least k, for all $k \in \mathbb{N}$. (The *arboricity* of a graph is defined in Chapter 3.5.)

10.⁻ A k-chromatic graph is called *critically k-chromatic*, or just *critical*, if $\chi(G - v) < k$ for every $v \in V(G)$. Show that every k-chromatic graph has a critical k-chromatic induced subgraph, and that any such subgraph has minimum degree at least $k - 1$.

11. Determine the critical 3-chromatic graphs.

12.⁺ Show that every critical k-chromatic graph is $(k - 1)$-edge-connected.

13. Given $k \in \mathbb{N}$, find a constant $c_k > 0$ such that every graph G with $|G| \geqslant 3k$ and $\alpha(G) \leqslant k$ contains a cycle of length at least $c_k |G|$.

14.⁻ Find a graph G for which Brooks's theorem yields a significantly weaker bound on $\chi(G)$ than Proposition 5.2.2.

15.⁺ Show that, in order to prove Brooks's theorem for a graph $G = (V, E)$, we may assume that $\kappa(G) \geqslant 2$ and $\Delta(G) \geqslant 3$. Prove the theorem under these assumptions, showing first the following two lemmas.

   (i) Let $v_1, \ldots, v_n$ be an enumeration of V. If every v_i $(i < n)$ has a neighbour v_j with $j > i$, and if $v_1 v_n, v_2 v_n \in E$ but $v_1 v_2 \notin E$, then the greedy algorithm uses at most $\Delta(G)$ colours.

   (ii) If G is not complete and v_n has maximum degree in G, then v_n has neighbours v_1, v_2 as in (i).

16. Given a graph G and $k \in \mathbb{N}$, let $P_G(k)$ denote the number of vertex colourings $V(G) \to \{1, \ldots, k\}$. Show that P_G is a polynomial in k of degree $n := |G|$, in which the coefficient of k^n is 1 and the coefficient of k^{n-1} is $-\|G\|$. (P_G is called the *chromatic polynomial* of G.)

   (Hint. Apply induction on $\|G\|$. In the induction step, compare the values of $P_G(k)$, $P_{G-e}(k)$ and $P_{G/e}(k)$.)

17.⁺ Determine the class of all graphs G for which $P_G(k) = k(k-1)^{n-1}$. (As in the previous exercise, let $n := |G|$, and let P_G denote the chromatic polynomial of G.)

18. In the definition of k-constructible graphs, replace the axiom (ii) by

   (ii)′ *Every supergraph of a k-constructible graph is k-constructible*;

   and the axiom (iii) by

   (iii)′ *If G is a graph with vertices x, y_1, y_2 such that $y_1 y_2 \in E(G)$ but $xy_1, xy_2 \notin E(G)$, and if both $G + xy_1$ and $G + xy_2$ are k-constructible, then G is k-constructible.*

   Show that a graph is k-constructible with respect to this new definition if and only if its chromatic number is at least k.

19. Determine the chromatic index of the Petersen graph (Fig. 6.6.1).

20.⁻ An $n \times n$-matrix with entries from $\{1, \ldots, n\}$ is called a *Latin square* if every element of $\{1, \ldots, n\}$ appears exactly once in each column and exactly once in each row. Recast the problem of constructing Latin squares as a colouring problem.

21. Without using Proposition 5.3.1, show that $\chi'(G) = k$ for every k-regular bipartite graph G.

(Hint. How are edge colourings related to matchings?)

22. Prove Proposition 5.3.1 from the statement of the previous exercise.

23.+ For every $k \in \mathbb{N}$, construct a triangle-free k-chromatic graph.

(Hint. Induction on k. In the induction step $k \to k + 1$, construct a triangle-free $(k + 1)$-chromatic graph from the disjoint union of triangle-free i-chromatic graphs for $i = 1, \ldots, k$.)

24. For every integer k, find a 2-chromatic graph whose choice number is at least k.

25.− Without using Theorem 5.4.2, show that every plane graph is 6-list-colourable.

26.− Find a general upper bound for ch$'(G)$ in terms of $\chi'(G)$.

27. A *total colouring* of a graph $G = (V, E)$ is a mapping defined on $V \cup E$ that assigns different values to adjacent or incident elements. Denote the smallest k such that G has a total colouring with k colours by $\chi''(G)$.

 (i) Find an upper bound for $\chi''(G)$ in terms of $\Delta(G)$.

 (ii) Show that the list-colouring conjecture implies $\chi''(G) \leqslant \Delta(G)+3$.

28.− Find a directed graph that has no kernel.

29.+ Prove *Richardson's theorem*: every directed graph without odd directed cycles has a kernel.

(Hint. Call a set S of vertices in a directed graph D a *core* if D contains a directed v–S path for every vertex $v \in D - S$. If, in addition, D contains no directed path between any two vertices of S, call S a *strong core*. Show first that every core contains a strong core. Next, define inductively a partition of $V(D)$ into 'levels' $L_0, \ldots, L_n$ such that, for even i, L_i is a suitable strong core in $D_i := D - (L_0 \cup \ldots \cup L_{i-1})$, while for odd i, L_i consists of the vertices of D_i that send an edge to L_{i-1}. Show that, if D has no directed odd cycle, the even levels together form a kernel of D.)

30. Show that every bipartite planar graph is 3-list-colourable.

(Hint. Apply the previous exercise and Lemma 5.4.3. Construct the desired orientation in steps: if, in the current orientation, there are still vertices v with $d^+(v) \geqslant 3$, then reverse the directions of all the edges along a directed path from v to some suitable vertex w. For a proof that w exists, use Euler's formula to determine the greatest possible average degree of a bipartite planar graph.)

31.− Show that perfection is closed neither under edge deletion nor under edge contraction.

32.− Deduce Theorem 5.5.5 from the perfect graph conjecture.

33. Use König's Theorem 2.1.1 to show that the complement of any bipartite graph is perfect.

   (Hint. Exercise 12, Chapter 3.)

34. Using the results of this chapter, find a one-line proof of the following theorem of König, the dual of Theorem 2.1.1: in any bipartite graph without isolated vertices, the minimum number of edges meeting all vertices equals the maximum number of independent vertices.

35. A graph is called a *comparability graph* if there exists a partial ordering of its vertex set such that two vertices are adjacent if and only if they are comparable. Show that every comparability graph is perfect.

   (Hint. Define the colour classes of a given induced subgraph $H \subseteq G$ inductively, starting with the class of all minimal elements.)

36. A graph G is called an *interval graph* if there exists a set $\{ I_v \mid v \in V(G) \}$ of real intervals such that $I_u \cap I_v \neq \emptyset$ if and only if $uv \in E(G)$.

   (i) Show that every interval graph is chordal.

   (ii) Show that the complement of any interval graph is a comparability graph.

   (Conversely, a chordal graph is an interval graph if its complement is a comparability graph; this is a theorem of Gilmore and Hoffman (1964).)

37. Show that $\chi(H) \in \{\omega(H), \omega(H) + 1\}$ for every line graph H of a graph G.

38.[+] Characterize the graphs whose line graphs are perfect.

39. Show that a graph G is perfect if and only if every non-empty induced subgraph H of G contains an independent set $A \subseteq V(H)$ such that $\omega(H - A) < \omega(H)$.

40.[+] Consider the graphs G for which every induced subgraph H has the property that every maximal complete subgraph of H meets every maximal independent vertex set in H. Show that these graphs G are precisely the graphs not containing an induced copy of P^3.

41.[+] Show that in every perfect graph G one can find a set $\mathcal{A}$ of independent vertex sets and a set $\mathcal{O}$ of vertex sets of complete subgraphs such that $\bigcup \mathcal{A} = V(G) = \bigcup \mathcal{O}$ and every set in $\mathcal{A}$ meets every set in $\mathcal{O}$.

   (Hint. Lemma 5.5.4.)

42.[+] Let G be a perfect graph. As in the proof of Theorem 5.5.3, replace every vertex x of G with a perfect graph G_x (not necessarily complete). Show that the resulting graph G' is again perfect.

   (Hint. Reduce the general case to the case that all but one of the G_x are trivial; then imitate the proof of Lemma 5.5.4.)

43. Let $\mathcal{H}_1$ and $\mathcal{H}_2$ be two sets of imperfect graphs, each minimal with the property that a graph is perfect if and only if it has no induced subgraph in $\mathcal{H}_i$ ($i = 1, 2$). Do $\mathcal{H}_1$ and $\mathcal{H}_2$ contain the same graphs, up to isomorphism?

Notes

The authoritative reference work on all questions of graph colouring is T.R. Jensen & B. Toft, *Graph Coloring Problems*, Wiley 1995. Starting with a brief survey of the most important results and areas of research in the field, this monograph gives a detailed account of over 200 open colouring problems, complete with extensive background surveys and references. Most of the remarks below are discussed comprehensively in this book, and all the references for this chapter can be found there.

The *four colour problem*, whether every map can be coloured with four colours so that adjacent countries are shown in different colours, was raised by a certain Francis Guthrie in 1852. He put the question to his brother Frederick, who was then a mathematics undergraduate in Cambridge. The problem was first brought to the attention of a wider public when Cayley presented it to the London Mathematical Society in 1878. A year later, Kempe published an incorrect proof, which was in 1890 modified by Heawood into a proof of the five colour theorem. In 1880, Tait announced 'further proofs' of the four colour conjecture, which never materialized; see the notes for Chapter 10.

The first widely accepted proof of the four colour theorem was published by Appel and Haken in 1977. The proof builds on ideas that can be traced back as far as Kempe's paper, and were developed largely by Birkhoff and Heesch. Very roughly, the proof sets out first to show that every plane triangulation must contain at least one of 1482 certain 'unavoidable configurations'. In a second step, a computer is used to show that each of those configurations is 'reducible', i.e., that any plane triangulation containing such a configuration can be 4-coloured by piecing together 4-colourings of smaller plane triangulations. Taken together, these two steps amount to an inductive proof that all plane triangulations, and hence all planar graphs, can be 4-coloured.

Appel & Haken's proof has not been immune to criticism, not only because of their use of a computer. The authors responded with a 741 page long algorithmic version of their proof, which addresses the various criticisms and corrects a number of errors (e.g. by adding more configurations to the 'unavoidable' list): K. Appel & W. Haken, *Every Planar Map is Four Colorable*, American Mathematical Society 1989. A much shorter proof, which is based on the same ideas (and, in particular, uses a computer in the same way) but can be more readily verified at least in its verbal part, has been given by N. Robertson, D. Sanders, P.D. Seymour & R. Thomas, The four-colour theorem, *J. Combin. Theory B* **70** (1997).

A relatively short proof of Grötzsch's theorem was found by C. Thomassen, Grötzsch's 3-color theorem and its counterparts for the torus and the projective plane, *J. Combin. Theory B* **62** (1994), 268–279. Although not touched upon in this chapter, colouring problems for graphs embedded in surfaces other than the plane form a substantial and interesting part of colouring theory; see B. Mohar & C. Thomassen, *Graphs on Surfaces*, Johns Hopkins University Press 1997.

The proof of Brooks's theorem indicated in Exercise 15, where the greedy algorithm is applied to a carefully chosen vertex ordering, is due to Lovász (1973). Lovász (1968) was also the first to *construct* graphs of arbitrarily large girth and chromatic number, graphs whose existence Erdős had proved by probabilistic methods ten years earlier.

Urquhart showed as recently as 1996 that not only do the graphs of chromatic number at least k each *contain* a k-constructible graph (as by Hajós's theorem); they are in fact all themselves k-constructible. Algebraic tools for showing that the chromatic number of a graph is large have been developed by Kleitman & Lovász (1982), and by Alon & Tarsi (1992); see Alon's paper cited below.

List colourings were first introduced in 1976 by Vizing. Among other things, Vizing proved the list-colouring equivalent of Brooks's theorem. Voigt (1993) constructed a plane graph of order 238 that is not 4-choosable; thus, Thomassen's list version of the five colour theorem is best possible. A stimulating survey on the list-chromatic number and how it relates to the more classical graph invariants (including a proof of Theorem 5.4.1) is given by N. Alon, Restricted colorings of graphs, in (K. Walker, ed.) *Surveys in Combinatorics*, LMS Lecture Notes **187**, Cambridge University Press 1993. Both the list colouring conjecture and Galvin's proof of the bipartite case are originally stated for multigraphs. Kahn (1994) proved that the conjecture is asymptotically correct, as follows: given any $\epsilon > 0$, every graph G with large enough maximum degree satisfies $\mathrm{ch}'(G) \leqslant (1 + \epsilon)\Delta(G)$.

A gentle introduction to the basic facts about perfect graphs and their applications is given by M.C. Golumbic, *Algorithmic Graph Theory and Perfect Graphs*, Academic Press 1980. Our first proof of the perfect graph theorem follows L. Lovász's survey on perfect graphs in (L.W. Beineke and R.J. Wilson, eds.) *Selected Topics in Graph Theory 2*, Academic Press 1983. The theorem was also proved independently, and only a little later, by Fulkerson. Our second proof, the proof of Theorem 5.5.5, is due to G.S. Gasparian, Minimal imperfect graphs: a simple approach, *Combinatorica* **16** (1996), 209–212. The approximate proof of the perfect graph conjecture is due to H.J. Prömel & A. Steger, Almost all Berge graphs are perfect, *Combinatorics, Probability and Computing* **1** (1992), 53–79.

6   Flows

Let us view a graph as a network: its edges carry some kind of flow—of
water, electricity, data or similar. How could we model this precisely?

For a start, we ought to know how much flow passes through each
edge $e = xy$, and in which direction. In our model, we could assign
a positive integer k to the pair (x, y) to express that a flow of k units
passes through e from x to y, or assign $-k$ to (x, y) to express that k
units of flow pass through e the other way, from y to x. For such an
assignment $f: V^2 \to \mathbb{Z}$ we would thus have $f(x, y) = -f(y, x)$ whenever
x and y are adjacent vertices of G.

Typically, a network will have only a few nodes where flow enters
or leaves the network; at all other nodes, the total amount of flow into
that node will equal the total amount of flow out of it. For our model
this means that, at most nodes x, the function f will satisfy *Kirchhoff's
law*

$$\sum_{y \in N(x)} f(x, y) = 0 \, .$$

Kirchhoff's
law

In this chapter, we call any map $f: V^2 \to \mathbb{Z}$ with the above two
properties a 'flow' on G. Sometimes, we shall replace $\mathbb{Z}$ with another
group, and as a rule we consider multigraphs rather than graphs.[1] As
it turns out, the theory of those 'flows' is not only useful as a model for
real flows: it blends so well with other parts of graph theory that some
deep and surprising connections become visible, connections particularly
with connectivity and colouring problems.

---

[1] For consistency, we shall phrase some of our proposition for graphs only: those
whose proofs rely on assertions proved (for graphs) earlier in the book. However, all
those results remain true for multigraphs.

6.1 Circulations

In the context of flows, we have to be able to speak about the 'directions'
$G = (V, E)$ of an edge. Since, in a multigraph $G = (V, E)$, an edge $e = xy$ is not
identified uniquely by the pair (x, y) or (y, x), we define directed edges as
triples:

$\vec{E}$
$$\vec{E} := \{ (e, x, y) \mid e \in E; \ x, y \in V; \ e = xy \}.$$

direction
(e, x, y)
Thus, an edge $e = xy$ with $x \neq y$ has the two *directions* (e, x, y) and
(e, y, x); a loop $e = xx$ has only one direction, the triple (e, x, x). For
$\overleftarrow{e}$
given $\vec{e} = (e, x, y) \in \vec{E}$, we set $\overleftarrow{e} := (e, y, x)$, and for an arbitrary set
$F \subseteq \vec{E}$ of edge directions we put

$\overleftarrow{F}$
$$\overleftarrow{F} := \{ \overleftarrow{e} \mid \vec{e} \in F \}.$$

Note that $\vec{E}$ itself is symmetrical: $\overleftarrow{E} = \vec{E}$. For $X, Y \subseteq V$ and $F \subseteq \vec{E}$,
define

$\vec{F}(X, Y)$
$$\vec{F}(X, Y) := \{ (e, x, y) \in F \mid x \in X; \ y \in Y; \ x \neq y \},$$

$\vec{F}(x, Y)$
abbreviate $\vec{F}(\{ x \}, Y)$ to $\vec{F}(x, Y)$ etc., and write

$\vec{F}(x)$
$$\vec{F}(x) := \vec{F}(x, V) = \vec{F}(\{ x \}, \overline{\{ x \}}).$$

$\overline{X}$
Here, as below, $\overline{X}$ denotes the complement $V \smallsetminus X$ of a vertex set $X \subseteq V$.
Note that any loops at vertices $x \in X \cap Y$ are disregarded in the defini-
tions of $\vec{F}(X, Y)$ and $\vec{F}(x)$.

0 Let H be an abelian semigroup,[2] written additively with zero 0.
f Given vertex sets $X, Y \subseteq V$ and a function $f \colon \vec{E} \to H$, let

$f(X, Y)$
$$f(X, Y) := \sum_{\vec{e} \, \in \, \vec{E}(X,Y)} f(\vec{e}).$$

$f(x, Y)$
Instead of $f(\{ x \}, Y)$ we again write $f(x, Y)$, etc.
circulation
From now on, we assume that H is a group. We call f a *circulation*
on G (with values in H), or an *H-circulation*, if f satisfies the following
two conditions:

(F1) $f(e, x, y) = -f(e, y, x)$ for all $(e, x, y) \in \vec{E}$ with $x \neq y$;

(F2) $f(v, V) = 0$ for all $v \in V$.

---

[2] This chapter contains no group theory. The only semigroups we ever consider
for H are the natural numbers, the integers, the reals, the cyclic groups $\mathbb{Z}_k$, and
(once) the Klein four-group.

If f satisfies (F1), then

$$f(X, X) = 0$$

for all $X \subseteq V$. If f satisfies (F2), then

$$f(X, V) = \sum_{x \in X} f(x, V) = 0.$$

Together, these two basic observations imply that, in a circulation, the net flow across any cut is zero:

[6.3.1]
[6.5.2]
[6.6.1]

Proposition 6.1.1. *If f is a circulation, then $f(X, \overline{X}) = 0$ for every set $X \subseteq V$.*

Proof. $f(X, \overline{X}) = f(X, V) - f(X, X) = 0 - 0 = 0.$ $\qquad\square$

Since bridges form cuts by themselves, Proposition 6.1.1 implies that circulations are always zero on bridges:

Corollary 6.1.2. *If f is a circulation and $e = xy$ is a bridge in G, then $f(e, x, y) = 0$.* $\qquad\square$

6.2 Flows in networks

In this section we give a brief introduction to the kind of network flow theory that is now a standard proof technique in areas such as matching and connectivity. By way of example, we shall prove a classic result of this theory, the so-called *max-flow min-cut* theorem of Ford and Fulkerson. This theorem alone implies Menger's theorem without much difficulty (Exercise 3), which indicates some of the natural power lying in this approach.

Consider the task of modelling a network with one source s and one sink t, in which the amount of flow through a given link between two nodes is subject to a certain capacity of that link. Our aim is to determine the maximum net amount of flow through the network from s to t. Somehow, this will depend both on the structure of the network and on the various capacities of its connections—how exactly, is what we wish to find out.

Let $G = (V, E)$ be a multigraph, $s, t \in V$ two fixed vertices, and $c \colon \vec{E} \to \mathbb{N}$ a map; we call c a *capacity function* on G, and the tuple $N := (G, s, t, c)$ a *network*. Note that c is defined independently for the two directions of an edge. A function $f \colon \vec{E} \to \mathbb{R}$ is a *flow* in N if it satisfies the following three conditions (Fig. 6.2.1):

$G = (V, E)$
s, t, c, N
network
flow

(F1) $f(e, x, y) = -f(e, y, x)$ for all $(e, x, y) \in \vec{E}$ with $x \neq y$;
(F2′) $f(v, V) = 0$ for all $v \in V \smallsetminus \{s, t\}$;
(F3) $f(\vec{e}) \leqslant c(\vec{e})$ for all $\vec{e} \in \vec{E}$.

integral   We call f *integral* if all its values are integers.

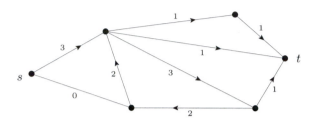

Fig. 6.2.1. A network flow in short notation: all values refer to
the direction indicated (capacities are not shown)

f
cut in N
capacity

Let f be a flow in N. If $S \subseteq V$ is such that $s \in S$ and $t \in \overline{S}$, we call
the pair $(S, \overline{S})$ a *cut in N*, and $c(S, \overline{S})$ the *capacity* of this cut.

Since f now has to satisfy only (F2′) rather than (F2), we no longer
have $f(X, \overline{X}) = 0$ for all $X \subseteq V$ (as in Proposition 6.1.1). However, the
value is the same for all cuts:

Proposition 6.2.1. *Every cut* $(S, \overline{S})$ *in* N *satisfies* $f(S, \overline{S}) = f(s, V)$.

Proof. As in the proof of Proposition 6.1.1, we have

$$
\begin{aligned}
f(S, \overline{S}) &= f(S, V) - f(S, S) \\
&\underset{(F1)}{=} f(s, V) + \sum_{v \in S \smallsetminus \{s\}} f(v, V) - 0 \\
&\underset{(F2')}{=} f(s, V).
\end{aligned}
$$

$\square$

total value
$|f|$

The common value of $f(S, \overline{S})$ in Proposition 6.2.1 will be called the *total
value* of f and denoted by $|f|$;[3] the flow shown in Figure 6.2.1 has total
value 3.

By (F3), we have

$$
|f| = f(S, \overline{S}) \leqslant c(S, \overline{S})
$$

for every cut $(S, \overline{S})$ in N. Hence the total value of a flow in N is never
larger than the smallest capacity of a cut. The following *max-flow min-
cut* theorem states that this upper bound is always attained by some
flow:

---

[3] Thus, formally, $|f|$ may be negative. In practice, however, we can change the
sign of $|f|$ simply by swapping the roles of s and t.

Theorem 6.2.2. (Ford & Fulkerson 1956)   max-flow
In every network, the maximum total value of a flow equals the minimum   min-cut
capacity of a cut.   theorem

Proof. Let $N = (G, s, t, c)$ be a network, and $G =: (V, E)$. We shall define
a sequence $f_0, f_1, f_2, \ldots$ of integral flows in N of strictly increasing total
value, i.e. with

$$|f_0| < |f_1| < |f_2| < \cdots$$

Clearly, the total value of an integral flow is again an integer, so in fact
$|f_{n+1}| \geqslant |f_n| + 1$ for all n. Since all these numbers are bounded above
by the capacity of any cut in N, our sequence will terminate with some
flow f_n. Corresponding to this flow, we shall find a cut of capacity
$c_n = |f_n|$. Since no flow can have a total value greater than c_n, and no
cut can have a capacity less than $|f_n|$, this number is simultaneously the
maximum and the minimum referred to in the theorem.

For f_0, we set $f_0(\vec{e}) := 0$ for all $\vec{e} \in \vec{E}$. Having defined an integral
flow f_n in N for some $n \in \mathbb{N}$, we denote by S_n the set of all vertices v S_n
such that G contains an s–v walk $x_0 e_0 \ldots e_{\ell-1} x_\ell$ with

$$f_n(\vec{e_i}) < c(\vec{e_i})$$

for all $i < \ell$; here, $\vec{e_i} := (e_i, x_i, x_{i+1})$ (and, of course, $x_0 = s$ and $x_\ell = v$).

If $t \in S_n$, let $W = x_0 e_0 \ldots e_{\ell-1} x_\ell$ be the corresponding s–t walk; W
without loss of generality we may assume that W does not repeat any
vertices. Let

$$\epsilon := \min \{ c(\vec{e_i}) - f_n(\vec{e_i}) \mid i < \ell \}.$$ ϵ

Then $\epsilon > 0$, and since f_n (like c) is integral by assumption, ϵ is an integer.
Let

$$f_{n+1} \colon \vec{e} \mapsto \begin{cases} f_n(\vec{e}) + \epsilon & \text{for } \vec{e} = \vec{e_i}, \ i = 0, \ldots, \ell-1; \\ f_n(\vec{e}) - \epsilon & \text{for } \vec{e} = \overleftarrow{e_i}, \ i = 0, \ldots, \ell-1; \\ f_n(\vec{e}) & \text{for } e \notin W. \end{cases}$$

Intuitively, f_{n+1} is obtained from f_n by sending additional flow of value ϵ
along W from s to t (Fig. 6.2.2).

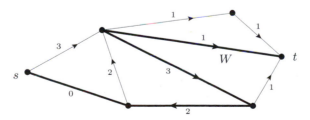

Fig. 6.2.2. An 'augmenting path' W with increment $\epsilon = 2$, for
constant flow $f_n = 0$ and capacities $c = 3$

Clearly, f_{n+1} is again an integral flow in N. Let us compute its total value $|f_{n+1}| = f_{n+1}(s, V)$. Since W contains the vertex s only once, $\vec{e}_0$ is the only triple (e, x, y) with $x = s$ and $y \in V$ whose f-value was changed. This value, and hence that of $f_{n+1}(s, V)$ was raised. Therefore $|f_{n+1}| > |f_n|$ as desired.

If $t \notin S_n$, then $(S_n, \overline{S_n})$ is a cut in N. By (F3) for f_n, and the definition of S_n, we have

$$f_n(\vec{e}) = c(\vec{e})$$

for all $\vec{e} \in \vec{E}(S_n, \overline{S_n})$, so

$$|f_n| = f_n(S_n, \overline{S_n}) = c(S_n, \overline{S_n})$$

as desired.   □

Since the flow constructed in the proof of Theorem 6.2.2 is integral, we have also proved the following:

Corollary 6.2.3. *In every network (with integral capacity function) there exists an integral flow of maximum total value.* □

6.3 Group-valued flows

Let $G = (V, E)$ be a multigraph and H an abelian group. If f and
$f + g$ g are two H-circulations then, clearly, $(f + g): \vec{e} \mapsto f(\vec{e}) + g(\vec{e})$ and
$-f$ $-f: \vec{e} \mapsto -f(\vec{e})$ are again H-circulations. The H-circulations on G thus form a group in a natural way.

nowhere   A function $f: \vec{E} \to H$ is *nowhere zero* if $f(\vec{e}) \neq 0$ for all $\vec{e} \in \vec{E}$. An
zero H-circulation that is nowhere zero is called an *H-flow*.[4] Note that the
H-flow   set of H-flows on G is not closed under addition: if two H-flows add up to zero on some edge $\vec{e}$, then their sum is no longer an H-flow. By Corollary 6.1.2, a graph with an H-flow cannot have a bridge.

For finite groups H, the number of H-flows on G—and, in particular, their existence—surprisingly depends only on the order of H, not on H itself:

Theorem 6.3.1. (Tutte 1954)
For every multigraph G there exists a polynomial P such that, for any finite abelian group H, the number of H-flows on G is $P(|H| - 1)$.

---

[4] This terminology seems simplest for our purposes but is not standard; see the notes.

Proof. Let $G =: (V, E)$; we use induction on $m := |E|$. Let us assume   (6.1.1)
first that all the edges of G are loops. Then, given any finite abelian
group H, every map $\vec{E} \to H \smallsetminus \{0\}$ is an H-flow on G. Since $|\vec{E}| = |E|$
when all edges are loops, there are $(|H| - 1)^m$ such maps, and $P := x^m$
is the polynomial sought.

Now assume there is an edge $e_0 = xy \in E$ that is not a loop; let $e_0 = xy$
$\vec{e}_0 := (e_0, x, y)$ and $E' := E \smallsetminus \{e_0\}$. We consider the multigraphs E'

$$G_1 := G - e_0 \quad \text{and} \quad G_2 := G/e_0 \,.$$

By the induction hypothesis, there are polynomials P_i for $i = 1, 2$ such P_1, P_2
that, for any finite abelian group H and $k := |H| - 1$, the number of k
H-flows on G_i is $P_i(k)$. We shall prove that the number of H-flows on
G equals $P_2(k) - P_1(k)$; then $P := P_2 - P_1$ is the desired polynomial.

Let H be given, and denote the set of all H-flows on G by F. We H
are trying to show that F

$$|F| = P_2(k) - P_1(k) \,. \tag{1}$$

The H-flows on G_1 are precisely the restrictions to $\vec{E}'$ of those H-circu-
lations on G that are zero on e_0 but nowhere else. Let us denote the set
of these circulations on G by F_1; then F_1

$$P_1(k) = |F_1| \,.$$

Our aim is to show that, likewise, the H-flows on G_2 correspond bijectively to those H-circulations on G that are nowhere zero except possibly
on e_0. The set F_2 of those circulations on G then satisfies F_2

$$P_2(k) = |F_2| \,,$$

and F_2 is the disjoint union of F_1 and F. This will prove (1), and hence
the theorem.

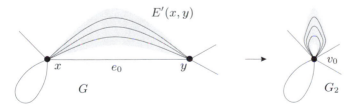

Fig. 6.3.1. Contracting the edge e_0

In G_2, let $v_0 := v_{e_0}$ be the vertex contracted from e_0 (Fig. 6.3.1; v_0
see Chapter 1.10). We are looking for a bijection $f \mapsto g$ between F_2

and the set of H-flows on G_2. Given f, let g be the restriction of f to $\vec{E'} \smallsetminus \vec{E'}(y, x)$. (As the x–y edges $e \in E'$ become loops in G_2, they have only the one direction (e, v_0, v_0) there; as its g-value, we choose $f(e, x, y)$.) Then g is indeed an H-flow on G_2; note that (F2) holds at v_0 by Proposition 6.1.1 for G, with $X := \{x, y\}$.

It remains to show that the map $f \mapsto g$ is a bijection. If we are given an H-flow g on G_2 and try to find an $f \in F_2$ with $f \mapsto g$, then $f(\vec{e})$ is already determined as $f(\vec{e}) = g(\vec{e})$ for all $\vec{e} \in \vec{E'} \smallsetminus \vec{E'}(y, x)$; by (F1), we further have $f(\vec{e}) = -f(\vec{e})$ for all $\vec{e} \in \vec{E'}(y, x)$. Thus our map $f \mapsto g$ is bijective if and only if for given g there is always a unique way to define the remaining values of $f(\vec{e_0})$ and $f(\bar{e_0})$ so that f satisfies (F1) in e_0 and (F2) in x and y.

V' This is indeed the case. Let $V' := V \smallsetminus \{x, y\}$. The f-values fixed already satisfy

$$f(x, V') + f(y, V') = g(v_0, V') = 0, \tag{2}$$

by (F2) for g. With

$$h := \sum_{\vec{e} \in \vec{E'}(x,y)} f(\vec{e}) \quad \left(= \sum_{e \in E'(x,y)} g(e, v_0, v_0) \right)$$

h we have

$$f(x, V) = f(\vec{e_0}) + h + f(x, V')$$

and

$$f(y, V) = f(\bar{e_0}) - h + f(y, V').$$

Hence, f will satisfy (F1) and (F2) if and only if we set

$$f(\vec{e_0}) := -f(x, V') - h \underset{(2)}{=} f(y, V') - h$$

and $f(\bar{e_0}) := -f(\vec{e_0})$. $\square$

flow
polynomial The polynomial P of Theorem 6.3.1 is known as the *flow polynomial* of G.

[6.4.5] **Corollary 6.3.2.** *If H and H' are two finite abelian groups of equal order, then G has an H-flow if and only if G has an H'-flow.* $\square$

Corollary 6.3.2 has fundamental implications for the theory of algebraic flows: it indicates that crucial difficulties in existence proofs of H-flows are unlikely to be of a group-theoretic nature. On the other hand, being able to choose a convenient group can be quite helpful; we shall see a pretty example for this in Proposition 6.4.5.

Let $k \geqslant 1$ be an integer and $G = (V, E)$ a multigraph. A $\mathbb{Z}$-flow f on G such that $0 < |f(\vec{e})| < k$ for all $\vec{e} \in \vec{E}$ is called a k-*flow*. Clearly, any k-flow is also an ℓ-flow for all $\ell > k$. Thus, we may ask which is the least integer k such that G admits a k-flow—assuming that such a k exists. We call this least k the *flow number* of G and denote it by $\varphi(G)$; if G has no k-flow for any k, we put $\varphi(G) := \infty$.

k

k-flow

flow number $\varphi(G)$

The task of determining flow numbers quickly leads to some of the deepest open problems in graph theory. We shall consider these later in the chapter. First, however, let us see how k-flows are related to the more general concept of H-flows.

There is an intimate connection between k-flows and $\mathbb{Z}_k$-flows. Let σ_k denote the natural homomorphism $i \mapsto \bar{i}$ from $\mathbb{Z}$ to $\mathbb{Z}_k$. By composition with σ_k, every k-flow defines a $\mathbb{Z}_k$-flow. As the following theorem shows, the converse holds too: from every $\mathbb{Z}_k$-flow on G we can construct a k-flow on G. In view of Corollary 6.3.2, this means that the general question about the existence of H-flows for arbitrary groups H reduces to the corresponding question for k-flows.

σ_k

Theorem 6.3.3. (Tutte 1950)
A multigraph admits a k-flow if and only if it admits a $\mathbb{Z}_k$-flow.

[6.4.1]
[6.4.2]
[6.4.3]
[6.4.5]

Proof. Let g be a $\mathbb{Z}_k$-flow on a multigraph $G = (V, E)$; we construct a k-flow f on G. We may assume without loss of generality that G has no loops. Let F be the set of all functions $f : \vec{E} \to \mathbb{Z}$ that satisfy (F1), $|f(\vec{e})| < k$ for all $\vec{e} \in \vec{E}$, and $\sigma_k \circ f = g$; note that, like g, any $f \in F$ is nowhere zero.

g

F

Let us show first that $F \neq \emptyset$. Since we can express every value $g(\vec{e}) \in \mathbb{Z}_k$ as $\bar{i}$ with $|i| < k$ and then put $f(\vec{e}) := i$, there is clearly a map $f : \vec{E} \to \mathbb{Z}$ such that $|f(\vec{e})| < k$ for all $\vec{e} \in \vec{E}$ and $\sigma_k \circ f = g$. For each edge $e \in E$, let us choose one of its two directions and denote this by $\vec{e}$. We may then define $f' : \vec{E} \to \mathbb{Z}$ by setting $f'(\vec{e}) := f(\vec{e})$ and $f'(\overleftarrow{e}) := -f(\vec{e})$ for every $e \in E$. Then f' is a function satisfying (F1) and with values in the desired range; it remains to show that $\sigma_k \circ f'$ and g agree not only on the chosen directions $\vec{e}$ but also on their inverses $\overleftarrow{e}$. Since σ_k is a homomorphism, this is indeed so:

$$(\sigma_k \circ f')(\overleftarrow{e}) = \sigma_k(-f(\vec{e})) = -(\sigma_k \circ f)(\vec{e}) = -g(\vec{e}) = g(\overleftarrow{e}).$$

Hence $f' \in F$, so F is indeed non-empty.

Our aim is to find an $f \in F$ that satisfies Kirchhoff's law (F2), and is thus a k-flow. As a candidate, let us consider an $f \in F$ for which the sum

f

$$K(f) := \sum_{x \in V} |f(x, V)|$$

K

of all deviations from Kirchhoff's law is minimum. We shall prove that $K(f) = 0$; then, clearly, $f(x, V) = 0$ for every x, as desired.

Suppose $K(f) \neq 0$. Since f satisfies (F1), and hence $\sum_{x \in V} f(x, V) = f(V, V) = 0$, there exists a vertex x with

x

$$f(x, V) > 0. \tag{1}$$

X

Let $X \subseteq V$ be the set of all vertices x' for which G contains a walk $x_0 e_0 \dots e_{\ell-1} x_\ell$ from x to x' such that $f(e_i, x_i, x_{i+1}) > 0$ for all $i < \ell$;

X'

furthermore, let $X' := X \setminus \{ x \}$.

We first show that X' contains a vertex x' with $f(x', V) < 0$. By definition of X, we have $f(e, x', y) \leqslant 0$ for all edges $e = x'y$ such that $x' \in X$ and $y \in \overline{X}$. In particular, this holds for $x' = x$. Thus, (1) implies $f(x, X') > 0$. Then $f(X', x) < 0$ by (F1), as well as $f(X', X') = 0$. Thus if $f(x', V) \geqslant 0$ for all $x' \in X'$, then

$$0 \leqslant \sum_{x' \in X'} f(x', V) = f(X', V) = f(X', \overline{X}) + f(X', x) + f(X', X') < 0,$$

a contradiction.

x'

So there exists an $x' \in X'$ with

$$f(x', V) < 0; \tag{2}$$

W

let $W = x_0 e_0 \dots e_{\ell-1} x_\ell$ be the corresponding x–x' walk from the defi-

f'

nition of X. Let us modify f by sending some flow back along W: we define $f' : \vec{E} \to \mathbb{Z}$ by

$$f' : \vec{e} \mapsto \begin{cases} f(\vec{e}) - k & \text{for } \vec{e} = (e_i, x_i, x_{i+1}), \ i = 0, \dots, \ell - 1; \\ f(\vec{e}) + k & \text{for } \vec{e} = (e_i, x_{i+1}, x_i), \ i = 0, \dots, \ell - 1; \\ f(\vec{e}) & \text{for } e \notin W. \end{cases}$$

By definition of W, we have $|f'(\vec{e})| < k$ for all $\vec{e} \in \vec{E}$. Hence f', like f, lies in F.

How does the modification of f affect K? At all inner vertices v of W, as well as outside W, the deviation from Kirchhoff's law remains unchanged:

$$f'(v, V) = f(v, V) \qquad \text{for all } v \in V \setminus \{ x, x' \}. \tag{3}$$

For x and x', on the other hand, we have

$$f'(x, V) = f(x, V) - k \quad \text{and} \quad f'(x', V) = f(x', V) + k. \tag{4}$$

Since g is a $\mathbb{Z}_k$-flow and hence

$$\sigma_k(f(x, V)) = g(x, V) = \bar{0} \in \mathbb{Z}_k$$

and

$$\sigma_k(f(x', V)) = g(x', V) = \bar{0} \in \mathbb{Z}_k,$$

$f(x, V)$ and $f(x', V)$ are both multiples of k. Thus $f(x, V) \geqslant k$ and $f(x', V) \leqslant -k$, by (1) and (2). But then (4) implies that

$$|f'(x, V)| < |f(x, V)| \quad \text{and} \quad |f'(x', V)| < |f(x', V)|.$$

Together with (3), this gives $K(f') < K(f)$, a contradiction to the choice of f.

Therefore $K(f) = 0$ as claimed, and f is indeed a k-flow. $\square$

Since the sum of two $\mathbb{Z}_k$-circulations is always another $\mathbb{Z}_k$-circulation, $\mathbb{Z}_k$-flows are often easier to construct (by summing over suitable partial flows) than k-flows. In this way, Theorem 6.3.3 may be of considerable help in determining whether or not some given graph has a k-flow. In the following sections we shall meet a number of examples for this.

6.4 k-Flows for small k

Trivially, a graph has a 1-flow (the empty set) if and only if it has no edges. In this section we collect a few simple examples of sufficient conditions under which a graph has a 2-, 3- or 4-flow. More examples can be found in the exercises.

Proposition 6.4.1. *A graph has a 2-flow if and only if all its degrees are even.* [6.6.1]

Proof. By Theorem 6.3.3, a graph $G = (V, E)$ has a 2-flow if and only if it has a $\mathbb{Z}_2$-flow, i.e. if and only if the constant map $\vec{E} \to \mathbb{Z}_2$ with value $\bar{1}$ satisfies (F2). This is the case if and only if all degrees are even. $\square$ (6.3.3)

For the remainder of this chapter, let us call a graph *even* if all its vertex degrees are even.   *even graph*

Proposition 6.4.2. *A cubic graph has a 3-flow if and only if it is bipartite.*

(1.6.1)
(6.3.3)
Proof. Let $G = (V, E)$ be a cubic graph. Let us assume first that G has a 3-flow, and hence also a $\mathbb{Z}_3$-flow f. We show that any cycle $C = x_0 \ldots x_\ell x_0$ in G has even length (cf. Proposition 1.6.1). Consider two consecutive edges on C, say $e_{i-1} := x_{i-1}x_i$ and $e_i := x_i x_{i+1}$. If f assigned the same value to these edges in the direction of the forward orientation of C, i.e. if $f(e_{i-1}, x_{i-1}, x_i) = f(e_i, x_i, x_{i+1})$, then f could not satisfy (F2) at x_i for any non-zero value of the third edge at x_i. Therefore f assigns the values $\bar{1}$ and $\bar{2}$ to the edges of C alternately, and in particular C has even length.

Conversely, let G be bipartite, with vertex bipartition $\{X, Y\}$. Since G is cubic, the map $\vec{E} \to \mathbb{Z}_3$ defined by $f(e, x, y) := \bar{1}$ and $f(e, y, x) := \bar{2}$ for all edges $e = xy$ with $x \in X$ and $y \in Y$ is a $\mathbb{Z}_3$-flow on G. By Theorem 6.3.3, then, G has a 3-flow. $\qquad\square$

What are the flow numbers of the complete graphs K^n? For odd $n > 1$, we have $\varphi(K^n) = 2$ by Proposition 6.4.1. Moreover, $\varphi(K^2) = \infty$, and $\varphi(K^4) = 4$; this is easy to see directly (and it follows from Propositions 6.4.2 and 6.4.5). Interestingly, K^4 is the only complete graph with flow number 4:

Proposition 6.4.3. *For all even $n > 4$, $\varphi(K^n) = 3$.*

(6.3.3)
Proof. Proposition 6.4.1 implies that $\varphi(K^n) \geqslant 3$ for even n. We show, by induction on n, that every $G = K^n$ with even $n > 4$ has a 3-flow.

For the induction start, let $n = 6$. Then G is the edge-disjoint union of three graphs G_1, G_2, G_3, with $G_1, G_2 = K^3$ and $G_3 = K_{3,3}$. By Proposition 6.4.1, G_1 and G_2 each have a 2-flow, while G_3 has a 3-flow by Proposition 6.4.2. The union of all these flows is a 3-flow on G.

Now let $n > 6$, and assume the assertion holds for $n - 2$. Clearly, G is the edge-disjoint union of a K^{n-2} and a graph $G' = (V', E')$ with $G' = \overline{K^{n-2}} * K^2$. The K^{n-2} has a 3-flow by induction. By Theorem 6.3.3, it thus suffices to find a $\mathbb{Z}_3$-flow on G'. For every vertex z of the $\overline{K^{n-2}} \subseteq G'$, let f_z be a $\mathbb{Z}_3$-flow on the triangle $zxyz \subseteq G'$, where $e = xy$ is the edge of the K^2 in G'. Let $f \colon \vec{E'} \to \mathbb{Z}_3$ be the sum of these flows. Clearly, f is nowhere zero, except possibly in (e, x, y) and (e, y, x). If $f(e, x, y) \neq \bar{0}$, then f is the desired $\mathbb{Z}_3$-flow on G'. If $f(e, x, y) = \bar{0}$, then $f + f_z$ (for any z) is a $\mathbb{Z}_3$-flow on G'. $\qquad\square$

Proposition 6.4.4. *Every 4-edge-connected graph has a 4-flow.*

(3.5.2)
Proof. Let G be a 4-edge-connected graph. By Corollary 3.5.2, G has two edge-disjoint spanning trees T_i, $i = 1, 2$. For each edge $e \notin T_i$, let
$f_{1,e}, f_{2,e}$
$C_{i,e}$ be the unique cycle in $T_i + e$, and let $f_{i,e}$ be a $\mathbb{Z}_4$-flow of value $\bar{i}$ around $C_{i,e}$—more precisely: a $\mathbb{Z}_4$-circulation on G with values $\bar{i}$ and $-\bar{i}$ on the edges of $C_{i,e}$ and zero otherwise.

Let $f_1 := \sum_{e \notin T_1} f_{1,e}$. Since each $e \notin T_1$ lies on only one cycle $C_{1,e'}$
(namely, for $e = e'$), f_1 takes only the values $\bar{1}$ and $-\bar{1} (= \bar{3})$ outside T_1.
Let f_1

$$ F := \{\, e \in E(T_1) \mid f_1(e) = \bar{0} \,\} $$

and $f_2 := \sum_{e \in F} f_{2,e}$. As above, $f_2(e) = \bar{2} = -\bar{2}$ for all $e \in F$. Now f_2
$f := f_1 + f_2$ is the sum of $\mathbb{Z}_4$-circulations, and hence itself a $\mathbb{Z}_4$-circula- f
tion. Moreover, f is nowhere zero: on edges in F it takes the value $\bar{2}$, on
edges of $T_1 - F$ it agrees with f_1 (and is hence non-zero by the choice
of F), and on all edges outside T_1 it takes one of the values $\bar{1}$ or $\bar{3}$. Hence,
f is a $\mathbb{Z}_4$-flow on G, and the assertion follows by Theorem 6.3.3. $\qquad \square$

The following proposition describes the graphs with a 4-flow in terms
of those with a 2-flow:

Proposition 6.4.5.

(i) *A graph has a 4-flow if and only if it is the union of two even
    subgraphs.*

(ii) *A cubic graph has a 4-flow if and only if it is 3-edge-colourable.*

Proof. Let $\mathbb{Z}_2^2 = \mathbb{Z}_2 \times \mathbb{Z}_2$ be the Klein four-group. (Thus, the elements of (6.3.2)
$\mathbb{Z}_2^2$ are the pairs (a, b) with $a, b \in \mathbb{Z}_2$, and $(a, b) + (a', b') = (a + a', b + b')$.) (6.3.3)
By Corollary 6.3.2 and Theorem 6.3.3, a graph has a 4-flow if and only
if it has a $\mathbb{Z}_2^2$-flow.

(i) now follows directly from Proposition 6.4.1.

(ii) Let $G = (V, E)$ be a cubic graph. We assume first that G has a
$\mathbb{Z}_2^2$-flow f, and define an edge colouring $E \to \mathbb{Z}_2^2 \smallsetminus \{\, 0 \,\}$. As $a = -a$ for
all $a \in \mathbb{Z}_2^2$, we have $f(\vec{e}) = f(\overleftarrow{e})$ for every $\vec{e} \in E$; let us colour the edge
e with this colour $f(\vec{e})$. Now if two edges with a common end v had
the same colour, then these two values of f would sum to zero; by (F2),
f would then assign zero to the third edge at v. As this contradicts the
definition of f, our edge colouring is correct.

Conversely, since the three non-zero elements of $\mathbb{Z}_2^2$ sum to zero,
every 3-edge-colouring $c : E \to \mathbb{Z}_2^2 \smallsetminus \{\, 0 \,\}$ defines a $\mathbb{Z}_2^2$-flow on G by letting
$f(\vec{e}) = f(\overleftarrow{e}) = c(e)$ for all $\vec{e} \in E$. $\qquad \square$

Corollary 6.4.6. *Every cubic 3-edge-colourable graph is bridgeless.*
$\qquad \square$

6.5 Flow-colouring duality

In this section we shall see a surprising connection between flows and colouring: every k-flow on a plane multigraph gives rise to a k-vertex-colouring of its dual, and vice versa. In this way, the investigation of k-flows appears as a natural generalization of the familiar map colouring problems in the plane.

$G = (V, E)$
G^*

Let $G = (V, E)$ and $G^* = (V^*, E^*)$ be dual plane multigraphs. For simplicity, let us assume that G and G^* have neither bridges nor loops and are non-trivial. For edge sets $F \subseteq E$, let us write

F^*

$$F^* := \{ e^* \in E^* \mid e \in F \} .$$

Conversely, if a subset of E^* is given, we shall usually write it immediately in the form F^*, and thus let $F \subseteq E$ be defined implicitly via the bijection $e \mapsto e^*$.

Suppose we are given a circulation g on G^*: how can we employ the duality between G and G^* to derive from g some information about G? The most general property of all circulations is Proposition 6.1.1, which says that $g(X, \overline{X}) = 0$ for all $X \subseteq V^*$. By Proposition 4.6.1, the minimal cuts $E^*(X, \overline{X})$ in G^* correspond precisely to the cycles in G. Thus if we take the composition f of the maps $e \mapsto e^*$ and g, and sum its values over the edges of a cycle in G, then this sum should again be zero.

Of course, there is still a technical hitch: since g takes its arguments not in E^* but in $\vec{E^*}$, we cannot simply define f as above: we first have to refine the bijection $e \mapsto e^*$ into one from $\vec{E}$ to $\vec{E^*}$, i.e. assign to every $\vec{e} \in \vec{E}$ canonically one of the two directions of e^*. This will be the purpose of our first lemma. After that, we shall show that f does indeed sum to zero along any cycle in G.

If $C = v_0 \ldots v_{\ell-1} v_0$ is a cycle with edges $e_i = v_i v_{i+1}$ (and $v_\ell := v_0$), we shall call

$\vec{C}$

$$\vec{C} := \{ (e_i, v_i, v_{i+1}) \mid i < \ell \}$$

cycle with orientation

a *cycle with orientation*. Note that this definition of $\vec{C}$ depends on the vertex enumeration chosen to denote C: every cycle has two orientations. Conversely, of course, C can be reconstructed from the set $\vec{C}$. In practice, we shall therefore speak about C freely even when, formally, only $\vec{C}$ has been defined.

Lemma 6.5.1. *There exists a bijection* $*: \vec{e} \mapsto \vec{e}^*$ *from* $\vec{E}$ *to* $\vec{E^*}$ *with the following properties.*

(i) *The underlying edge of* $\vec{e}^*$ *is always* e^*, *i.e.* $\vec{e}^*$ *is one of the two directions* $\vec{e^*}, \overleftarrow{e^*}$ *of* e^*.

(ii) *If* $C \subseteq G$ *is a cycle,* $F := E(C)$, *and if* $X \subseteq V^*$ *is such that* $F^* = E^*(X, \overline{X})$, *then there exists an orientation* $\vec{C}$ *of* C *with* $\{ \vec{e}^* \mid \vec{e} \in \vec{C} \} = \vec{E^*}(X, \overline{X})$.

The proof of Lemma 6.5.1 is not entirely trivial: it is based on the so-called *orientability* of the plane, and we cannot give it here. Still, the assertion of the lemma is intuitively plausible. Indeed if we define for $e = vw$ and $e^* = xy$ the assignment $(e, v, w) \mapsto (e, v, w)^* \in \{ (e^*, x, y), (e^*, y, x) \}$ simply by turning e and its ends clockwise onto e^* (Fig. 6.5.1), then the resulting map $\vec{e} \mapsto \vec{e}^*$ satisfies the two assertions of the lemma.

Fig. 6.5.1. Oriented cycle-cut duality

Given an abelian group H, let $f \colon \vec{E} \to H$ and $g \colon \vec{E}^* \to H$ be two maps such that

$$f(\vec{e}) = g(\vec{e}^*)$$

for all $\vec{e} \in \vec{E}$. For $\vec{F} \subseteq \vec{E}$, we set

$$f(\vec{F}) := \sum_{\vec{e} \in \vec{F}} f(\vec{e}).$$

f, g

$f(\vec{C})$ etc.

Lemma 6.5.2.

(i) *The map g satisfies (F1) if and only if f does.*

(ii) *The map g is a circulation on G^* if and only if f satisfies (F1) and $f(\vec{C}) = 0$ for every cycle $\vec{C}$ with orientation.*

Proof. Assertion (i) follows from Lemma 6.5.1 (i) and the fact that $\vec{e} \mapsto \vec{e}^*$ is bijective.

For the forward implication of (ii), let us assume that g is a circulation on G^*, and consider a cycle $C \subseteq G$ with some given orientation. Let $F := E(C)$. By Proposition 4.6.1, F^* is a minimal cut in G^*, i.e. $F^* = E^*(X, \overline{X})$ for some suitable $X \subseteq V^*$. By definition of f and g, Lemma 6.5.1 (ii) and Proposition 6.1.1 give

$$f(\vec{C}) = \sum_{\vec{e} \in \vec{C}} f(\vec{e}) = \sum_{\vec{d} \in \vec{E}^*(X, \overline{X})} g(\vec{d}) = g(X, \overline{X}) = 0$$

for one of the two orientations $\vec{C}$ of C. Then, by $f(\overleftarrow{C}) = -f(\vec{C})$, also

(4.6.1)
(6.1.1)

the corresponding value for our given orientation of C must be zero.

For the backward implication it suffices by (i) to show that g satisfies (F2), i.e. that $g(x, V^*) = 0$ for every $x \in V^*$. We shall prove that $g(x, V(B)) = 0$ for every block B of G^* containing x; since every edge of G^* at x lies in exactly one such block, this will imply $g(x, V^*) = 0$.

B

So let $x \in V^*$ be given, and let B be any block of G^* containing x. Since G^* is a non-trivial plane dual, and hence connected, we have $B - x \neq \emptyset$. Let F^* be the set of all edges of B at x (Fig. 6.5.2),

F^*, F

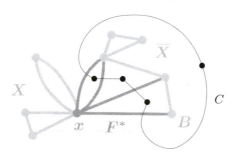

Fig. 6.5.2. The cut F^* in G^*

X

and let X be the vertex set of the component of $G^* - F^*$ containing x. Then $\emptyset \neq V(B - x) \subseteq \overline{X}$, by the maximality of B as a cutvertex-free subgraph. Hence

$$F^* = E^*(X, \overline{X}) \tag{1}$$

by definition of X, i.e. F^* is a cut in G^*. As a dual, G^* is connected, so $G^*[\overline{X}]$ too is connected. Indeed, every vertex of $\overline{X}$ is linked to x by a path $P \subseteq G^*$ whose last edge lies in F^*. Then $P - x$ is a path in $G^*[\overline{X}]$ meeting B. Since x does not separate B, this shows that $G^*[\overline{X}]$ is connected.

C

Thus, X and $\overline{X}$ are both connected in G^*, so F^* is even a minimal cut in G^*. Let $C \subseteq G$ be the cycle with $E(C) = F$ that exists by Proposition 4.6.1. By Lemma 6.5.1 (ii), C has an orientation $\vec{C}$ such that $\{ \vec{e}^* \mid \vec{e} \in \vec{C} \} = \vec{E}^*(X, \overline{X})$. By (1), however, $\vec{E}^*(X, \overline{X}) = \vec{E}^*(x, V(B))$, so

$$g(x, V(B)) = g(X, \overline{X}) = f(\vec{C}) = 0$$

by definition of f and g.   □

With the help of Lemma 6.5.2, we can now prove our colouring-flow duality theorem for plane multigraphs. If $P = v_0 \dots v_\ell$ is a path with edges $e_i = v_i v_{i+1}$ $(i < \ell)$, we set (depending on our vertex enumeration of P)

$\vec{P}$

$$\vec{P} := \{ (e_i, v_i, v_{i+1}) \mid i < \ell \}$$

$v_0 \to v_\ell$
path

and call $\vec{P}$ a $v_0 \to v_\ell$ path. Again, P may be given implicitly by $\vec{P}$.

Theorem 6.5.3. (Tutte 1954)
For every dual pair G, G^* of plane multigraphs,

$$\chi(G) = \varphi(G^*) \,.$$

Proof. Let $G =: (V, E)$ and $G^* =: (V^*, E^*)$. For $|G| \in \{1, 2\}$ the assertion is easily checked; we shall assume that $|G| \geqslant 3$, and apply induction on the number of bridges in G. If $e \in G$ is a bridge, then e^* is a loop, and $G^* - e^*$ is a plane dual of G/e (why?). Hence, by the induction hypothesis,

$$\chi(G) = \chi(G/e) = \varphi(G^* - e^*) = \varphi(G^*) \,;$$

for the first and the last equality we use that, by $|G| \geqslant 3$, e is not the only edge of G.

  So all that remains to be checked is the induction start: let us assume that G has no bridge. If G has a loop, then G^* has a bridge, and $\chi(G) = \infty = \varphi(G^*)$ by convention. So we may also assume that G has no loop. Then $\chi(G)$ is finite; we shall prove for given $k \geqslant 2$ that G is k-colourable if and only if G^* has a k-flow. As G—and hence G^*—has neither loops nor bridges, we may apply Lemmas 6.5.1 and 6.5.2 to G and G^*. Let $\vec{e} \mapsto \vec{e}^*$ be the bijection between $\vec{E}$ and $\vec{E}^*$ from Lemma 6.5.1.

  We first assume that G^* has a k-flow. Then G^* also has a $\mathbb{Z}_k$-flow g. As before, let $f \colon \vec{E} \to \mathbb{Z}_k$ be defined by $f(\vec{e}) := g(\vec{e}^*)$. We shall use f to define a vertex colouring $c \colon V \to \mathbb{Z}_k$ of G.

  Let us start with an auxiliary observation for vertices $v, w \in G$:

$$\text{Any two } v \to w \text{ paths } \vec{P}, \vec{P}' \text{ in } G \text{ satisfy } f(\vec{P}) = f(\vec{P}'). \tag{1}$$

This follows easily by induction on $|P| + |P'|$. Indeed, if P and P' have no inner vertex in common, then either $P = P' \simeq K^1$, or $\vec{P} \cup \vec{P}'$ is a cycle with orientation, and

$$\vec{0} = f(\vec{P} \cup \vec{P}') = f(\vec{P}) - f(\vec{P}')$$

by Lemma 6.5.2 (ii); in either case, the assertion follows. On the other hand, if u is a vertex in $\mathring{P} \cap \mathring{P}'$, then by the induction hypothesis

$$f(\vec{P}) = f(\vec{Pu}) + f(\vec{uP}) = f(\vec{P'u}) + f(\vec{uP'}) = f(\vec{P'}) \,.$$

This completes the proof of (1).

  Let us fix a vertex $v_0 \in G$. For every vertex $w \in V$, we choose a $v_0 \to w$ path $\vec{P}$ in G and set $c(w) := f(\vec{P})$; by (1), this makes c well defined on all of V.

It remains to show that c is a proper colouring, i.e. that $c(w) \neq c(w')$ for adjacent vertices w, w'. If $w, w' \in V$ are adjacent then, without loss of generality, G contains a $v_0 \to w'$ path $\vec{P}$ ending with the edge $e := ww'$; if not, just swap the roles of w and w'. Since g is nowhere zero, we have $f(e, w, w') \neq \bar{0}$ by definition of f. Then by (1)

$$c(w') - c(w) = f(\vec{P}) - f(\vec{Pw}) = f(e, w, w') \neq \bar{0} \,,$$

so $c(w) \neq c(w')$ as desired.

c

Conversely, we now assume that G has a k-colouring c. Let us define $f \colon \vec{E} \to \mathbb{Z}$ by

$$f(e, v, w) := c(w) - c(v) \,,$$

g

and $g \colon \vec{E}^* \to \mathbb{Z}$ by $g(\vec{e}^*) := f(\vec{e})$. Clearly, f satisfies (F1) and takes values in $\{ \pm 1, \ldots, \pm(k-1) \}$, so by Lemma 6.5.2 (i) the same holds for g. By definition of f, we further have $f(\vec{C}) = 0$ for every cycle $\vec{C}$ with orientation. By Lemma 6.5.2 (ii), therefore, g is a k-flow. $\qquad \square$

6.6 Tutte's flow conjectures

How can we determine the flow number of a graph? Indeed, does every (bridgeless) graph have a flow number, a k-flow for some k? Can flow numbers, like chromatic numbers, become arbitrarily large? Can we characterize the graphs admitting a k-flow, for given k?

Of these four questions, we shall answer the second and third in this section: we prove that every bridgeless graph has a 6-flow. In particular, a graph has a flow number if and only if it has no bridge. The question asking for a characterization of the graphs with a k-flow remains interesting for $k = 3, 4, 5$. Partial answers are suggested by the following three conjectures of Tutte, who initiated algebraic flow theory.

The oldest and best known of the Tutte conjectures is his *5-flow conjecture*:

Five-Flow Conjecture. (Tutte 1954)
Every bridgeless multigraph has a 5-flow.

Which graphs have a 4-flow? By Proposition 6.4.4, the 4-edge-connected graphs are among them. The Petersen graph (Fig. 6.6.1), on the other hand, is an example of a bridgeless graph without a 4-flow: since it is cubic but not 3-edge-colourable (Ex. 19, Ch. 5), it cannot have a 4-flow by Proposition 6.4.5 (ii).

Tutte's *4-flow conjecture* states that the Petersen graph must be present in every graph without a 4-flow:

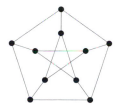

Fig. 6.6.1. The Petersen graph

Four-Flow Conjecture. (Tutte 1966)
Every bridgeless multigraph not containing the Petersen graph as a minor has a 4-flow.

By Proposition 1.7.2, we may replace the word 'minor' in the 4-flow conjecture by 'topological minor'.

Even if true, the 4-flow conjecture will not be best possible: a K^{11}, for example, contains the Petersen graph as a minor but has a 4-flow, even a 2-flow. The conjecture appears more natural for sparser graphs, and indeed the cubic graphs form an important special case.

A cubic bridgeless graph or multigraph without a 4-flow (equivalently, without a 3-edge-colouring) is called a *snark*. The 4-flow conjecture for cubic graphs says that every snark contains the Petersen graph as a minor; in this sense, the Petersen graph would then be the smallest snark. Snarks form the hard core both of the four colour theorem and of the 5-flow conjecture: the four colour theorem is equivalent to the assertion that no snark is planar (exercise), and it is not difficult to reduce the 5-flow conjecture to the case of snarks.[5] However, although the snarks form a very special class of graphs, none of the problems mentioned seems to become much easier by this reduction.[6]

snark

Three-Flow Conjecture. (Tutte 1972)
Every multigraph without a cut consisting of exactly one or exactly three edges has a 3-flow.

Again, the 3-flow conjecture will not be best possible: it is easy to construct graphs with three-edge cuts that have a 3-flow (exercise).

By our duality theorem (6.5.3), all three flow conjectures are true for planar graphs and thus motivated: the 3-flow conjecture translates to Grötzsch's theorem (5.1.3), the 4-flow conjecture to the four colour theorem (since the Petersen graph is not planar, it is not a minor of a planar graph), the 5-flow conjecture to the five colour theorem.

---

[5] The same applies to another well-known conjecture, the *cycle double cover conjecture*; see Exercise 13.

[6] That snarks are elusive has been known to mathematicians for some time; cf. Lewis Carroll, *The Hunting of the Snark*, Macmillan 1876.

We finish this section with the main result of the chapter:

Theorem 6.6.1. (Seymour 1981)
Every bridgeless graph has a 6-flow.

(3.3.5)
(6.1.1)
(6.4.1)

Proof. Let $G = (V, E)$ be a bridgeless graph. Since 6-flows on the components of G will add up to a 6-flow on G, we may assume that G is connected; as G is bridgeless, it is then 2-edge-connected. Note that any two vertices in a 2-edge-connected graph lie in some common even connected subgraph—for example, in the union of two edge-disjoint paths linking these vertices by Menger's theorem (3.3.5 (ii)). We shall use this fact repeatedly.

$H_0, \ldots, H_n$
$F_1, \ldots, F_n$

We shall construct a sequence $H_0, \ldots, H_n$ of disjoint connected and even subgraphs of G, together with a sequence $F_1, \ldots, F_n$ of non-empty sets of edges between them. The sets F_i will each contain only one or two edges, between H_i and $H_0 \cup \ldots \cup H_{i-1}$. We write $H_i =: (V_i, E_i)$,

V_i, E_i

H^i

$$H^i := (H_0 \cup \ldots \cup H_i) + (F_1 \cup \ldots \cup F_i)$$

V^i, E^i

and $H^i =: (V^i, E^i)$. Note that each $H^i = (H^{i-1} \cup H_i) + F_i$ is connected (induction on i). Our assumption that H_i is even implies by Proposition 6.4.1 (or directly by Proposition 1.2.1) that H_i has no bridge.

As H_0 we choose any K^1 in G. Now assume that $H_0, \ldots, H_{i-1}$ and $F_1, \ldots, F_{i-1}$ have been defined for some $i > 0$. If $V^{i-1} = V$, we terminate the construction and set $i - 1 =: n$. Otherwise, we let $X_i \subseteq \overline{V^{i-1}}$ be minimal such that $X_i \neq \emptyset$ and

n
X_i

$$\left| E(X_i, \overline{V^{i-1}} \setminus X_i) \right| \leqslant 1 \tag{1}$$

(Fig. 6.6.2); such an X_i exists, because $\overline{V^{i-1}}$ is a candidate. Since G is 2-edge-connected, (1) implies that $E(X_i, V^{i-1}) \neq \emptyset$. By the minimality of X_i, the graph $G[X_i]$ is connected and bridgeless, i.e. 2-edge-connected or a K^1. As the elements of F_i we pick one or two edges from $E(X_i, V^{i-1})$, if possible two. As H_i we choose any connected even subgraph of $G[X_i]$ containing the ends in X_i of the edges in F_i.

F_i

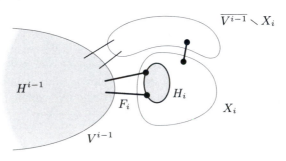

Fig. 6.6.2. Constructing the H_i and F_i

When our construction is complete, we set $H^n =: H$ and $E' :=$ $E \setminus E(H)$. By definition of n, H is a spanning connected subgraph of G. $\hfill H \\ E'$

We now define, by 'reverse' induction, a sequence $f_n, \ldots, f_0$ of $\mathbb{Z}_3$-circulations on G. For every edge $e \in E'$, let $\vec{C}_e$ be a cycle (with orientation) in $H + e$ containing e, and f_e a positive flow around $\vec{C}_e$; formally, we let f_e be a $\mathbb{Z}_3$-circulation on G such that $f_e^{-1}(\vec{0}) = \vec{E} \setminus (\vec{C}_e \cup \tilde{C}_e)$. Let f_n be the sum of all these f_e. Since each $e' \in E'$ lies on just one of the cycles C_e (namely, on $C_{e'}$), we have $f_n(\vec{e}) \neq \vec{0}$ for all $\vec{e} \in \vec{E'}$. $\hfill f_n, \ldots, f_0 \\ \vec{C}_e \\ f_e \\ f_n$

Assume now that $\mathbb{Z}_3$-circulations $f_n, \ldots, f_i$ on G have been defined for some $i \leqslant n$, and that $\hfill f_i$

$$f_i(\vec{e}) \neq \vec{0} \quad \text{for all} \quad \vec{e} \in \vec{E'} \cup \bigcup_{j > i} \vec{F}_j, \qquad (2)$$

where $\vec{F}_j := \{ \vec{e} \in \vec{E} \mid e \in F_j \}$. Our aim is to define f_{i-1} in such a way that (2) also holds for $i - 1$. $\hfill \vec{F}_j$

We first consider the case that $|F_i| = 1$, say $F_i = \{e\}$. We then let $f_{i-1} := f_i$, and thus have to show that f_i is non-zero on (the two directions of) e. Our assumption of $|F_i| = 1$ implies by the choice of F_i that G contains no X_i–V^{i-1} edge other than e. Since G is 2-edge-connected, it therefore has at least—and thus, by (1), exactly—one edge e' between X_i and $\overline{V^{i-1}} \setminus X_i$. We show that f_i is non-zero on e'; as $\{e, e'\}$ is a cut in G, this implies by Proposition 6.1.1 that f_i is also non-zero on e. $\hfill e \\ e'$

To show that f_i is non-zero on e', we use (2): we show that $e' \in E' \cup \bigcup_{j > i} F_j$, i.e. that e' lies in no H_k and in no F_j with $j \leqslant i$. Since e' has both ends in $\overline{V^{i-1}}$, it clearly lies in no F_j with $j \leqslant i$ and in no H_k with $k < i$. But every H_k with $k \geqslant i$ is a subgraph of $G[\overline{V^{i-1}}]$. Since e' is a bridge of $G[\overline{V^{i-1}}]$ but H_k has no bridge, this means that $e' \notin H_k$. Hence, f_{i-1} does indeed satisfy (2) for $i - 1$ in the case considered.

It remains to consider the case that $|F_i| = 2$, say $F_i = \{e_1, e_2\}$. Since H_i and H^{i-1} are both connected, we can find a cycle C in $H^i = (H_i \cup H^{i-1}) + F_i$ that contains e_1 and e_2. If f_i is non-zero on both these edges, we again let $f_{i-1} := f_i$. Otherwise, there are directions $\vec{e_1}$ and $\vec{e_2}$ of e_1 and e_2 such that, without loss of generality, $f_i(\vec{e_1}) = \vec{0}$ and $f_i(\vec{e_2}) \in \{\vec{0}, \vec{1}\}$. Let $\vec{C}$ be the orientation of C with $\vec{e_2} \in \vec{C}$, and let g be a flow of value $\vec{1}$ around $\vec{C}$ (formally: let g be a $\mathbb{Z}_3$-circulation on G such that $g(\vec{e_2}) = \vec{1}$ and $g^{-1}(\vec{0}) = \vec{E} \setminus (\vec{C} \cup \tilde{C})$). We then let $f_{i-1} := f_i + g$. By choice of the directions $\vec{e_1}$ and $\vec{e_2}$, f_{i-1} is non-zero on both edges. Since f_{i-1} agrees with f_i on all of $\vec{E'} \cup \bigcup_{j > i} \vec{F}_j$ and (2) holds for i, we again have (2) also for $i - 1$. $\hfill e_1, e_2 \\ C$

Eventually, f_0 will be a $\mathbb{Z}_3$-circulation on G that is nowhere zero except possibly on edges of $H_0 \cup \ldots \cup H_n$. Composing f_0 with the map $\overline{h} \mapsto \overline{2h}$ from $\mathbb{Z}_3$ to $\mathbb{Z}_6$ ($h \in \{1, 2\}$), we obtain a $\mathbb{Z}_6$-circulation f on G $\hfill f$

with values in $\{\bar{0}, \bar{2}, \bar{4}\}$ for all edges lying in some H_i, and with values in $\{\bar{2}, \bar{4}\}$ for all other edges. Adding to f a 2-flow on each H_i (formally: a $\mathbb{Z}_6$-circulation on G with values in $\{\bar{1}, -\bar{1}\}$ on the edges of H_i and $\bar{0}$ otherwise; this exists by Proposition 6.4.1), we obtain a $\mathbb{Z}_6$-circulation on G that is nowhere zero. Hence, G has a 6-flow by Theorem 6.3.3.

$\square$

Exercises

1.⁻ Prove Proposition 6.2.1 by induction on $|S|$.

2. (i)⁻ Given $n \in \mathbb{N}$, find a capacity function for the network below such that the algorithm from the proof of the max-flow min-cut theorem will need more than n augmenting paths W if these are badly chosen.

   (ii)⁺ Show that, if all augmenting paths are chosen as short as possible, their number is bounded by a function of the size of the network.

3. Derive Menger's Theorem 3.3.4 from the max-flow min-cut theorem.

   (Hint. The edge version is easy. For the vertex version, split every vertex x into two vertices x^-, x^+. Define the edges of the arising graph, and their capacities, in such a way that positive flow through an edge $x^- x^+$ describes a piece of a path in G through x.)

4.⁻ Let f be an H-circulation on G and $g\colon H \to H'$ a group homomorphism. Show that $g \circ f$ is an H'-circulation on G. Is $g \circ f$ an H'-flow if f is an H-flow?

5.⁻ Given $k \geqslant 1$, show that a graph has a k-flow if and only if each of its blocks has a k-flow.

6.⁻ Show that $\varphi(G/e) \leqslant \varphi(G)$ whenever G is a multigraph and e an edge of G. Does this imply that, for every k, the class of all multigraphs admitting a k-flow is closed under taking minors?

7.⁻ Work out the flow number of K^4 directly, without using any results from the text.

8. Let H be a finite abelian group, G a graph, and T a spanning tree of G. Show that every mapping from the directions of $E(G) \setminus E(T)$ to H that satisfies (F1) extends uniquely to an H-circulation on G.

Do not use the 6-flow Theorem 6.6.1 for the following three exercises.

9. Show that $\varphi(G) < \infty$ for every bridgeless multigraph G.

10. Assume that a graph G has m spanning trees such that no edge of G lies in all of these trees. Show that $\varphi(G) \leqslant 2^m$.

11.[+] Let G be a bridgeless connected graph with n vertices and m edges. By considering a depth-first search tree of G (see Ex. 14, Ch. 1), show that $\varphi(G) \leqslant m - n + 2$.

12. Show that every graph with a Hamilton cycle has a 4-flow. (A *Hamilton cycle* of G is a cycle in G that contains all the vertices of G.)

13. A family of (not necessarily distinct) cycles in a graph G is called a *cycle double cover* of G if every edge of G lies on exactly two of these cycles. The *cycle double cover conjecture* asserts that every bridgeless multigraph has a cycle double cover. Prove the conjecture for graphs with a 4-flow.

14.[-] Determine the flow number of $C^5 * K^1$, the wheel with 5 spokes.

15. Find bridgeless graphs G and $H = G - e$ such that $2 < \varphi(G) < \varphi(H)$.

16.[-] Prove Proposition 6.4.1 from Euler's Theorem 1.8.1.

17.[+] Prove *Heawood's theorem* that a plane triangulation is 3-colourable if and only if all its vertices have even degree.

18.[-] Find a bridgeless graph that has both a 3-flow and a cut of exactly three edges.

19. Show that the 3-flow conjecture for planar multigraphs is equivalent to Grötzsch's Theorem 5.1.3.

20. (i)[-] Show that the four colour theorem is equivalent to the non-existence of a planar snark, i.e. to the statement that every cubic bridgeless planar multigraph has a 4-flow.

(ii) Can 'bridgeless' in (i) be replaced by '3-connected'?

21.[+] Show that a graph $G = (V, E)$ has a k-flow if and only if it admits an orientation D that directs, for every $X \subseteq V$, at least $1/k$ of the edges in $E(X, \overline{X})$ from X towards $\overline{X}$.

(Hint. For the 'if' implication, consider a circulation f on G, with values in $\{ 0, \pm 1, \ldots, \pm(k-1) \}$, that respects the given orientation (i.e. is positive or zero on the edge directions assigned by D) and is zero on as few edges as possible. Then show that f is nowhere zero, as follows. If f is zero on $e = st \in E$ and D directs e from t to s, define a network $N = (G, s, t, c)$ such that any flow in N of positive total value contradicts the choice of f, but any cut in N of zero capacity contradicts the property assumed for D.)

22.[-] Generalize the 6-flow Theorem 6.6.1 to multigraphs.

Notes

Network flow theory is an application of graph theory that has had a major and lasting impact on its development over decades. As is illustrated already by the fact that Menger's theorem can be deduced easily from the max-flow min-cut theorem (Exercise 3), the interaction between graphs and networks may go either way: while 'pure' results in areas such as connectivity, matching and random graphs have found applications in network flows, the intuitive power of the latter has boosted the development of proof techniques that have in turn brought about theoretic advances.

The standard reference for network flows is L.R. Ford & D.R. Fulkerson, *Flows in Networks*, Princeton University Press 1962. A more recent and comprehensive account is given by R.K. Ahuja, T.L. Magnanti & J.B. Orlin, *Network flows*, Prentice-Hall 1993. For more theoretical aspects, see A. Frank's chapter in the *Handbook of Combinatorics* (R.L. Graham, M. Grötschel & L. Lovász, eds.), North-Holland 1995. A general introduction to graph algorithms is given in A. Gibbons, *Algorithmic Graph Theory*, Cambridge University Press 1985.

If one recasts the maximum flow problem in linear programming terms, one can derive the max-flow min-cut theorem from the linear programming duality theorem; see A. Schrijver, *Theory of integer and linear programming*, Wiley 1986.

The more algebraic theory of group-valued flows and k-flows has been developed largely by Tutte; he gives a thorough account in his monograph W.T. Tutte, *Graph Theory*, Addison-Wesley 1984. Tutte's flow conjectures are covered also in F. Jaeger's survey, Nowhere-zero[7] flow problems, in (L.W. Beineke & R.J. Wilson, eds.) *Selected Topics in Graph Theory 3*, Academic Press 1988. For the flow conjectures, see also T.R. Jensen & B. Toft, *Graph Coloring Problems*, Wiley 1995. Seymour's 6-flow theorem is proved in P.D. Seymour, Nowhere-zero 6-flows, *J. Combin. Theory B* **30** (1981), 130–135. This paper also indicates how Tutte's 5-flow conjecture reduces to snarks.

Finally, Tutte discovered a 2-variable polynomial associated with a graph, which generalizes both its chromatic polynomial and its flow polynomial. What little is known about this *Tutte polynomial* can hardly be more than the tip of the iceberg: it has far-reaching, and largely unexplored, connections to areas as diverse as knot theory and statistical physics. See D.J.A. Welsh, *Complexity: knots, colourings and counting* (LMS Lecture Notes **186**), Cambridge University Press 1993.

---

[7] In the literature, the term 'flow' is often used to mean what we have called 'circulation', i.e. flows are not required to be nowhere zero unless this is stated explicitly.

7 Substructures in Dense Graphs

In this chapter and the next, we study how global parameters of a graph, such as its edge density or chromatic number, have a bearing on the existence of certain local substructures. How many edges, for instance, do we have to give a graph on n vertices to be sure that, no matter how these edges happen to be arranged, the graph will contain a K^r subgraph for some given r? Or at least a K^r minor? Or a topological K^r minor? Will some sufficiently high average degree or chromatic number ensure that one of these substructures occurs?

Questions of this type are among the most natural ones in graph theory, and there is a host of deep and interesting results. Collectively, these are known as *extremal graph theory*.

Extremal graph problems in this sense fall neatly into two categories, as follows. If we are looking for ways to ensure by global assumptions that a graph G contains some given graph H as a *minor* (or topological minor), it will suffice to raise $\|G\|$ above the value of some linear function of $|G|$ (depending on H), i.e. to make $\varepsilon(G)$ large enough. The existence of such a function was already established in Theorem 3.6.1. The precise growth rate needed will be investigated in Chapter 8, where we study substructures of such 'sparse' graphs. Since a large enough value of ε gives rise to an H minor for any given graph H, its occurrence could be forced alternatively by raising some other global invariants (such as κ or χ) which, in turn, force up the value of ε, at least in some subgraph. This, too, will be a topic for Chapter 8.

On the other hand, if we ask what global assumptions might imply the existence of some given graph H as a *subgraph*, it will not help to raise any of the invariants ε, κ or χ, let alone any of the other invariants discussed in Chapter 1. Indeed, as mentioned in Chapter 5.2, given any graph H that contains at least one cycle, there are graphs of arbitrarily large chromatic number not containing H as a subgraph (Theorem 11.2.2). By Corollary 5.2.3 and Theorem 1.4.2, such graphs have subgraphs of arbitrarily large average degree and connectivity, so these invariants too can be large without the presence of an H subgraph.

Thus, unless H is a forest, the only way to force the presence of an H subgraph in an arbitrary graph G by global assumptions on G is to raise $\|G\|$ substantially above any value implied by large values of the above invariants. If H is not bipartite, then any function f such that $f(n)$ edges on n vertices force an H subgraph must even grow quadratically with n: since complete bipartite graphs can have $\frac{1}{4}n^2$ edges, $f(n)$ must exceed $\frac{1}{4}n^2$.

dense

edge density

Graphs with a number of edges roughly[1] quadratic in their number of vertices are usually called *dense*; the number $\|G\|/\binom{|G|}{2}$—the proportion of its potential edges that G actually has—is the *edge density* of G. The question of exactly which edge density is needed to force a given subgraph is the archetypal extremal graph problem in its original (narrower) sense; it is the topic of this chapter. Rather than attempting to survey the wide field of (dense) extremal graph theory, however, we shall concentrate on its two most important results and portray one powerful general proof technique.

The two results are Turán's classic extremal graph theorem for $H = K^r$, a result that has served as a model for countless similar theorems for other graphs H, and the fundamental Erdős-Stone theorem, which gives precise asymptotic information for all H at once (Section 7.1). The proof technique, one of increasing importance in the extremal theory of dense graphs, is the use of the Szemerédi *regularity lemma*. This lemma is presented and proved in Section 7.2. In Section 7.3, we outline a general method for applying the regularity lemma, and illustrate this in the proof of the Erdős-Stone theorem postponed from Section 7.1. Another application of the regularity lemma will be given in Chapter 9.2.

7.1 Subgraphs

Let H be a graph and $n \geqslant |H|$. How many edges will suffice to force an H subgraph in any graph on n vertices, no matter how these edges are arranged? Or, to rephrase the problem: which is the greatest possible number of edges that a graph on n vertices can have *without* containing a copy of H as a subgraph? What will such a graph look like? Will it be unique?

---

[1] Note that, formally, the notions of sparse and dense make sense only for families of graphs whose order tends to infinity, not for individual graphs.

A graph $G \not\supseteq H$ on n vertices with the largest possible number of
edges is called *extremal* for n and H; its number of edges is denoted by
$\mathrm{ex}(n, H)$. Clearly, any graph G that is extremal for some n and H will
also be edge-maximal with $H \not\subseteq G$. Conversely, though, edge-maximality
does not imply extremality: G may well be edge-maximal with $H \not\subseteq G$
while having fewer than $\mathrm{ex}(n, H)$ edges (Fig. 7.1.1).

extremal

$\mathrm{ex}(n, H)$

Fig. 7.1.1. Two graphs that are edge-maximal with $P^3 \not\subseteq G$; is
the right one extremal?

As a case in point, we consider our problem for $H = K^r$ (with $r > 1$).
A moment's thought suggests some obvious candidates for extremality
here: all complete $(r-1)$-partite graphs are edge-maximal without containing K^r. But which among these have the greatest number of edges?
Clearly those whose partition sets are as equal as possible, i.e. differ in
size by at most 1: if V_1, V_2 are two partition sets with $|V_1| - |V_2| \geqslant 2$, we
may increase the number of edges in our complete $(r-1)$-partite graph
by moving a vertex from V_1 across to V_2.

The unique complete $(r-1)$-partite graphs on $n \geqslant r-1$ vertices
whose partition sets differ in size by at most 1 are called *Turán graphs*;
we denote them by $T^{r-1}(n)$ and their number of edges by $t_{r-1}(n)$
(Fig. 7.1.2). For $n < r-1$ we shall formally continue to use these
definitions, with the proviso that—contrary to our usual terminology—
the partition sets may now be empty; then, clearly, $T^{r-1}(n) = K^n$ for
all $n \leqslant r-1$.

$T^{r-1}(n)$

$t_{r-1}(n)$

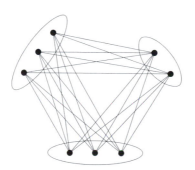

Fig. 7.1.2. The Turán graph $T^3(8)$

The following theorem tells us that $T^{r-1}(n)$ is indeed extremal for
n and K^r, and as such unique; in particular, $\mathrm{ex}(n, K^r) = t_{r-1}(n)$.

[7.1.2]
[9.2.2]

Theorem 7.1.1. (Turán 1941)
For all integers r, n with $r > 1$, every graph $G \not\supseteq K^r$ with n vertices and $\mathrm{ex}(n, K^r)$ edges is a $T^{r-1}(n)$.

Proof. We apply induction on n. For $n \leqslant r - 1$ we have $G = K^n = T^{r-1}(n)$ as claimed. For the induction step, let now $n \geqslant r$.

K Since G is edge-maximal without a K^r subgraph, G has a subgraph $K = K^{r-1}$. By the induction hypothesis, $G - K$ has at most $t_{r-1}(n - r + 1)$ edges, and each vertex of $G - K$ has at most $r - 2$ neighbours in K. Hence,

$$\|G\| \leqslant t_{r-1}(n - r + 1) + (n - r + 1)(r - 2) + \binom{r-1}{2} = t_{r-1}(n); \quad (1)$$

the equality on the right follows by inspection of the Turán graph $T^{r-1}(n)$ (Fig. 7.1.3).

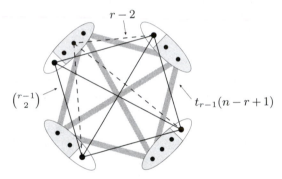

Fig. 7.1.3. The equation from (1) for $r = 5$ and $n = 14$

Since G is extremal for K^r (and $T^{r-1}(n) \not\supseteq K^r$), we have equality in (1). Thus, every vertex of $G - K$ has *exactly* $r - 2$ neighbours in K—

$x_1, \ldots, x_{r-1}$ just like the vertices $x_1, \ldots, x_{r-1}$ of K itself. For $i = 1, \ldots, r - 1$ let

$V_1, \ldots, V_{r-1}$
$$V_i := \{\, v \in V(G) \mid vx_i \notin E(G) \,\}$$

be the set of all vertices of G whose $r - 2$ neighbours in K are precisely the vertices other than x_i. Since $K^r \not\subseteq G$, each of the sets V_i is independent, and they partition $V(G)$. Hence, G is $(r - 1)$-partite. As $T^{r-1}(n)$ is the unique $(r - 1)$-partite graph with n vertices and the maximum number of edges, our claim that $G = T^{r-1}(n)$ follows from the assumed extremality of G.   □

The Turán graphs $T^{r-1}(n)$ are dense: in order of magnitude, they have about n^2 edges. More exactly, for every n and r we have

$$t_{r-1}(n) \leqslant \tfrac{1}{2} n^2 \, \frac{r-2}{r-1},$$

with equality whenever $r - 1$ divides n (Exercise 8). It is therefore remarkable that just ϵn^2 more edges (for any fixed $\epsilon > 0$ and n large) give us not only a K^r subgraph (as does Turán's theorem) but a K_s^r for any given integer s—a graph itself teeming with K^r subgraphs:

Theorem 7.1.2. (Erdős & Stone 1946)
For all integers $r \geqslant 2$ and $s \geqslant 1$, and every $\epsilon > 0$, there exists an integer n_0 such that every graph with $n \geqslant n_0$ vertices and at least

$$t_{r-1}(n) + \epsilon n^2$$

edges contains K_s^r as a subgraph.

We shall prove this theorem in Section 7.3.

The Erdős-Stone theorem is interesting not only in its own right: it also has a most interesting corollary. In fact, it was this entirely unexpected corollary that established the theorem as a kind of meta-theorem for the extremal theory of dense graphs, and thus made it famous.

Given a graph H and an integer n, consider the number $h_n := \mathrm{ex}(n, H)/\binom{n}{2}$: the maximum edge density that an n-vertex graph can have without containing a copy of H. Could it be that this critical density is essentially just a function of H, that h_n converges as $n \to \infty$? Theorem 7.1.2 implies this, and more: the limit of h_n is determined by a very simple function of a natural invariant of H—its chromatic number!

Corollary 7.1.3. *For every graph H,*

$$\lim_{n \to \infty} \mathrm{ex}(n, H) \binom{n}{2}^{-1} = \frac{\chi(H) - 2}{\chi(H) - 1}.$$

For the proof of Corollary 7.1.3 we need as a lemma that $t_{r-1}(n)$ never deviates much from the value it takes when $r - 1$ divides n (see above), and that $t_{r-1}(n)/\binom{n}{2}$ converges accordingly. The proof of the lemma is left as an easy exercise with hint (Exercise 9).

Lemma 7.1.4.

[7.1.2]

$$\lim_{n \to \infty} t_{r-1}(n) \binom{n}{2}^{-1} = \frac{r - 2}{r - 1}.$$

□

Proof of Corollary 7.1.3. Let $r := \chi(H)$. Since H cannot be coloured with $r - 1$ colours, we have $H \not\subseteq T^{r-1}(n)$ for all $n \in \mathbb{N}$, and hence

$$t_{r-1}(n) \leqslant \mathrm{ex}(n, H).$$

On the other hand, $H \subseteq K_s^r$ for all sufficiently large s, so

$$\mathrm{ex}(n, H) \leqslant \mathrm{ex}(n, K_s^r)$$

for all those s. Let us fix such an s. For every $\epsilon > 0$, Theorem 7.1.2 implies that eventually (i.e. for large enough n)

$$\mathrm{ex}(n, K_s^r) < t_{r-1}(n) + \epsilon n^2.$$

Hence for n large,

$$
\begin{aligned}
t_{r-1}(n)/\tbinom{n}{2} &\leqslant \mathrm{ex}(n, H)/\tbinom{n}{2} \\
&\leqslant \mathrm{ex}(n, K_s^r)/\tbinom{n}{2} \\
&< t_{r-1}(n)/\tbinom{n}{2} + \epsilon n^2/\tbinom{n}{2} \\
&= t_{r-1}(n)/\tbinom{n}{2} + 2\epsilon/(1 - \tfrac{1}{n}) \\
&\leqslant t_{r-1}(n)/\tbinom{n}{2} + 4\epsilon \qquad \text{(assume } n \geqslant 2\text{)}.
\end{aligned}
$$

Therefore, since $t_{r-1}(n)/\tbinom{n}{2}$ converges to $\frac{r-2}{r-1}$ (Lemma 7.1.4), so does $\mathrm{ex}(n, H)/\tbinom{n}{2}$. Thus

$$\lim_{n \to \infty} \mathrm{ex}(n, H)\binom{n}{2}^{-1} = \frac{r-2}{r-1}$$

as claimed. $\qquad\square$

For bipartite graphs H, Corollary 7.1.3 says that substantially fewer than $\binom{n}{2}$ edges suffice to force an H subgraph. It turns out that

$$c_1 n^{2-2/r} \leqslant \mathrm{ex}(n, K_{r,r}) \leqslant c_2 n^{2-1/r}$$

for suitable constants c_1, c_2 depending on r; the lower bound is obtained by random graphs,[2] the upper bound is calculated in Exercise 12. If H is a forest, then $H \subseteq G$ as soon as $\varepsilon(G)$ is large enough, so $\mathrm{ex}(n, H)$ is at most linear in n (Exercise 6).

---

[2] see Chapter 11

7.2 Szemerédi's regularity lemma

More than 20 years ago, in the course of the proof of a major result on
the Ramsey properties of arithmetic progressions, Szemerédi developed a
graph theoretical tool whose fundamental importance has been realized
more and more in recent years: his so-called *regularity* or *uniformity
lemma*. Very roughly, the lemma says that all graphs can be approx-
imated by random graphs in the following sense: every graph can be
partitioned, into a bounded number of equal parts, so that most of its
edges run between different parts and the edges between any two parts
are distributed fairly uniformly—just as we would expect it if they had
been generated at random.

   In order to state the regularity lemma precisely, we need some definitions. Let $G = (V, E)$ be a graph, and let $X, Y \subseteq V$ be disjoint. Then
we denote by $\|X, Y\|$ the number of X–Y edges of G, and call $\|X, Y\|$

$$d(X, Y) := \frac{\|X, Y\|}{|X|\,|Y|}$$

$d(X, Y)$

the *density* of the pair (X, Y). (This is a real number between 0 and 1.)   *density*
Given some $\epsilon > 0$, we call a pair (A, B) of disjoint sets $A, B \subseteq V$ ϵ-*regular* ϵ-*regular*
if all $X \subseteq A$ and $Y \subseteq B$ with   *pair*

$$|X| \geqslant \epsilon|A| \quad \text{and} \quad |Y| \geqslant \epsilon|B|$$

satisfy

$$\bigl|d(X, Y) - d(A, B)\bigr| \leqslant \epsilon\,.$$

The edges in an ϵ-regular pair are thus distributed fairly uniformly: the
smaller ϵ, the more uniform their distribution.

   Consider a partition $\{\,V_0, V_1, \ldots, V_k\,\}$ of V in which one set V_0 has   *exceptional*
been singled out as an *exceptional set*. (This exceptional set V_0 may   *set*
be empty.[3]) We call such a partition an ϵ-*regular* partition of G if it
satisfies the following three conditions:

(i) $|V_0| \leqslant \epsilon|V|$;   *$\epsilon$-regular*
                                                                                  partition
(ii) $|V_1| = \ldots = |V_k|$;

(iii) all but at most ϵk^2 of the pairs (V_i, V_j) with $1 \leqslant i < j \leqslant k$ are
      ϵ-regular.

   The role of the exceptional set V_0 is one of pure convenience: it
makes it possible to require that all the other partition sets have exactly
the same size. Since condition (iii) affects only the sets $V_1, \ldots, V_k$, we

---

[3] So V_0 may be an exception also to our terminological rule that partition sets
are not normally empty.

may think of V_0 as a kind of bin: its vertices are disregarded when the uniformity of the partition is assessed, but there are only few such vertices.

Lemma 7.2.1. (Regularity Lemma)

[9.2.2]   *For every $\epsilon > 0$ and every integer $m \geqslant 1$ there exists an integer M such that every graph of order at least m admits an ϵ-regular partition $\{V_0, V_1, \ldots, V_k\}$ with $m \leqslant k \leqslant M$.*

The regularity lemma thus says that, given any $\epsilon > 0$, every graph has an ϵ-regular partition into a bounded number of sets. The upper bound M on the number of partition sets ensures that for large graphs the partition sets are large too; note that ϵ-regularity is trivial when the partition sets are singletons, and a powerful property when they are large. In addition, the lemma allows us to specify a lower bound m on the number of partition sets; by choosing m large, we may increase the proportion of edges running between different partition sets (rather than inside one), i.e. the proportion of edges that are subject to the regularity assertion.

Note that the regularity lemma is designed for use with dense graphs:[4] for sparse graphs it becomes trivial, because all densities of pairs—and hence their differences—tend to zero (Exercise 16).

The remainder of this section is devoted to the proof of the regularity lemma. Although the proof is not difficult, a reader meeting the regularity lemma here for the first time is likely to draw more insight from seeing how the lemma is typically applied than from studying the technicalities of its proof. Any such reader is encouraged to skip to the start of Section 7.3 now and come back to the proof at his or her leisure.

We shall need the following inequality for reals $\mu_1, \ldots, \mu_k > 0$ and $e_1, \ldots, e_k \geqslant 0$:

$$\sum \frac{e_i^2}{\mu_i} \geqslant \frac{\left(\sum e_i\right)^2}{\sum \mu_i}. \tag{1}$$

This follows from the Cauchy-Schwartz inequality $\sum a_i^2 \sum b_i^2 \geqslant \left(\sum a_i b_i\right)^2$ by taking $a_i := \sqrt{\mu_i}$ and $b_i := e_i/\sqrt{\mu_i}$.

$G = (V, E)$ Let $G = (V, E)$ be a graph and $n := |V|$. For disjoint sets $A, B \subseteq V$
n we define

$q(A, B)$
$$q(A, B) := \frac{|A|\,|B|}{n^2}\, d^2(A, B) = \frac{\|A, B\|^2}{|A|\,|B|\,n^2}.$$

For partitions $\mathcal{A}$ of A and $\mathcal{B}$ of B we set

$q(\mathcal{A}, \mathcal{B})$
$$q(\mathcal{A}, \mathcal{B}) := \sum_{A' \in \mathcal{A};\ B' \in \mathcal{B}} q(A', B'),$$

---

[4]   Sparse versions do exist, though; see the notes.

and for a partition $\mathcal{P} = \{C_1, \ldots, C_k\}$ of V we let

$$q(\mathcal{P}) := \sum_{i<j} q(C_i, C_j).$$ $q(\mathcal{P})$

However, if $\mathcal{P} = \{C_0, C_1, \ldots, C_k\}$ is a partition of V *with exceptional set* C_0, we treat C_0 as a set of singletons and define

$$q(\mathcal{P}) := q(\tilde{\mathcal{P}}),$$

where $\tilde{\mathcal{P}} := \{C_1, \ldots, C_k\} \cup \{\{v\} : v \in C_0\}$. $\tilde{\mathcal{P}}$

The function $q(\mathcal{P})$ plays a pivotal role in the proof of the regularity lemma. On the one hand, it measures the uniformity of the partition $\mathcal{P}$: if $\mathcal{P}$ has too many irregular pairs (A, B), we may take the pairs (X, Y) of subsets violating the regularity of the pairs (A, B) and make those sets X and Y into partition sets of their own; as we shall prove, this refines $\mathcal{P}$ into a partition for which q is substantially greater than for $\mathcal{P}$. Here, 'substantial' means that the increase of $q(\mathcal{P})$ is bounded below by some constant depending only on ϵ. On the other hand,

$$\begin{aligned}
q(\mathcal{P}) &= \sum_{i<j} q(C_i, C_j) \\
&= \sum_{i<j} \frac{|C_i|\,|C_j|}{n^2} \, d^2(C_i, C_j) \\
&\leqslant \frac{1}{n^2} \sum_{i<j} |C_i|\,|C_j| \\
&\leqslant 1.
\end{aligned}$$

The number of times that $q(\mathcal{P})$ can be increased by a constant is thus also bounded by a constant—in other words, after some bounded number of refinements our partition will be ϵ-regular! To complete the proof of the regularity lemma, all we have to do then is to note how many sets that last partition can possibly have if we start with a partition into m sets, and to choose this number as our desired bound M.

Let us make all this precise. We begin by showing that, when we refine a partition, the value of q will not decrease:

Lemma 7.2.2.

(i) *Let* $C, D \subseteq V$ *be disjoint. If* $\mathcal{C}$ *is a partition of* C *and* $\mathcal{D}$ *is a partition of* D, *then* $q(\mathcal{C}, \mathcal{D}) \geqslant q(C, D)$.

(ii) *If* $\mathcal{P}, \mathcal{P}'$ *are partitions of* V *and* $\mathcal{P}'$ *refines* $\mathcal{P}$, *then* $q(\mathcal{P}') \geqslant q(\mathcal{P})$.

Proof. (i) Let $\mathcal{C} =: \{C_1, \ldots, C_k\}$ and $\mathcal{D} =: \{D_1, \ldots, D_\ell\}$. Then

$$q(\mathcal{C}, \mathcal{D}) = \sum_{i,j} q(C_i, D_j)$$

$$= \frac{1}{n^2} \sum_{i,j} \frac{\|C_i, D_j\|^2}{|C_i| |D_j|}$$

$$\underset{(1)}{\geq} \frac{1}{n^2} \frac{\left(\sum_{i,j} \|C_i, D_j\|\right)^2}{\sum_{i,j} |C_i| |D_j|}$$

$$= \frac{1}{n^2} \frac{\|C, D\|^2}{\left(\sum_i |C_i|\right)\left(\sum_j |D_j|\right)}$$

$$= q(C, D).$$

(ii) Let $\mathcal{P} =: \{C_1, \ldots, C_k\}$, and for $i = 1, \ldots, k$ let $\mathcal{C}_i$ be the partition of C_i induced by $\mathcal{P}'$. Then

$$q(\mathcal{P}) = \sum_{i<j} q(C_i, C_j)$$

$$\underset{(i)}{\leq} \sum_{i<j} q(\mathcal{C}_i, \mathcal{C}_j)$$

$$\leq q(\mathcal{P}'),$$

since $q(\mathcal{P}') = \sum_i q(\mathcal{C}_i) + \sum_{i<j} q(\mathcal{C}_i, \mathcal{C}_j)$. $\qquad\square$

Next, we show that refining a partition by subpartitioning an irregular pair of partition sets increases the value of q a little; since we are dealing here with a single pair only, the amount of this increase will still be less than any constant.

Lemma 7.2.3. *Let $\epsilon > 0$, and let $C, D \subseteq V$ be disjoint. If (C, D) is not ϵ-regular, then there are partitions $\mathcal{C} = (C_1, C_2)$ of C and $\mathcal{D} = (D_1, D_2)$ of D such that*

$$q(\mathcal{C}, \mathcal{D}) \geq q(C, D) + \epsilon^4 \frac{|C| |D|}{n^2}.$$

Proof. Suppose (C, D) is not ϵ-regular. Then there are sets $C_1 \subseteq C$ and $D_1 \subseteq D$ with $|C_1| > \epsilon |C|$ and $|D_1| > \epsilon |D|$ such that

$$|\eta| > \epsilon \qquad\qquad (2)$$

η

for $\eta := d(C_1, D_1) - d(C, D)$. Let $\mathcal{C} := \{C_1, C_2\}$ and $\mathcal{D} := \{D_1, D_2\}$, where $C_2 := C \setminus C_1$ and $D_2 := D \setminus D_1$.

Let us show that C and D satisfy the conclusion of the lemma. We shall write $c_i := |C_i|$, $d_i := |D_i|$, $e_{ij} := \|C_i, D_j\|$, $c := |C|$, $d := |D|$ c_i, d_i, e_{ij}
and $e := \|C, D\|$. As in the proof of Lemma 7.2.2, c, d, e

$$q(C, D) = \frac{1}{n^2} \sum_{i,j} \frac{e_{ij}^2}{c_i d_j}$$

$$= \frac{1}{n^2} \left(\frac{e_{11}^2}{c_1 d_1} + \sum_{i+j>2} \frac{e_{ij}^2}{c_i d_j} \right)$$

$$\underset{(1)}{\geqslant} \frac{1}{n^2} \left(\frac{e_{11}^2}{c_1 d_1} + \frac{(e - e_{11})^2}{cd - c_1 d_1} \right).$$

By definition of η, we have $e_{11} = c_1 d_1 e / cd + \eta c_1 d_1$, so

$$n^2 q(C, D) \geqslant \frac{1}{c_1 d_1} \left(\frac{c_1 d_1 e}{cd} + \eta c_1 d_1 \right)^2$$

$$+ \frac{1}{cd - c_1 d_1} \left(\frac{cd - c_1 d_1}{cd} e - \eta c_1 d_1 \right)^2$$

$$= \frac{c_1 d_1 e^2}{c^2 d^2} + \frac{2 e \eta c_1 d_1}{cd} + \eta^2 c_1 d_1$$

$$+ \frac{cd - c_1 d_1}{c^2 d^2} e^2 - \frac{2 e \eta c_1 d_1}{cd} + \frac{\eta^2 c_1^2 d_1^2}{cd - c_1 d_1}$$

$$\geqslant \frac{e^2}{cd} + \eta^2 c_1 d_1$$

$$\underset{(2)}{\geqslant} \frac{e^2}{cd} + \epsilon^4 cd$$

since $c_1 \geqslant \epsilon c$ and $d_1 \geqslant \epsilon d$ by the choice of C_1 and D_1.   □

Finally, we show that if a partition has enough irregular pairs of partition sets to fall short of the definition of an ϵ-regular partition, then subpartitioning all those pairs at once results in an increase of q by a constant:

Lemma 7.2.4. Let $0 < \epsilon \leqslant 1/4$, and let $\mathcal{P} = \{C_0, C_1, \ldots, C_k\}$ be a partition of V, with exceptional set C_0 of size $|C_0| \leqslant \epsilon n$ and $|C_1| = \ldots = |C_k| =: c$. If $\mathcal{P}$ is not ϵ-regular, then there is a partition c
$\mathcal{P}' = \{C_0', C_1', \ldots, C_\ell'\}$ of V with exceptional set C_0', where $k \leqslant \ell \leqslant k4^k$, such that $|C_0'| \leqslant |C_0| + n/2^k$, all other sets C_i' have equal size, and

$$q(\mathcal{P}') \geqslant q(\mathcal{P}) + \epsilon^5/2.$$

C_{ij}

Proof. For all $1 \leqslant i < j \leqslant k$, let us define a partition $\mathcal{C}_{ij}$ of C_i and a partition $\mathcal{C}_{ji}$ of C_j, as follows. If the pair (C_i, C_j) is ϵ-regular, we let $\mathcal{C}_{ij} := \{C_i\}$ and $\mathcal{C}_{ji} := \{C_j\}$. If not, then by Lemma 7.2.3 there are partitions $\mathcal{C}_{ij}$ of C_i and $\mathcal{C}_{ji}$ of C_j with $|\mathcal{C}_{ij}| = |\mathcal{C}_{ji}| = 2$ and

$$q(\mathcal{C}_{ij}, \mathcal{C}_{ji}) \geqslant q(C_i, C_j) + \epsilon^4 \frac{|C_i||C_j|}{n^2} = q(C_i, C_j) + \frac{\epsilon^4 c^2}{n^2} . \qquad (3)$$

$\mathcal{C}_i$

For each $i = 1, \ldots, k$, let $\mathcal{C}_i$ be the unique minimal partition of C_i that refines every partition $\mathcal{C}_{ij}$ with $j \neq i$. (In other words, if we consider two elements of C_i as equivalent whenever they lie in the same partition set of $\mathcal{C}_{ij}$ for every $j \neq i$, then $\mathcal{C}_i$ is the set of equivalence classes.) Thus, $|\mathcal{C}_i| \leqslant 2^{k-1}$. Now consider the partition

$\mathcal{C}$

$$\mathcal{C} := \{C_0\} \cup \bigcup_{i=1}^{k} \mathcal{C}_i$$

of V, with C_0 as exceptional set. Then $\mathcal{C}$ refines $\mathcal{P}$, and

$$k \leqslant |\mathcal{C}| \leqslant k2^k. \qquad (4)$$

C_0

Let $\mathcal{C}_0 := \{\{v\} : v \in C_0\}$. Now if $\mathcal{P}$ is not ϵ-regular, then for more than ϵk^2 of the pairs (C_i, C_j) with $1 \leqslant i < j \leqslant k$ the partition $\mathcal{C}_{ij}$ is non-trivial. Hence, by our definition of q for partitions with exceptional set, and Lemma 7.2.2 (i),

$$
\begin{aligned}
q(\mathcal{C}) &= \sum_{1 \leqslant i < j} q(\mathcal{C}_i, \mathcal{C}_j) + \sum_{1 \leqslant i} q(\mathcal{C}_0, \mathcal{C}_i) + \sum_{0 \leqslant i} q(\mathcal{C}_i) \\
&\geqslant \sum_{1 \leqslant i < j} q(\mathcal{C}_{ij}, \mathcal{C}_{ji}) + \sum_{1 \leqslant i} q(\mathcal{C}_0, \{C_i\}) + q(\mathcal{C}_0) \\
&\underset{(3)}{\geqslant} \sum_{1 \leqslant i < j} q(C_i, C_j) + \epsilon k^2 \frac{\epsilon^4 c^2}{n^2} + \sum_{1 \leqslant i} q(\mathcal{C}_0, \{C_i\}) + q(\mathcal{C}_0) \\
&= q(\mathcal{P}) + \epsilon^5 \left(\frac{kc}{n}\right)^2 \\
&\geqslant q(\mathcal{P}) + \epsilon^5/2 .
\end{aligned}
$$

(For the last inequality, recall that $|C_0| \leqslant \epsilon n \leqslant \frac{1}{4}n$, so $kc \geqslant \frac{3}{4}n$.)

In order to turn $\mathcal{C}$ into our desired partition $\mathcal{P}'$, all that remains to do is to cut its sets up into pieces of some common size, small enough that all remaining vertices can be collected into the exceptional set without making this too large. Let $C'_1, \ldots, C'_\ell$ be a maximal collection of disjoint sets of size $d := \lfloor c/4^k \rfloor$ such that each C'_i is contained in some

d

$C \in \mathcal{C} \smallsetminus \{C_0\}$, and put $C_0' := V \smallsetminus \bigcup C_i'$. Then $\mathcal{P}' = \{C_0', C_1', \ldots, C_\ell'\}$ $\qquad$ $\mathcal{P}'$
is indeed a partition of V. Moreover, $\tilde{\mathcal{P}}'$ refines $\tilde{\mathcal{C}}$, so

$$q(\mathcal{P}') \geqslant q(\mathcal{C}) \geqslant q(\mathcal{P}) + \epsilon^5/2$$

by Lemma 7.2.2 (ii). Since each set $C_i' \neq C_0'$ is also contained in one
of the sets $C_1, \ldots, C_k$, but no more than 4^k sets C_i' can lie inside the
same C_j (by the choice of d), we also have $k \leqslant \ell \leqslant k4^k$ as required.
Finally, the sets $C_1', \ldots, C_\ell'$ use all but at most d vertices from each set
$C \neq C_0$ of $\mathcal{C}$. Hence,

$$
\begin{aligned}
|C_0'| &\leqslant |C_0| + d\,|\mathcal{C}| \\
&\underset{(4)}{\leqslant} |C_0| + \frac{c}{4^k} k 2^k \\
&= |C_0| + ck/2^k \\
&\leqslant |C_0| + n/2^k.
\end{aligned}
$$

$\qquad\square$

The proof of the regularity lemma now follows easily by repeated
application of Lemma 7.2.4:

Proof of Lemma 7.2.1. Let $\epsilon > 0$ and $m \geqslant 1$ be given; without loss $\qquad$ ϵ, m
of generality, $\epsilon \leqslant 1/4$. Let $s := 2/\epsilon^5$. This number s is an upper bound $\qquad$ s
on the number of iterations of Lemma 7.2.4 that can be applied to a
partition of a graph before it becomes ϵ-regular; recall that $q(\mathcal{P}) \leqslant 1$ for
all partitions $\mathcal{P}$.

There is one formal requirement which a partition $\{C_0, C_1, \ldots, C_k\}$
with $|C_1| = \ldots = |C_k|$ has to satisfy before Lemma 7.2.4 can be (re-)
applied: the size $|C_0|$ of its exceptional set must not exceed ϵn. With
each iteration of the lemma, however, the size of the exceptional set can
grow by up to $n/2^k$. (More precisely, by up to $n/2^\ell$, where ℓ is the
number of other sets in the current partition; but $\ell \geqslant k$ by the lemma,
so $n/2^k$ is certainly an upper bound for the increase.) We thus want
to choose k large enough that even s increments of $n/2^k$ add up to at
most $\frac{1}{2}\epsilon n$, and n large enough that, for any initial value of $|C_0| < k$, we
have $|C_0| \leqslant \frac{1}{2}\epsilon n$. (If we give our starting partition k non-exceptional
sets $C_1, \ldots, C_k$, we should allow an initial size of up to k for C_0, to be
able to achieve $|C_1| = \ldots = |C_k|$.)

So let $k \geqslant m$ be large enough that $2^{k-1} \geqslant s/\epsilon$. Then $s/2^k \leqslant \epsilon/2$, $\qquad$ k
and hence

$$k + \frac{s}{2^k} n \leqslant \epsilon n \tag{5}$$

whenever $k/n \leqslant \epsilon/2$, i.e. for all $n \geqslant 2k/\epsilon$.

Let us now choose M. This should be an upper bound on the
number of (non-exceptional) sets in our partition after up to s iterations

M

n

of Lemma 7.2.4, where in each iteration this number may grow from its current value r to at most $r4^r$. So let f be the function $x \mapsto x4^x$, and take $M := \max\{ f^s(k), 2k/\epsilon \}$; the second term in the maximum ensures that any $n \geqslant M$ is large enough to satisfy (5).

We finally have to show that every graph $G = (V, E)$ of order at least m has an ϵ-regular partition $\{ V_0, V_1, \ldots, V_k \}$ with $m \leqslant k \leqslant M$. So let G be given, and let $n := |G|$. If $n \leqslant M$, we partition G into $k := n$ singletons, choosing $V_0 := \emptyset$ and $|V_1| = \ldots = |V_k| = 1$. This partition of G is clearly ϵ-regular. Suppose now that $n > M$. Let $C_0 \subseteq V$ be minimal such that k divides $|V \smallsetminus C_0|$, and let $\{ C_1, \ldots, C_k \}$ be any partition of $V \smallsetminus C_0$ into sets of equal size. Then $|C_0| < k$, and hence $|C_0| \leqslant \epsilon n$ by (5). Starting with $\{ C_0, C_1, \ldots, C_k \}$ we apply Lemma 7.2.4 again and again, until the partition of G obtained is ϵ-regular; this will happen after at most s iterations, since by (5) the size of the exceptional set in the partitions stays below ϵn, so the lemma could indeed be reapplied up to the theoretical maximum of s times.   □

7.3 Applying the regularity lemma

The purpose of this section is to illustrate how the regularity lemma is typically applied in the context of (dense) extremal graph theory. Suppose we are trying to prove that a certain edge density of a graph G suffices to force the occurrence of some given subgraph H, and that we have an ϵ-regular partition of G. The edges between each pair (V_i, V_j) of partition sets are distributed uniformly, although their density may depend on the pair. But since G has many edges, this density cannot be zero for all the pairs: some sizeable proportion of the pairs will have positive density. Now if G is large, then so are the pairs: recall that the number of partition sets is bounded, and they have equal size. But any large enough bipartite graph with equal partition sets, fixed positive edge density (however small!) and a uniform distribution of edges will contain any given bipartite subgraph[5]—this will be made precise below. Thus if enough pairs in our partition of G have positive density that H can be written as the union of bipartite graphs each arising in one of those pairs, we may hope that $H \subseteq G$ as desired.

These ideas will be formalized by Lemma 7.3.2 below. We shall then use this and the regularity lemma to prove the Erdős-Stone theorem from Section 7.1; another application will be given later, in the proof of Theorem 9.2.2.

Before we state Lemma 7.3.2, let us note a simple consequence of the ϵ-regularity of a pair (A, B): for any subset $Y \subseteq B$ that is not too

---

[5] Readers already acquainted with random graphs may find it instructive to compare this statement with Proposition 11.3.1.

small, most vertices of A have about the expected number of neighbours in Y:

Lemma 7.3.1. *Let (A, B) be an ϵ-regular pair, of density d say, and let $Y \subseteq B$ have size $|Y| \geqslant \epsilon |B|$. Then all but at most $\epsilon |A|$ of the vertices in A have (each) at least $(d - \epsilon)|Y|$ neighbours in Y.*

Proof. Let $X \subseteq A$ be the set of vertices with fewer than $(d - \epsilon)|Y|$ neighbours in Y. Then $\|X, Y\| < |X|(d - \epsilon)|Y|$, so

$$d(X, Y) = \frac{\|X, Y\|}{|X| \, |Y|} < d - \epsilon = d(A, B) - \epsilon \,.$$

Since (A, B) is ϵ-regular, this implies that $|X| < \epsilon |A|$. $\qquad \square$

Let G be a graph with an ϵ-regular partition $\{ V_0, V_1, \ldots, V_k \}$, with exceptional set V_0 and $|V_1| = \ldots = |V_k| =: \ell$. Given $d \in (0, 1]$, let R be the graph with vertices $V_1, \ldots, V_k$ in which two vertices are adjacent if and only if they form an ϵ-regular pair in G of density $\geqslant d$. We shall call R a *regularity graph* of G with parameters ϵ, ℓ and d. Given $s \in \mathbb{N}$, let us now replace every vertex V_i of R by a set V_i^s of s vertices, and every edge by a complete bipartite graph between the corresponding s-sets. The resulting graph will be denoted by R_s. (For $R = K^r$, for example, we have $R_s = K_s^r$.) $\qquad$

The following lemma says that subgraphs of R_s can also be found in G, provided that ϵ is small enough and the V_i are large enough. In fact, the values of ϵ and ℓ required depend only on (d and) the maximum degree of the subgraph:

R

regularity
graph
V_i^s

R_s

Lemma 7.3.2. *For all $d \in (0, 1]$ and $\Delta \geqslant 1$ there exists an $\epsilon_0 > 0$ with the following property: if G is any graph, H is a graph with $\Delta(H) \leqslant \Delta$, $s \in \mathbb{N}$, and R is any regularity graph of G with parameters $\epsilon \leqslant \epsilon_0$, $\ell \geqslant s/\epsilon_0$ and d, then*

[9.2.2]

$$H \subseteq R_s \;\; \Rightarrow \;\; H \subseteq G \,.$$

Proof. Given d and Δ, choose $\epsilon_0 < d$ small enough that

d, Δ

$$\frac{\Delta + 1}{(d - \epsilon_0)^\Delta} \epsilon_0 \leqslant 1 \,; \qquad\qquad (1)$$

ϵ_0

such a choice is possible, since $(\Delta + 1)\epsilon/(d - \epsilon)^\Delta \to 0$ as $\epsilon \to 0$. Now let G, H, s and R be given as stated. Let $\{ V_0, V_1, \ldots, V_k \}$ be the ϵ-regular partition of G that gave rise to R; thus, $\epsilon \leqslant \epsilon_0$, $V(R) = \{ V_1, \ldots, V_k \}$ and $|V_1| = \ldots = |V_k| = \ell$. Let us assume that H is actually a subgraph

G, H, R, R_s
V_i
ϵ, k, ℓ

u_i, h

σ

v_i

of R_s (not just isomorphic to one), with vertices $u_1, \ldots, u_h$ say. Each vertex u_i lies in one of the s-sets V_j^s of R_s; this defines a map $\sigma : i \mapsto j$. Our aim is to define an embedding $u_i \mapsto v_i \in V_{\sigma(i)}$ of H in G; thus, $v_1, \ldots, v_h$ will be distinct, and $v_i v_j$ will be an edge of G whenever $u_i u_j$ is an edge of H.

Our plan is to choose the vertices $v_1, \ldots, v_h$ inductively. Throughout the induction, we shall have a 'target set' $Y_i \subseteq V_{\sigma(i)}$ assigned to each i; this contains the vertices that are still candidates for the choice of v_i. Initially, Y_i is the entire set $V_{\sigma(i)}$. As the embedding proceeds, Y_i will get smaller and smaller (until it collapses to $\{ v_i \}$ when v_i is chosen): whenever we choose a vertex v_j with $j < i$ and $u_j u_i \in E(H)$, we delete all those vertices from Y_i that are not adjacent to v_j. The set Y_i thus evolves as

$$V_{\sigma(i)} = Y_i^0 \supseteq \ldots \supseteq Y_i^i = \{ v_i \},$$

where Y_i^j denotes the version of Y_i current after the definition of v_j (and any corresponding deletion of vertices from Y_i^{j-1}).

In order to make this approach work, we have to ensure that the target sets Y_i do not get too small. When we come to embed a vertex u_j, we consider all the indices $i > j$ with $u_j u_i \in E(H)$; there are at most Δ such i. For each of these i, we wish to select v_j so that

$$Y_i^j = N(v_j) \cap Y_i^{j-1} \tag{2}$$

is large, i.e. not much smaller than Y_i^{j-1}. Now this can be done by Lemma 7.3.1 (with $A = V_{\sigma(j)}$, $B = V_{\sigma(i)}$ and $Y = Y_i^{j-1}$): unless Y_i^{j-1} is tiny (of size less than $\epsilon \ell$), all but at most $\epsilon \ell$ choices of v_j will be such that (2) implies

$$|Y_i^j| \geq (d - \epsilon)|Y_i^{j-1}|. \tag{3}$$

Doing this simultaneously for all of the at most Δ values of i considered, we find that all but at most $\Delta \epsilon \ell$ choices of v_j from $V_{\sigma(j)}$, and in particular from $Y_j^{j-1} \subseteq V_{\sigma(j)}$, satisfy (3) for all i.

It remains to show that the sets Y considered for Lemma 7.3.1 above are indeed never tiny, and that $|Y_j^{j-1}| - \Delta \epsilon \ell \geq s$ to ensure that a suitable choice for v_j exists: since $\sigma(j') = \sigma(j)$ for at most $s - 1$ of the vertices $u_{j'}$ with $j' < j$, a choice between s suitable candidates for v_j will suffice to keep v_j distinct from $v_1, \ldots, v_{j-1}$. But all this follows from our choice of ϵ_0. Indeed, the initial target sets Y_i^0 have size ℓ, and each Y_i has vertices deleted from it only when some v_j with $j < i$ and $u_j u_i \in E(H)$ is defined, which happens at most Δ times. Thus,

$$|Y_i^j| - \Delta \epsilon \ell \underset{(3)}{\geq} (d - \epsilon)^\Delta \ell - \Delta \epsilon \ell \geq (d - \epsilon_0)^\Delta \ell - \Delta \epsilon_0 \ell \underset{(1)}{\geq} \epsilon_0 \ell \geq s$$

whenever $j < i$, so in particular $|Y_i^j| \geq \epsilon_0 \ell \geq \epsilon \ell$ and $|Y_j^{j-1}| - \Delta \epsilon \ell \geq s$. $\qquad \square$

We are now ready to prove the Erdős-Stone theorem.

Proof of Theorem 7.1.2. Let $r \geqslant 2$ and $s \geqslant 1$ be given as in the (7.1.1)
(7.1.4)
statement of the theorem. For $s = 1$ the assertion follows from Turán's r, s
theorem, so we assume that $s \geqslant 2$. Let $\gamma > 0$ be given; this γ will play γ
the role of the ϵ of the theorem. Let G be a graph with $|G| =: n$ and

$$\|G\| \geqslant t_{r-1}(n) + \gamma n^2.$$ $\|G\|$

(Thus, $\gamma < 1$.) We want to show that $K_s^r \subseteq G$ if n is large enough.

Our plan is to use the regularity lemma to show that G has a regularity graph R dense enough to contain a K^r by Turán's theorem. Then R_s contains a K_s^r, so we may hope to use Lemma 7.3.2 to deduce that $K_s^r \subseteq G$.

On input $d := \gamma$ and $\Delta := \Delta(K_s^r)$, Lemma 7.3.2 returns an $\epsilon_0 > 0$; d, Δ
since the lemma's assertion about ϵ_0 becomes weaker when ϵ_0 is made ϵ_0
smaller, we may assume that $\epsilon_0 < 1$. To apply the regularity lemma, let
$m > 1/\gamma$ and choose $\epsilon > 0$ small enough that $\epsilon \leqslant \epsilon_0$ and m, ϵ

$$\delta := 2\gamma - \epsilon^2 - 4\epsilon - d - \frac{1}{m} > 0;$$ δ

this is possible, since $2\gamma - d - \frac{1}{m} > 0$. On input ϵ and m, the regularity
lemma returns an integer M. Let us assume that M

$$n \geqslant \frac{Ms}{\epsilon_0(1 - \epsilon)}.$$ n

Since this number is at least m, the regularity lemma provides us with
an ϵ-regular partition $\{V_0, V_1, \ldots, V_k\}$ of G, where $m \leqslant k \leqslant M$; let k
$|V_1| = \ldots = |V_k| =: \ell$. Then ℓ

$$n \geqslant k\ell, \tag{1}$$

and

$$\ell = \frac{n - |V_0|}{k} \geqslant \frac{n - \epsilon n}{M} = n\frac{1 - \epsilon}{M} \geqslant \frac{s}{\epsilon_0}$$

by the choice of n. Let R be the regularity graph of G with parameters R
ϵ, ℓ, d corresponding to the above partition. Since $\epsilon \leqslant \epsilon_0$ and $\ell \geqslant s/\epsilon_0$, the
regularity graph R satisfies the premise of Lemma 7.3.2, and by definition
of Δ we have $\Delta(K_s^r) = \Delta$. Thus in order to conclude by Lemma 7.3.2
that $K_s^r \subseteq G$, all that remains to be checked is that $K^r \subseteq R$ (and hence
$K_s^r \subseteq R_s$).

Our plan was to show $K^r \subseteq R$ by Turán's theorem. We thus have to
check that R has enough edges, i.e. that enough ϵ-regular pairs (V_i, V_j)

have density at least d. This should follow from our assumption that G has at least $t_{r-1}(n) + \gamma n^2$ edges, i.e. an edge density of about $\frac{r-2}{r-1} + 2\gamma$: this lies substantially above the approximate edge density $\frac{r-2}{r-1}$ of the Turán graph $T^{r-1}(k)$, and hence substantially above any density that G could have if no more than $t_{r-1}(k)$ of the pairs (V_i, V_j) had density $\geqslant d$—even if all those pairs had density 1!

Let us then estimate $\|R\|$ more precisely. How many edges of G lie outside ϵ-regular pairs? At most $\binom{|V_0|}{2}$ edges lie inside V_0, and by condition (i) in the definition of ϵ-regularity these are at most $\frac{1}{2}(\epsilon n)^2$ edges. At most $|V_0| k\ell \leqslant \epsilon n k\ell$ edges join V_0 to other partition sets. The at most ϵk^2 other pairs (V_i, V_j) that are not ϵ-regular contain at most ℓ^2 edges each, together at most $\epsilon k^2 \ell^2$. The ϵ-regular pairs of insufficient density $(< d)$ each contain no more than $d\ell^2$ edges, altogether at most $\frac{1}{2} k^2 d\ell^2$ edges. Finally, there are at most $\binom{\ell}{2}$ edges inside each of the partition sets $V_1, \ldots, V_k$, together at most $\frac{1}{2}\ell^2 k$ edges. All *other* edges of G lie in ϵ-regular pairs of density at least d, and thus contribute to edges of R. Since each edge of R corresponds to at most ℓ^2 edges of G, we thus have in total

$$\|G\| \leq \tfrac{1}{2}\epsilon^2 n^2 + \epsilon n k\ell + \epsilon k^2\ell^2 + \tfrac{1}{2}k^2 d\ell^2 + \tfrac{1}{2}\ell^2 k + \|R\|\,\ell^2.$$

Hence, for all sufficiently large n,

$$\|R\| \geq \tfrac{1}{2}k^2 \frac{\|G\| - \tfrac{1}{2}\epsilon^2 n^2 - \epsilon n k\ell - \epsilon k^2\ell^2 - \tfrac{1}{2}dk^2\ell^2 - \tfrac{1}{2}k\ell^2}{\tfrac{1}{2}k^2\ell^2}$$

$$\underset{(1)}{\geq} \tfrac{1}{2}k^2 \left(\frac{\|G\|}{n^2/2} - \epsilon^2 - 2\epsilon - 2\epsilon - d - \frac{1}{k} \right)$$

$$\geqslant \tfrac{1}{2}k^2 \left(\frac{t_{r-1}(n)}{n^2/2} + 2\gamma - \epsilon^2 - 4\epsilon - d - \frac{1}{m} \right)$$

$$= \tfrac{1}{2}k^2 \left(t_{r-1}(n) \binom{n}{2}^{-1} \left(1 - \frac{1}{n}\right) + \delta \right)$$

$$> \tfrac{1}{2}k^2 \frac{r-2}{r-1}$$

$$\geqslant t_{r-1}(k).$$

(The strict inequality follows from Lemma 7.1.4.) Therefore $K^r \subseteq R$ by Theorem 7.1.1, as desired. □

Exercises

1.⁻ Show that $K_{1,3}$ is extremal without a P^3.

2.⁻ Given $k > 0$, determine the extremal graphs of chromatic number at most k.

3. Determine the value of $\mathrm{ex}(n, K_{1,r})$ for all $r, n \in \mathbb{N}$.

4. Is there a graph that is edge-maximal without a K^3 minor but not extremal?

5. Given $k > 0$, determine the extremal graphs without a matching of size k.

    (Hint. Theorem 2.2.3 and Ex. 10, Ch. 2.)

6. Show that, for every forest F, the value of $\mathrm{ex}(n, F)$ is bounded above by a linear function of n.

    (Hint. Proposition 1.2.2 and Corollary 1.5.4.)

7. Without using Turán's theorem, show that the maximum number of edges in a triangle-free graph of order $n > 1$ is $\lfloor n^2/4 \rfloor$.

    (Hint. Consider a vertex $x \in G$ of maximum degree, and count the edges in $G - x$.)

8. Show that

    $$t_{r-1}(n) \leqslant \tfrac{1}{2}n^2 \, \frac{r-2}{r-1} \, ,$$

    with equality whenever $r - 1$ divides n.

    (Hint. Choose k and i so that $n = (r-1)k + i$ with $0 \leqslant i < r-1$. Treat the case of $i = 0$ first, and then show for the general case that $t_{r-1}(n) = \frac{1}{2}\frac{r-2}{r-1}(n^2 - i^2) + \binom{i}{2}$.)

9. Show that $t_{r-1}(n)/\binom{n}{2}$ converges to $(r-2)/(r-1)$ as $n \to \infty$.

    (Hint. $t_{r-1}\bigl((r-1)\lfloor\frac{n}{r-1}\rfloor\bigr) \leqslant t_{r-1}(n) \leqslant t_{r-1}\bigl((r-1)\lceil\frac{n}{r-1}\rceil\bigr)$.)

10.⁺ Given non-adjacent vertices u, v in a graph G, denote by $G[u \to v]$ the graph obtained from G by first deleting all the edges at u and then joining u to all the neighbours of v. Show that $K^r \not\subseteq G[u \to v]$ if $K^r \not\subseteq G$. Applying this operation repeatedly to a given extremal graph for n and K^r, prove that $\mathrm{ex}(n, K^r) = t_{r-1}(n)$: in each iteration step, choose u and v so that the number of edges will not decrease, and so that eventually a complete multipartite graph is obtained.

11.⁺ For $0 < s \leqslant t \leqslant n$ let $z(n, s, t)$ denote the maximum number of edges in a bipartite graph whose partition sets both have size n, and which does not contain a $K_{s,t}$. Show that $2\,\mathrm{ex}(n, K_{s,t}) \leq z(n, s, t) \leq \mathrm{ex}(2n, K_{s,t})$.

12.[+] Let $1 \leqslant r \leqslant n$ be integers. Let G be a bipartite graph with bipartition $\{A, B\}$, where $|A| = |B| = n$, and assume that $K_{r,r} \not\subseteq G$. Show that

$$\sum_{x \in A} \binom{d(x)}{r} \leqslant (r-1)\binom{n}{r}.$$

Using the previous exercise, deduce that $\mathrm{ex}(n, K_{r,r}) \leqslant cn^{2-1/r}$ for some constant c depending only on r.

(Hint. For the displayed inequality, count the pairs (x, Y) such that $x \in A$ and $Y \subseteq B$, with $|Y| = r$ and x adjacent to all of Y. For the bound on $\mathrm{ex}(n, K_{r,r})$, use the estimate $(s/t)^t \leq \binom{s}{t} \leq s^t$ and the fact that the function $z \mapsto z^r$ is convex.)

13. The *upper density* of an infinite graph G is the supremum of $\|H\| \binom{|H|}{2}^{-1}$, taken over all non-empty finite subgraphs H of G. Show that this number always takes one of the countably many values $0, 1, \frac{1}{2}, \frac{2}{3}, \frac{3}{4}, \dots$.

(Hint. Erdős-Stone.)

14.[-] In the definition of an ϵ-regular pair, what is the purpose of the requirement that $|X| > \epsilon |A|$ and $|Y| > \epsilon |B|$?

15.[-] Show that any ϵ-regular pair in G is also ϵ-regular in $\overline{G}$.

16. Prove the regularity lemma for sparse graphs, that is, for every sequence $(G_n)_{n \in \mathbb{N}}$ of graphs such that $\|G_n\|/n^2 \to 0$ as $n \to \infty$.

Notes

The standard reference work for results and open problems in extremal graph theory (in a very broad sense) is still B. Bollobás, *Extremal Graph Theory*, Academic Press 1978. A kind of update on the book is given by its author in his chapter of the *Handbook of Combinatorics* (R.L. Graham, M. Grötschel & L. Lovász, eds.), North-Holland 1995. An instructive survey of extremal graph theory in the narrower sense of our chapter is given by M. Simonovits in (L.W. Beineke & R.J. Wilson, eds.) *Selected Topics in Graph Theory 2*, Academic Press 1983. This paper focuses among other things on the particular role played by the Turán graphs. A more recent survey by the same author can be found in (R.L. Graham & J. Nešetřil, eds.) *The Mathematics of Paul Erdős*, Vol. 2, Springer 1996.

Turán's theorem is not merely one extremal result among others: it is the result that sparked off the entire line of research. Our proof of Turán's theorem is essentially the original one; the proof indicated in Exercise 10 is due to Zykov.

Our version of the Erdős-Stone theorem is a slight simplification of the original. A direct proof, not using the regularity lemma, is given in L. Lovász, *Combinatorial Problems and Exercises* (2nd edn.), North-Holland 1993. Its most fundamental application, Corollary 7.1.3, was only found 20 years after the theorem, by Erdős and Simonovits (1966).

The regularity lemma is proved in E. Szemerédi, Regular partitions of graphs, *Colloques Internationaux CNRS* **260**—*Problèmes Combinatoires et*

Théorie des Graphes, Orsay (1976), 399–401. Our rendering follows an account by Scott (personal communication). A broad survey on the regularity lemma and its applications is given by J. Komlós & M. Simonovits in (D. Miklos, V.T. Sós & T. Szőnyi, eds.) *Paul Erdős is 80*, Vol. 2, Proc. Colloq. Math. Soc. János Bolyai (1996); the concept of a regularity graph and Lemma 7.3.2 are taken from this paper. An adaptation of the regularity lemma for use with sparse graphs was developed independently by Kohayakawa and by Rödl; see Y. Kohayakawa, Szemerédi's regularity lemma for sparse graphs, in (F. Cucker & M. Shub, eds.) *Foundations of Computational Mathematics*, Selected papers of a conference held at IMPA in Rio de Janeiro, January 1997, Springer 1997.

8 Substructures in Sparse Graphs

In this chapter we study how global assumptions about a graph—on its average degree, chromatic number, or even (large) girth—can force it to contain a given graph H as a minor or topological minor. As we know already from Mader's theorem 3.6.1, there exists a function h such that an average degree of $d(G) \geqslant h(r)$ suffices to create a TK^r subgraph in G, and hence a (topological) H minor if $r \geqslant |H|$. Since a graph with n vertices and average degree d has $\frac{1}{2}dn$ edges this shows that, for every H, there is a 'constant' c (depending on H but not on n) such that a topological H minor occurs in every graph with n vertices and at least cn edges. Such graphs with a number of edges about linear[1] in their order are called *sparse*—so this is a chapter about substructures in sparse graphs.

sparse

The first question, then, will be the analogue of Turán's theorem: given a positive integer r, what is the minimum value of the above 'constant' c for $H = K^r$, i.e. the smallest growth rate of a function $h(r)$ as in Theorem 3.6.1? This was a major open problem until very recently; we present its solution, which builds on some fascinating methods the problem has inspired over time, in Section 8.1.

If raising the average degree suffices to force the occurrence of a certain minor, then so does raising any other invariant which in turn forces up the average degree. For example, if $d(G) \geqslant c$ implies $H \preccurlyeq G$, then so will $\chi(G) \geqslant c+1$ (by Corollary 5.2.3). However, is this best possible? Even if the value of c above is least possible for $d(G) \geqslant c$ to imply $H \preccurlyeq G$, it need not be so for $\chi(G) \geqslant c+1$ to imply $H \preccurlyeq G$. One of the most famous conjectures in graph theory, the *Hadwiger conjecture*,

---

[1] Compare the footnote at the beginning of Chapter 7.

suggests that there is indeed a gap here: while a value of $c = c'r\sqrt{\log r}$ (where c' is independent of both n and r) is best possible for $d(G) \geqslant c$ to imply $H \preccurlyeq G$ (Section 8.2), the conjecture says that $\chi(G) \geqslant r$ will do the same! Thus, if true, then Hadwiger's conjecture shows that the effect of a large chromatic number on the occurrence of minors somehow goes beyond that part which is well-understood: its effect via mere edge density. We shall consider Hadwiger's conjecture in Section 8.3.

8.1 Topological minors

In this section we prove that an average degree of cr^2 suffices to force the occurrence of a topological K^r minor in a graph; complete bipartite graphs show that, up to the constant c, this is best possible (Exercise 4).

    The following theorem was proved independently by Bollobás & Thomason and by Komlós & Szemerédi (1996).

Theorem 8.1.1. *There exists a $c \in \mathbb{R}$ such that, for every $r \in \mathbb{N}$, every graph G of average degree $d(G) \geqslant cr^2$ contains K^r as a topological minor.*

    The proof of this theorem, in which we follow Bollobás & Thomason, will occupy us for the remainder of this section. A set $U \subseteq V(G)$ will *linked* be called *linked* (in G) if for any distinct vertices $u_1, \ldots, u_{2h} \in U$ there are h disjoint paths $P_i = u_{2i-1} \ldots u_{2i}$ in G, $i = 1, \ldots, h$.[2] The graph G (k, ℓ)-*linked* itself is (k, ℓ)-*linked* if every k-set of its vertices contains a linked ℓ-set.

    How can we hope to find the TK^r in G claimed to exist by Theorem 8.1.1? Our basic approach will be to identify first some r-set X as a set of branch vertices, and to choose for each $x \in X$ a set Y_x of $r - 1$ neighbours, one for every edge incident with x in the K^r. If the constant c from the theorem is large enough, the $r + r(r-1) = r^2$ vertices of $X \cup \bigcup Y_x$ can be chosen distinct: by Proposition 1.2.2, G has a subgraph of minimum degree at least $\varepsilon(G) = \frac{1}{2}d(G) \geqslant \frac{1}{2}cr^2$, so we can choose X and its neighbours inside this subgraph. Having fixed X and the sets Y_x, we then have to link up the correct pairs of vertices in $Y := \bigcup Y_x$ by disjoint paths in $G - X$, to obtain the desired TK^r.

    This would be possible at once if Y were linked in $G - X$. Unfortunately, this is unrealistic to hope for: no average degree, however large, will force *every* $r(r-1)$-set to be linked. (Why not?) However, if we pick for X significantly more than the r vertices needed eventually, and for each $x \in X$ significantly more than $r - 1$ neighbours as Y_x, then Y might become so large that the high average degree of G guarantees the

---

    [2] Thus, in a k-linked graph—see Chapter 3.6—every set of up to $2k + 1$ vertices is linked.

existence of some large linked subset $Z \subseteq Y$. This would be the case if G were (k, ℓ)-linked for some $k \leqslant |Y|$ and $\ell \geqslant |Z|$.

As above, a large enough constant c will easily ensure that X and Y can be chosen with many vertices to spare. Another problem, however, is more serious: it will not be enough to make ℓ (and hence Z) large in absolute terms. Indeed, if k (and Y) is much larger still, it might happen that Z, although large, consists of neighbours of only a few vertices in X! We thus have to ensure that ℓ is large also relative to k. This will be the purpose of our first lemma (8.1.2): it establishes a sufficient condition for G to be $(k, \lceil k/2 \rceil)$-linked.

What is this sufficient condition? It is the assumption that G has a particularly dense minor H, one whose minimum degree exceeds $\frac{1}{2}|H|$ by a positive fraction of k. (In particular, H will be dense in the sense of Chapter 7.) In view of Theorem 3.6.2, it is not surprising that such a dense graph H is highly linked. Given sufficiently high connectivity of G (which again follows easily if c is large enough), we may then try to link up the vertices of any Y as above to distinct branch sets of H by disjoint paths in G avoiding most of the other branch sets, and thus to transfer the linking properties of H to a $\lceil k/2 \rceil$-set $Z \subseteq Y$ (Fig. 8.1.1).

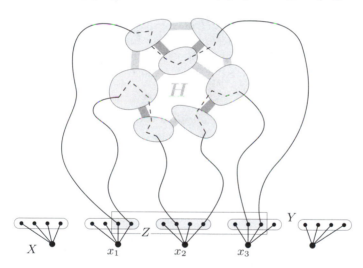

Fig. 8.1.1. Finding a TK^3 in G with branch vertices x_1, x_2, x_3

What is all the more surprising, however, is that the existence of such a dense minor H can be deduced from our assumption of $d(G) \geqslant cr^2$. This will be shown in another lemma (8.1.3); the assertion of the theorem itself will then follow easily.

Lemma 8.1.2. If G is k-connected and has a minor H with $2\delta(H) \geqslant |H| + \frac{3}{2}k$, then G is $(k, \lceil k/2 \rceil)$-linked.

Proof. Let $\mathcal{V} := \{\, V_x \mid x \in V(H) \,\}$ be the set of branch sets in G corresponding to the vertices of H. For our proof that G is $(k, \lceil k/2 \rceil)$-linked, let k distinct vertices $v_1, \dots, v_k \in G$ be given. Let us call a sequence $P_1, \dots, P_k$ of disjoint paths in G a *linkage* if the P_i each start in v_i and end in pairwise distinct sets $V \in \mathcal{V}$; the paths P_i themselves will be called *links*. Since our assumptions about H imply that $|H| \geqslant k$, and G is k-connected, such linkages exist: just pick k vertices from pairwise distinct sets $V \in \mathcal{V}$, and link them disjointly to $\{\, v_1, \dots, v_k \,\}$ by Menger's theorem.

Now let $\mathcal{P} = (P_1, \dots, P_k)$ be a linkage whose total number of edges outside $\bigcup_{V \in \mathcal{V}} G[V]$ is minimum. Thus, if $f(P)$ denotes the number of edges of P not lying in any $G[V_x]$, we choose $\mathcal{P}$ so as to minimize $\sum_{i=1}^{k} f(P_i)$. Then for every $V \in \mathcal{V}$ that meets a path $P_i \in \mathcal{P}$ there exists one such path that ends in V: if not, we could terminate P_i in V and reduce $f(P_i)$. Thus, exactly k of the branch sets of H meet a link. Let us divide these sets into two classes:

$$\mathcal{U} := \{\, V \in \mathcal{V} \mid V \text{ meets exactly one link} \,\}$$
$$\mathcal{W} := \{\, V \in \mathcal{V} \mid V \text{ meets more than one link} \,\}.$$

Since H is dense and each $U \in \mathcal{U}$ meets only one link, it will be easy to show that the starting vertices v_i of those links form a linked set in G. Hence, our aim is to show that $|\mathcal{U}| \geqslant \lceil k/2 \rceil$, i.e. that $\mathcal{U}$ is no smaller than $\mathcal{W}$. (Recall that $|\mathcal{U}| + |\mathcal{W}| = k$.) To this end, we first prove the following:

Every $V \in \mathcal{W}$ is met by some link which leaves V again
and next meets a set from $\mathcal{U}$ (where it ends).   (1)

Suppose $V_x \in \mathcal{W}$ is a counterexample to (1). Since

$$2\delta(H) \geqslant |H| + \tfrac{3}{2}k \geqslant \delta(H) + \tfrac{3}{2}k,$$

we have $\delta(H) \geqslant \tfrac{3}{2}k$. As $|\mathcal{U} \cup \mathcal{W}| = k$, this implies that x has a neighbour y in H with $V_y \in \mathcal{V} \smallsetminus (\mathcal{U} \cup \mathcal{W})$; let $w_x w_y$ be an edge of G with $w_x \in V_x$ and $w_y \in V_y$. Let $Q = w \dots w_x w_y$ be a path in $G[V_x \cup \{\, w_y \,\}]$ of whose vertices only w lies on any link, say on P_i (Fig. 8.1.2). Replacing P_i in $\mathcal{P}$ by $P_i' := P_i w Q$ then yields another linkage.

If P_i is not the link ending in V_x, then $f(P_i') \leqslant f(P_i)$. The choice of $\mathcal{P}$ then implies that $f(P_i') = f(P_i)$, i.e. that P_i ends in the branch set W it enters immediately after V_x. Since V_x is a counterexample to (1) we have $W \notin \mathcal{U}$, i.e. $W \in \mathcal{W}$. Let $P \neq P_i$ be another link meeting W. Then P does not end in W (because P_i ends there); let $P' \subseteq P$ be the (minimal) initial segment of P that ends in W. If we now replace P_i and P by P_i' and P' in $\mathcal{P}$, we obtain a linkage contradicting the choice of $\mathcal{P}$.

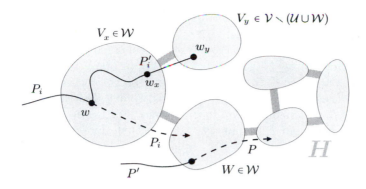

Fig. 8.1.2. If P_i does not end in V_x, we replace P_i and P by P_i' and P'

We now assume that P_i does end in V_x; then $f(P_i') = f(P_i) + 1$. As $V_x \in W$, there exists a link P_j that meets V_x and leaves it again; let P_j' be the initial segment of P_j ending in V_x (Fig 8.1.3). Then $f(P_j') \leqslant f(P_j) - 1$. In fact, since replacing P_i and P_j with P_i' and P_j' in $\mathcal{P}$ yields another linkage, the choice of $\mathcal{P}$ implies that $f(P_j') = f(P_j) - 1$, so P_j ends in the branch set W it enters immediately after V_x. Then $W \in \mathcal{W}$ as before, so we may define P and P' as before. Replacing P_i, P_j and P by P_i', P_j' and P' in $\mathcal{P}$, we finally obtain a linkage that contradicts the choice of $\mathcal{P}$. This completes the proof of (1).

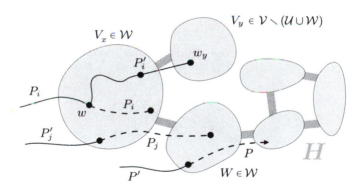

Fig. 8.1.3. If P_i ends in V_x, we replace P_i, P_j, P by P_i', P_j', P'

With the help of (1) we may define an injection $\mathcal{W} \to \mathcal{U}$ as follows: given $W \in \mathcal{W}$, choose a link that passes through W and next meets a set $U \in \mathcal{U}$, and map $W \mapsto U$. (This is indeed an injection, because different links end in different branch sets.) Thus $|\mathcal{U}| \geqslant |\mathcal{W}|$, and hence $|\mathcal{U}| \geqslant \lceil k/2 \rceil$.

Let us assume the enumeration of $v_1, \ldots, v_k$ to be such that the first $u := |\mathcal{U}|$ of the links $P_1, \ldots, P_k$ end in sets from $\mathcal{U}$. Since $2\delta(H) \geqslant$

u

$|H| + \frac{3}{2}k$, we can find for any two sets $V_x, V_y \in \mathcal{U}$ at least $\frac{3}{2}k$ sets V_z such that $xz, yz \in E(H)$. At least $k/2$ of these sets V_z do not lie in $\mathcal{U} \cup \mathcal{W}$. Thus whenever $U_1, \ldots, U_{2h}$ are distinct sets in $\mathcal{U}$ (so $h \leqslant u/2 \leqslant k/2$), we may find inductively h distinct sets $V^i \in \mathcal{V} \setminus (\mathcal{U} \cup \mathcal{W})$ $(i = 1, \ldots, h)$ such that V^i is joined in G to both U_{2i-1} and U_{2i}. For each i, any vertex of U_{2i-1} can be linked by a path through V^i to any desired vertex of U_{2i}, and these paths will be disjoint for different i. Joining up the appropriate pairs of paths from $\mathcal{P}$ in this way, we see that the set $\{v_1, \ldots, v_u\}$ is linked in G, and the lemma is proved. $\qquad\qquad\square$

Lemma 8.1.3. *Let $k \geqslant 6$ be an integer. Then every graph G with $\varepsilon(G) \geqslant k$ has a minor H such that $2\delta(H) \geqslant |H| + \frac{1}{6}k$.*

G_0

Proof. We begin by choosing a $(\preccurlyeq\text{-})$minimal minor G_0 of G with $\varepsilon(G_0) \geqslant k$. The minimality of G_0 implies that $\delta(G_0) > k$ and $\varepsilon(G_0) = k$ (otherwise we could delete a vertex or edge of G_0), and hence

$$k + 1 \leqslant \delta(G_0) \leqslant d(G_0) = 2k.$$

x_0

Let $x_0 \in G_0$ be a vertex of minimum degree.
    If k is odd, let $m := (k+1)/2$ and

$$G_1 := G_0 \left[\{x_0\} \cup N_{G_0}(x_0) \right].$$

Then $|G_1| = \delta(G_0) + 1 \leqslant 2k + 1 \leqslant 2(k+1) = 4m$. By the minimality of G_0, contracting any edge $x_0 y$ of G_0 will result in the loss of at least $k + 1$ edges. The vertices x_0 and y thus have at least k common neighbours, so $\delta(G_1) \geqslant k + 1 = 2m$ (Fig. 8.1.4).

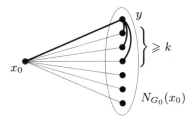

Fig. 8.1.4. The graph $G_1 \preccurlyeq G$: a first approximation to the desired minor H

If k is even, we let $m := k/2$ and

$$G_1 := G_0 \left[N_{G_0}(x_0) \right].$$

Then $|G_1| = \delta(G_0) \leqslant 2k = 4m$, and $\delta(G_1) \geqslant k = 2m$ as before.

Thus in either case we have found an integer m and a graph $G_1 \preccurlyeq G$ such that

$$m \geqslant k/2 \geqslant 3 \tag{1}$$

$$|G_1| \leqslant 4m \tag{2}$$

and $\delta(G_1) \geqslant 2m$, so

$$\varepsilon(G_1) \geqslant m. \tag{3}$$

As $2\delta(G_1) \geqslant 4m \geqslant |G_1|$, our graph G_1 is already quite a good candidate for the desired minor H of G. In order to jack up its value of 2δ by another $\frac{1}{6}k$ (as required for H), we shall reapply the above contraction process to G_1, and a little more rigorously than before: step by step, we shall contract edges as long as this results in a loss of no more than $\frac{7}{6}m$ edges per vertex. In other words, we permit a loss of edges slightly greater than maintaining $\varepsilon \geqslant m$ seems to allow. (Recall that, when we contracted G to G_0, we put this threshold at $\varepsilon(G) = k$.) If this second contraction process terminates with a non-empty graph H_0, then $\varepsilon(H_0)$ will be at least $\frac{7}{6}m$, higher than for G_1! The $\frac{1}{6}m$ thus gained will suffice to give the graph H_1, obtained from H_0 just as G_1 was obtained from G_0, the desired high minimum degree.

But how can we be sure that this second contraction process will indeed end with a non-empty graph? Paradoxical though it may seem, the reason is that even a permitted loss of up to $\frac{7}{6}m$ edges (and one vertex) per contraction step cannot destroy the $m|G_1|$ or more edges of G_1 in the $|G_1|$ steps possible: the graphs with fewer than m vertices towards the end of the process would simply be too small to be able to shed their allowance of $\frac{7}{6}m$ edges—and, by (2), these small graphs would account for about a quarter of the process!

Formally, we shall control the graphs H in the contraction process not by specifying an upper bound on the number of edges to be discarded at each step, but by fixing a lower bound for $\|H\|$. Let

$$f(n) := \tfrac{1}{6}m(n - m - 5) \qquad\qquad f$$

and

$$\mathcal{H} := \left\{ H \preccurlyeq G_1 : \|H\| \geqslant m|H| + f(|H|) - \binom{m}{2} \right\}. \qquad \mathcal{H}$$

By (2),

$$f(|G_1|) \leqslant f(4m) = \tfrac{1}{2}m^2 - \tfrac{5}{6}m < \binom{m}{2},$$

so $G_1 \in \mathcal{H}$ by (3).

For every $H \in \mathcal{H}$, any graph obtained from H by one of the following three operations will again be in $\mathcal{H}$:

(i) deletion of an edge, if $\|H\| \geqslant m\,|H| + f(|H|) - \binom{m}{2} + 1$;

(ii) deletion of a vertex of degree at most $\frac{7}{6}m$;

(iii) contraction of an edge $xy \in H$ such that x and y have at most $\frac{7}{6}m - 1$ common neighbours in H.

Starting with G_1, let us apply these operations as often as possible, and let $H_0 \in \mathcal{H}$ be the graph obtained eventually. Since

$$\|K^m\| = m\,|K^m| - m - \binom{m}{2}$$

and

$$f(m) = -\tfrac{5}{6}m > -m\,,$$

K^m does not have enough edges to be in $\mathcal{H}$; thus, $\mathcal{H}$ contains no graph on m vertices. Hence $|H_0| > m$, and in particular $H_0 \neq \emptyset$. Let $x_1 \in H_0$ be a vertex of minimum degree, and put

$$H_1 := H_0\left[\{x_1\} \cup N_{H_0}(x_1)\right].$$

We shall prove that the minimum degree of $H := H_1$ is as large as claimed in the lemma.

Note first that

$$\delta(H_1) > \tfrac{7}{6}m\,. \tag{4}$$

Indeed, since H_0 is minimal with respect to (ii) and (iii), we have $d(x_1) > \frac{7}{6}m$ in H_0 (and hence in H_1), and every vertex $y \neq x_1$ of H_1 has more than $\frac{7}{6}m - 1$ common neighbours with x_1 (and hence more than $\frac{7}{6}m$ neighbours in H_1 altogether). In order to convert (4) into the desired inequality of the form

$$2\delta(H_1) \geqslant |H_1| + \alpha m\,,$$

we need an upper bound for $|H_1|$ in terms of m. Since H_0 lies in $\mathcal{H}$ but is minimal with respect to (i), we have

$$\|H_0\| < m\,|H_0| + \left(\tfrac{1}{6}m\,|H_0| - \tfrac{1}{6}m^2 - \tfrac{5}{6}m\right) - \binom{m}{2} + 1$$
$$= \tfrac{7}{6}m\,|H_0| - \tfrac{4}{6}m^2 - \tfrac{1}{3}m + 1$$
$$\underset{(1)}{\leqslant} \tfrac{7}{6}m\,|H_0| - \tfrac{4}{6}m^2. \tag{5}$$

By the choice of x_1 and definition of H_1, therefore,

$$|H_1| - 1 = \delta(H_0)$$
$$\leqslant 2\,\varepsilon(H_0)$$

$$\underset{(5)}{\leqslant} \tfrac{7}{3}m - \tfrac{4}{3}m^2/|H_0|$$

$$\underset{(2)}{\leqslant} \tfrac{7}{3}m - \tfrac{1}{3}m$$

$$= 2m\,,$$

so $|H_1| \leqslant 2m$. Hence,

$$\underset{(4)}{2\delta(H_1) >} 2m + \tfrac{1}{3}m$$

$$\geqslant |H_1| + \tfrac{1}{3}m$$

$$\underset{(1)}{\geqslant} |H_1| + \tfrac{1}{6}k$$

as claimed.   □

Proof of Theorem 8.1.1. We prove the assertion for $c := 1116$. Let $(1.4.2)$
G be a graph with $d(G) \geqslant 1116r^2$. By Theorem 1.4.2, G has a subgraph
G_0 such that G_0

$$\kappa(G_0) \geqslant 279r^2 \geqslant 276r^2 + 3r\,.$$

Pick a set $X := \{\,x_1, \ldots, x_{3r}\,\}$ of $3r$ vertices in G_0, and let $G_1 := G_0 - X$. X
For each $i = 1, \ldots, 3r$ choose a set Y_i of $5r$ neighbours of x_i in G_1; let G_1, Y_i
these sets Y_i be disjoint for different i. (This is possible since $\delta(G_0) \geqslant$
$\kappa(G_0) \geqslant 15r^2 + |X|$.)
        As

$$\delta(G_1) \geqslant \kappa(G_1) \geqslant \kappa(G_0) - |X| \geqslant 276r^2,$$

we have $\varepsilon(G_1) \geqslant 138r^2$. By Lemma 8.1.3, G_1 has a minor H with
$2\delta(H) \geqslant |H| + 23r^2$ and is therefore $(15r^2, 7r^2)$-linked by Lemma 8.1.2;
let $Z \subseteq \bigcup_{i=1}^{3r} Y_i$ be a set of $7r^2$ vertices that is linked in G_1. Z
        For all $i = 1, \ldots, 3r$ let $Z_i := Z \cap Y_i$. Since Z is linked, it suffices Z_i
to find r indices i with $|Z_i| \geqslant r - 1$: then the corresponding x_i will be
the branch vertices of a TK^r in G_0. If r such i cannot be found, then
$|Z_i| \leqslant r - 2$ for all but at most $r - 1$ indices i. But then

$$|Z| = \sum_{i=1}^{3r} |Z_i| \leqslant (r-1)\,5r + (2r+1)(r-2) < 7r^2 = |Z|\,,$$

a contradiction.   □

        The following almost counter-intuitive result of Mader implies that
the existence of a topological K^r minor can be forced essentially by large
girth. In the next section, we shall prove the analogue of this for ordinary
minors.

Theorem 8.1.4. (Mader 1997)
*For every graph H of maximum degree $d \geqslant 3$ there exists an integer k
such that every graph G of minimum degree at least d and girth at least k
contains H as a topological minor.*

As discussed already in Chapter 5.2 and the introduction to Chapter 7, no constant average degree, however large, will force an arbitrary graph to contain a given graph H as a subgraph—as long as H contains at least one cycle. By Proposition 1.2.2 and Corollary 1.5.4, on the other hand, any graph G contains all trees on up to $\varepsilon(G) + 2$ vertices. Large average degree therefore does ensure the occurrence of any fixed tree T as a subgraph. What can we say, however, if we would like T to occur as an *induced* subgraph?

Here, a large average degree appears to do as much harm as good, even for graphs of bounded clique number. (Consider, for example, complete bipartite graphs.) It is all the more remarkable, then, that the assumption of a large chromatic number rather than a large average degree seems to make a difference here: according to a conjecture of Gyárfás, any graph of large enough chromatic number contains either a large complete graph or any given tree as an induced subgraph. (Formally: for every integer r and every tree T, there exists an integer k such that every graph G with $\chi(G) \geqslant k$ and $\omega(G) < r$ contains an induced copy of T.)

The weaker topological version of this is indeed true:

Theorem 8.1.5. (Scott 1997)
*For every integer r and every tree T there exists an integer k such that
every graph with $\chi(G) \geqslant k$ and $\omega(G) < r$ contains an induced copy of
some subdivision of T.*

8.2 Minors

According to Theorem 8.1.1, an average degree of cr^2 suffices to force the existence of a topological K^r minor in a given graph. If we are content with any minor, topological or not, an even smaller average degree will do: in a pioneering paper of 1968, Mader proved that every graph with an average degree of at least $cr \log r$ has a K^r minor. The following result, the analogue to Theorems 7.1.1 and 8.1.1 for general minors, determines the precise average degree needed as a function of r, up to a constant c:

Theorem 8.2.1. (Kostochka 1982; Thomason 1984)
*There exists a $c \in \mathbb{R}$ such that, for every $r \in \mathbb{N}$, every graph G of average
degree $d(G) \geqslant cr\sqrt{\log r}$ has a K^r minor. Up to the value of c, this
bound is best possible as a function of r.*

The easier implication of the theorem, the fact that in general an average degree of $cr\sqrt{\log r}$ is needed to force a K^r minor, follows from considering random graphs, to be introduced in Chapter 11. The converse implication, the fact that this average degree suffices, is proved by methods similar to those described in Section 8.1.

Rather than proving Theorem 8.2.1, we therefore devote the remainder of this section to another striking result on forcing minors. At first glance, this result is so surprising that it seems almost paradoxical: as long as we do not merely subdivide edges, we can force a K^r minor in a graph simply by raising its girth (Corollary 8.2.3)!

Theorem 8.2.2. (Thomassen 1983)
Given an integer k, every graph G with girth $g(G) \geqslant 4k-3$ and $\delta(G) \geqslant 3$ has a minor H with $\delta(H) \geqslant k$.

Proof. As $\delta(G) \geqslant 3$, every component of G contains a cycle. In particular, the assertion is trivial for $k \leqslant 2$; so let $k \geqslant 3$. Consider the vertex set V of a component of G, together with a partition $\{V_1, \ldots, V_m\}$ of V into as many connected sets V_i with at least $2k-2$ vertices each as possible. (Such a partition exists, since $|V| \geqslant g(G) > 2k-2$ and V is connected in G.)

We first show that every $G[V_i]$ is a tree. To this end, let T_i be a spanning tree of $G[V_i]$. If $G[V_i]$ has an edge $e \notin T_i$, then $T_i + e$ contains a cycle C; by assumption, C has length at least $4k-3$. The edge (about) opposite e on C therefore separates the path $C - e$, and hence also T_i, into two components with at least $2k-2$ vertices each. Together with the sets V_j for $j \neq i$, these two components form a partition of V into $m+1$ sets that contradicts the maximality of m.

So each $G[V_i]$ is indeed a tree, i.e. $G[V_i] = T_i$. As $\delta(G) \geqslant 3$, the degrees in G of the vertices in V_i sum to at least $3|V_i|$, while the edges of T_i account for only $2|V_i| - 2$ in this sum. Hence for each i, G has at least $|V_i| + 2 \geqslant 2k$ edges joining V_i to $V \smallsetminus V_i$. We shall prove that every V_i sends at most two edges to each of the other V_j; then V_i must send edges to at least k of those V_j, so the V_i are the branch sets of an $MH \subseteq G$ with $\delta(H) \geqslant k$.

Suppose, without loss of generality, that G has three V_1–V_2 edges. Then there are vertices $v_1 \in V_1$ and $v_2 \in V_2$ such that $G[V_1 \cup V_2]$ contains three independent v_1–v_2 paths P_1, P_2, P_3 (Fig. 8.2.1). At most one of these paths can have length less than $\frac{1}{2}g(G)$; we assume that P_1 and P_2 have length at least $\lceil \frac{1}{2}g(G) \rceil \geqslant 2k-1$. For $i = 1, 2$ let $P_i' \subseteq \mathring{P}_i$ be a subpath with exactly $2k-2$ vertices, and put $P_3' := (P_1 \cup P_2 \cup P_3) - (P_1' \cup P_2')$ and $C := P_1 \cup P_3$. Then

$$|P_3'| \geqslant |C - P_1'| \geqslant g(G) - (2k-2) \geqslant 2k-1,$$

(1.5.3)

V, V_i
m

T_i

P_1, P_2, P_3
P_1', P_2'
P_3'

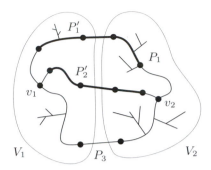

Fig. 8.2.1. Three edges between V_1 and V_2

and P_3' (as well as P_1' and P_2') is connected. Since $G[V_1 \cup V_2]$ is connect-ed, there exists a partition of $V_1 \cup V_2$ into three connected sets V_1', V_2', V_3' such that $V(P_i') \subseteq V_i'$ for $i = 1, 2, 3$. Replacing the two sets V_1, V_2 in our partition of V with the three sets V_1', V_2', V_3', we obtain a partition of V that contradicts the maximality of m. $\square$

The following combination of Theorems 8.2.1 and 8.2.2 brings out the paradoxical character of the latter particularly well:

Corollary 8.2.3. *There exists a $c \in \mathbb{R}$ such that, for every $r \in \mathbb{N}$, every graph G with girth $g(G) \geqslant c\, r\sqrt{\log r}$ and $\delta(G) \geqslant 3$ has a K^r minor.*

Proof. We prove the corollary for $c := 4c'$, where c' is the constant from Theorem 8.2.1. Let G be given as stated. By Theorem 8.2.2, G has a minor H with $\delta(H) \geqslant c'r\sqrt{\log r}$. By Theorem 8.2.1, H (and hence G) has a K^r minor. $\square$

8.3 Hadwiger's conjecture

As we saw in the preceding two sections, an average degree of $c\, r\sqrt{\log r}$ suffices to force an arbitrary graph to have a K^r minor, and an average degree of cr^2 forces it to have a topological K^r minor. If we replace 'average degree' above with 'chromatic number' then, with almost the same constants c, the two assertions remain true: this is because every graph with chromatic number k has a subgraph of average degree at least $k-1$ (Corollary 5.2.3).

Although both functions above, $c\,r\sqrt{\log r}$ and cr^2, are best possible (up to the constant c) for the said implications with 'average degree', the question arises whether they are still best possible with 'chromat-ic number'—or whether some slower-growing function would do in that case. What is lurking behind this problem about growth rates, of course, is a fundamental question about the nature of the invariant χ: can this invariant have some direct *structural* effect on a graph in terms of forcing concrete substructures, or is its effect no greater than that of the 'unstructural' property of having lots of edges somewhere, which it implies trivially?

Neither for general nor for topological minors is the answer to this question known. For general minors, however, the following conjecture of Hadwiger suggests a positive answer; the conjecture is considered by many as one of the deepest open problems in graph theory.

Conjecture. (Hadwiger 1943)
The following implication holds for every integer $r > 0$ and every graph G:

$$\chi(G) \geqslant r \;\Rightarrow\; G \succcurlyeq K^r.$$

Hadwiger's conjecture is trivial for $r \leqslant 2$, easy for $r = 3$ and $r = 4$ (exercises), and equivalent to the four colour theorem for $r = 5$ and $r = 6$. For $r \geqslant 7$, the conjecture is open. Rephrased as $G \succcurlyeq K^{\chi(G)}$, it is true for almost all graphs.[3] In general, the conjecture for $r + 1$ implies it for r (exercise).

The Hadwiger conjecture for any fixed r is equivalent to the assertion that every graph without a K^r minor has an $(r - 1)$-colouring. In this reformulation, the conjecture raises the question of what the graphs without a K^r minor look like: any sufficiently detailed structural description of those graphs should enable us to decide whether or not they can be $(r - 1)$-coloured.

For $r = 3$, for example, the graphs without a K^r minor are precisely the forests (why?), and these are indeed 2-colourable. For $r = 4$, there is also a simple structural characterization of the graphs without a K^r minor:

Proposition 8.3.1. *A graph with at least three vertices is edge-maximal without a K^4 minor if and only if it can be constructed recursively from triangles by pasting[4] along K^2s.*   [12.4.3]

Proof. Recall first that every MK^4 contains a TK^4, because $\Delta(K^4) = 3$ (1.7.2)
(Proposition 1.7.2); the graphs without a K^4 minor thus coincide with   (4.4.4)
those without a topological K^4 minor. The proof that any graph constructible as described is edge-maximal without a K^4 minor is left as an easy exercise; in order to deduce Hadwiger's conjecture for $r = 4$, we only need the converse implication anyhow. We prove this by induction on $|G|$.

---

[3] See Chapter 11 for the notion of 'almost all'.

[4] This was defined formally in Chapter 5.5.

Let G be given, edge-maximal without a K^4 minor. If $|G| = 3$ then G is itself a triangle, so let $|G| \geqslant 4$ for the induction step. Then G is not complete; let $S \subseteq V(G)$ be a separating set with $|S| = \kappa(G)$, and let C_1, C_2 be distinct components of $G - S$. Since S is a minimal separator, every vertex in S has a neighbour in C_1 and another in C_2. If $|S| \geqslant 3$, this implies that G contains three independent paths P_1, P_2, P_3 between a vertex $v_1 \in C_1$ and a vertex $v_2 \in C_2$. Since $\kappa(G) = |S| \geqslant 3$, the graph $G - \{v_1, v_2\}$ is connected and contains a (shortest) path P between two different P_i. Then $P \cup P_1 \cup P_2 \cup P_3 = TK^4$, a contradiction.

Hence $\kappa(G) \leqslant 2$, and the assertion follows from Lemma 4.4.4[5] and the induction hypothesis.   □

One of the interesting consequences of Proposition 8.3.1 is that all the edge-maximal graphs without a K^4 minor have the same number of edges, and are thus all 'extremal':

Corollary 8.3.2. *Every edge-maximal graph G without a K^4 minor has $2\,|G| - 3$ edges.*

Proof. Induction on $|G|$.   □

Corollary 8.3.3. *Hadwiger's conjecture holds for $r = 4$.*

Proof. If G arises from G_1 and G_2 by pasting along a complete graph, then $\chi(G) = \max\{\chi(G_1), \chi(G_2)\}$ (see the proof of Proposition 5.5.2). Hence, Proposition 8.3.1 implies by induction on $|G|$ that all edge-maximal (and hence all) graphs without a K^4 minor can be 3-coloured.   □

It is also possible to prove Corollary 8.3.3 by a simple direct argument (Exercise 11).

By the four colour theorem, Hadwiger's conjecture for $r = 5$ follows from the following structure theorem for the graphs without a K^5 minor, just as it follows from Proposition 8.3.1 for $r = 4$. The proof of Theorem 8.3.4 is similar to that of Proposition 8.3.1, but considerably longer. We therefore state the theorem without proof:

Theorem 8.3.4. (Wagner 1937)
Let G be an edge-maximal graph without a K^5 minor. If $|G| \geqslant 4$ then G can be constructed recursively, by pasting along triangles and K^2s, from plane triangulations and copies of the graph W (Fig. 8.3.1).

Using Corollary 4.2.8, one can easily compute which of the graphs constructed as in Theorem 8.3.4 have the most edges. It turns out that

---

[5] The proof of this lemma is elementary and can be read independently of the rest of Chapter 4.

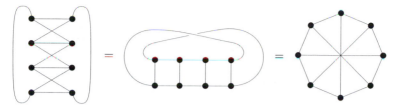

Fig. 8.3.1. Three representations of the *Wagner graph W*

these *extremal* graphs without a K^5 minor have no more edges than those
that are extremal with respect to $\{MK^5, MK_{3,3}\}$, i.e. the maximal
planar graphs:

Corollary 8.3.5. *A graph with n vertices and no K^5 minor has at most $3n - 6$ edges.* $\square$

Since $\chi(W) = 3$, Theorem 8.3.4 and the four colour theorem imply
Hadwiger's conjecture for $r = 5$:

Corollary 8.3.6. *Hadwiger's conjecture holds for $r = 5$.* $\square$

The Hadwiger conjecture for $r = 6$ is again substantially more difficult than the case $r = 5$, and again it relies on the four colour theorem. The proof shows (without using the four colour theorem) that any
$\preccurlyeq$-minimal counterexample arises from a planar graph by adding one
vertex—so by the four colour theorem it is not a counterexample after all.

Theorem 8.3.7. (Robertson, Seymour & Thomas 1993)
Hadwiger's conjecture holds for $r = 6$.

By Corollary 8.3.5, any graph with n vertices and more than $3n - 6$
edges contains an MK^5. In fact, it even contains a TK^5. This inconspicuous improvement is another deep result that had been conjectured
for over 30 years:

Theorem 8.3.8. (Mader 1997)
*Every graph with n vertices and more than $3n - 6$ edges contains K^5 as
a topological minor.*

No structure theorem for the graphs without a TK^5, analogous to
Proposition 8.3.1 and Theorem 8.3.4, is known. However, Mader has
recently characterized those with the greatest possible number of edges:

Theorem 8.3.9. (Mader 1997)
*A graph is extremal without a TK^5 if and only if it can be constructed
recursively from maximal planar graphs by pasting along triangles.*

Exercises

1. Prove, from first principles, the theorem of Wagner (1964) that every
    graph of chromatic number at least 2^r contains K^r as a minor.

    (Hint. Apply induction on r. For the induction step, use the fact
    that a graph of chromatic number at least 2^r cannot be coloured using
    $2^{r-1} - 1$ colours for each of the two vertex classes at even resp. odd
    distance from some fixed vertex x.)

2. Prove, from first principles, the result of Mader (1967) that every graph
    of average degree at least 2^{r-2} contains K^r as a minor.

    (Hint. Induction on r. Imitate the start of the proof of Lemma 8.1.3.)

3.$^-$ Derive Wagner's theorem (Ex. 1) from Mader's theorem (Ex. 2).

4.$^+$ Show that any function h as in Theorem 3.6.1 satisfies the inequality
    $h(r) \geqslant \frac{1}{8}r^2$, and hence that Theorem 8.1.1 is best possible up to the
    value of the constant c.

    (Hint. How large does s have to get before $K_{s,s}$ contains a TK^{2^r}?)

5. Prove the statement of Lemma 8.1.3 for $k < 6$.

6.$^+$ Explain how exactly the number $\frac{7}{6}$ in the proof of Lemma 8.1.3 was
    arrived at. Could it be replaced by $\frac{3}{2}$?

7.$^+$ For which trees T is there a function $f : \mathbb{N} \to \mathbb{N}$ tending to infinity, such
    that every graph G with $\chi(G) < f(d(G))$ contains an induced copy of T?
    (In other words: can we force the chromatic number up by raising the
    average degree, as long as T does not occur as an induced subgraph?
    Or, as in Gyárfás's conjecture: will a large average degree force an
    induced copy of T if the chromatic number is kept small?)

8.$^-$ Derive the four colour theorem from Hadwiger's conjecture for $r = 5$.

9.$^-$ Show that Hadwiger's conjecture for $r + 1$ implies the conjecture for r.

10.$^-$ Using the results from this chapter, prove or disprove the following
    approximate version of Hadwiger's conjecture: given any $\epsilon > 0$, every
    graph of chromatic number at least $r^{1+\epsilon}$ has a K^r minor, provided that
    r is large enough.

11. Prove Hadwiger's conjecture for $r = 4$ from first principles.

    (Hint. Show by induction on $|G|$ that any 3-colouring of a cycle in
    $G \not\supseteq K^4$ extends to all of G.)

12. Prove Hadwiger's conjecture for line graphs.

    (Hint. Reduce the statement to critical k-chromatic graphs and apply
    Vizing's theorem.)

13. (i)$^-$ Show that Hadwiger's conjecture is equivalent to the statement
    that $G \succcurlyeq K^{\chi(G)}$ for all graphs G.

    (ii) Show that any minimum-order counterexample G to Hadwiger's
    conjecture (as rephrased above) satisfies $K^{\chi(G)-1} \not\subseteq G$ and has a con-
    nected complement.

14.   Show that any graph constructed as in Theorem 8.3.1 is edge-maximal without a K^4 minor.

15.   Prove the implication $\delta(G) \geqslant 3 \Rightarrow G \supseteq TK^4$.

      (Hint. Theorem 8.3.1.)

16.   A multigraph is called *series-parallel* if it can be constructed recursively from a K^2 by the operations of subdividing and of doubling edges. Show that a 2-connected multigraph is series-parallel if and only if it has no (topological) K^4 minor.

      (Hint. To prove 'if', use induction and consider a cycle $C \subseteq G$. How are the neighbours of the components of $G - C$ arranged on C?)

17.   Prove Corollary 8.3.5.

18.   Characterize the graphs with n vertices and more than $3n - 6$ edges that contain no $TK_{3,3}$. In particular, determine $\mathrm{ex}(n, TK_{3,3})$.

      (Hint. By a theorem of Wagner, every edge-maximal graph without a $K_{3,3}$ minor can be constructed recursively from maximal planar graphs and copies of K^5 by pasting along K^2s.)

19.   By a theorem of Pelikán, every graph of minimum degree at least 4 contains a subdivision of K^5_-, a K^5 minus an edge. Using this theorem, prove Thomassen's 1974 result that every graph with $n \geqslant 5$ vertices and at least $4n - 10$ edges contains a TK^5.

      (Hint. Show by induction on $|G|$ that if $\|G\| \geqslant 4n - 10$ then for every vertex $x \in G$ there is a $TK^5 \subseteq G$ in which x is not a branch vertex.)

Notes

The investigation of graphs not containing a given graph as a minor, or topological minor, has a long history. It probably started with Wagner's 1935 PhD thesis, in which he sought to 'detopologize' the four colour problem by classifying the graphs without a K^5 minor. His hope was to be able to show abstractly that all those graphs were 4-colourable; since the graphs without a K^5 minor include the planar graphs, this would amount to a proof of the four colour conjecture involving no topology whatsoever. The result of Wagner's efforts, Theorem 8.3.4, falls tantalizingly short of this goal: although it succeeds in classifying the graphs without a K^5 minor in structural terms, planarity re-emerges as one of the criteria used in the classification. From this point of view, it is instructive to compare Wagner's K^5 theorem with similar classification theorems, such as his analogue for K^4 (Proposition 8.3.1), where the graphs are decomposed into parts from a *finite* set of irreducible graphs. See R. Diestel, *Graph Decompositions*, Oxford University Press 1990, for more such classification theorems.

    Despite its failure to resolve the four colour problem, Wagner's K^5 structure theorem had consequences for the development of graph theory like few others. To mention just two: it prompted Hadwiger to make his famous conjecture; and it inspired the notion of a tree-decomposition, which is fundamental to the work of Robertson and Seymour on minors (see Chapter 12). Wagner himself responded to Hadwiger's conjecture with a proof that, in order to force a K^r minor, it does suffice to raise the chromatic number of a graph to *some* value depending only on r (Exercise 1). This theorem then, along with its analogue for topological minors proved independently by Dirac and by Jung, prompted the question of which average degree suffices to force the desired minor.

The deepest contribution in this field of research was no doubt made by Mader, in a series of papers from the late sixties. Our proof of Lemma 8.1.3 is presented intentionally in a step-by-step fashion, to bring out some of Mader's ideas. Mader's own proof—not to mention that of Thomason's best possible version of the lemma, as used in the original proof of Theorem 8.1.1— is wrapped up so elegantly that it becomes hard to see the ideas behind it. Except for this lemma, our proof of Theorem 8.1.1 follows B. Bollobás & A.G. Thomason, Proof of a conjecture of Mader, Erdős and Hajnal on topological complete subgraphs, manuscript 1994. The constant c from the theorem was shown by J. Komlós & E. Szemerédi, Topological cliques in graphs II, *Combinatorics, Probability and Computing* **5** (1996), 79–90, to be no greater than about $\frac{1}{2}$, which is not far from the lower bound of $\frac{1}{8}$ given in Exercise 4.

Theorem 8.1.5 is due to A.D. Scott, Induced trees in graphs of large chromatic number, *J. Graph Theory* (1997). Theorem 8.2.1 was proved independently by Kostochka (1982; English translation: A.V. Kostochka, Lower bounds of the Hadwiger number of graphs by their average degree, *Combinatorica* **4** (1984), 307–316) and by A.G. Thomason, An extremal function for contractions of graphs, *Math. Proc. Camb. Phil. Soc.* **95** (1984), 261–265. Theorem 8.2.2 was taken from Thomassen's survey, Paths, Circuits and Subdivisions, in (L.W. Beineke & R.J. Wilson, eds.) *Selected Topics in Graph Theory 3*, Academic Press 1988. Theorem 8.1.4 is from W. Mader, Topological subgraphs in graphs of large girth, submitted to *Combinatorica*.

The proof of Hadwiger's conjecture for $r = 4$, hinted at in Exercise 11, is given by Hadwiger himself in the 1943 paper containing his conjecture. For a while, there was a counterpart to Hadwiger's conjecture for topological minors, the conjecture of Hajós that $\chi(G) \geqslant r$ even implies $G \supseteq TK^r$. A counterexample to this conjecture was found in 1979 by Catlin; a little later, Erdős und Fajtlowicz even proved that Hajós's conjecture is false for *almost all* graphs (see Chapter 11).

Mader's Theorem 8.3.8 that $3n - 5$ edges force a topological K^5 minor had been conjectured by Dirac in 1964. Its proof will comprise at least two papers, the proof of Theorem 8.3.9 another paper.

9 Ramsey Theory for Graphs

In this chapter we consider a type of problem which, on the face of it, appears to be similar to the theme of the last two chapters: what kind of substructures are necessarily present in *every* large enough graph? For this question to make sense, there ought to be at least two substructures that count as 'success', maybe of complementary kind; we may then ask whether every large enough graph contains at least one of them. For example: given an integer r, does every large enough graph contain either a K^r or an induced $\overline{K^r}$?

Despite its similarity to extremal problems in that we are looking for local implications of global assumptions, the above type of question leads to a kind of mathematics with a distinctive flavour of its own. Indeed, the theorems and proofs in this chapter have more in common with similar results in algebra or geometry, say, than with most other areas of graph theory. The study of their underlying methods, therefore, is generally regarded as a combinatorial subject in its own right: the discipline of *Ramsey theory*.

In line with the subject of this book, we shall focus on results that are naturally expressed in terms of graphs. Even from the viewpoint of general Ramsey theory, however, this is not as much of a limitation as it might seem: graphs are a natural setting for Ramsey problems, and the material in this chapter brings out a sufficient variety of ideas and methods to convey some of the fascination of the theory as a whole.

9.1 Ramsey's original theorems

In its simplest version, Ramsey's original theorem says that, given an
integer $r \geqslant 0$, every large enough graph G contains either K^r or $\overline{K^r}$ as
an induced subgraph.

How could we go about proving this? Let us try to build a K^r or
$\overline{K^r}$ in G inductively, starting with an arbitrary vertex $v_1 \in V_1 := V(G)$.
If $|G|$ is large, there will be a large set $V_2 \subseteq V_1 \smallsetminus \{ v_1 \}$ of vertices that are
either all adjacent to v_1 or all non-adjacent to v_1. Accordingly, we may
think of v_1 as the first vertex of a K^r or $\overline{K^r}$ whose other vertices all lie
in V_2. Let us then choose another vertex $v_2 \in V_2$ for our K^r or $\overline{K^r}$. Since
V_2 is large, it will have a subset V_3, still fairly large, of vertices that are
all 'of the same type' with respect to v_2 as well: either all adjacent or all
non-adjacent to it. We then continue our search for vertices inside V_3,
and so on (Fig. 9.1.1).

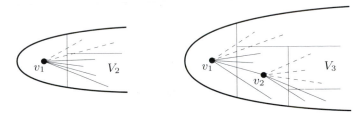

Fig. 9.1.1. Choosing the sequence $v_1, v_2, \dots$

How long can we go on in this way? This depends on the size of
our initial set V_1: each set V_i has at least half the size of its predeces-
sor V_{i-1}, so we shall be able to complete s construction steps if G has
order about 2^s. As the following proof shows, the choice of $s = 2r - 3$
vertices v_i suffices in order to find among them the vertices of a K^r
or $\overline{K^r}$.

Theorem 9.1.1. (Ramsey 1930)

[9.2.2]
*For every $r \in \mathbb{N}$ there exists an $n \in \mathbb{N}$ such that every graph of order at
least n contains either K^r or $\overline{K^r}$ as an induced subgraph.*

Proof. The assertion is trivial for $r \leqslant 1$; we assume that $r \geqslant 2$. Let
$n := 2^{2r-3}$, and let G be a graph of order at least n. We shall define
a sequence $V_1, \dots, V_{2r-2}$ of sets and choose vertices $v_i \in V_i$ with the
following properties:

    (i) $|V_i| = 2^{2r-2-i}$ $(i = 1, \dots, 2r-2)$;

    (ii) $V_i \subseteq V_{i-1} \smallsetminus \{ v_{i-1} \}$ $(i = 2, \dots, 2r-2)$;

    (iii) v_{i-1} is adjacent either to all vertices in V_i or to no vertex in V_i
        $(i = 2, \dots, 2r-2)$.

Let $V_1 \subseteq V(G)$ be any set of 2^{2r-3} vertices, and pick $v_1 \in V_1$ arbitrarily. Then (i) holds for $i = 1$, while (ii) and (iii) hold trivially. Suppose now that V_{i-1} and $v_{i-1} \in V_{i-1}$ have been chosen so as to satisfy (i)–(iii) for $i - 1$, where $1 < i \leqslant 2r - 2$. Since

$$|V_{i-1} \smallsetminus \{ v_{i-1} \}| = 2^{2r-1-i} - 1$$

is odd, V_{i-1} has a subset V_i satisfying (i)–(iii); we pick $v_i \in V_i$ arbitrarily.

Among the $2r - 3$ vertices $v_1, \dots, v_{2r-3}$, there are $r - 1$ vertices that show the same behaviour when viewed as v_{i-1} in (iii), being adjacent either to all the vertices in V_i or to none. Accordingly, these $r - 1$ vertices and v_{2r-2} induce either a K^r or a $\overline{K^r}$ in G, because $v_i, \dots, v_{2r-2} \in V_i$ for all i. $\qquad\square$

The least integer n associated with r as in Theorem 9.1.1 is the *Ramsey number $R(r)$* of r; our proof shows that $R(r) \leqslant 2^{2r-3}$. In Chapter 11 we shall use a simple probabilistic argument to show that $R(r)$ is bounded below by $2^{r/2}$ (Theorem 11.1.3).

Ramsey
number
$R(r)$

It is customary in Ramsey theory to think of partitions as colourings: a *colouring* of (the elements of) a set X *with c colours*, or *c-colouring* for short, is simply a partition of X into c classes (indexed by the 'colours'). In particular, these colourings need not satisfy any non-adjacency requirements as in Chapter 5. Given a c-colouring of $[X]^k$, the set of all k-subsets of X, we call a set $Y \subseteq X$ *monochromatic* if all the elements of $[Y]^k$ have the same colour,[1] i.e. belong to the same of the c partition classes of $[X]^k$. Similarly, if $G = (V, E)$ is a graph and and all the edges of $H \subseteq G$ have the same colour in some colouring of E, we call H a *monochromatic subgraph* of G, speak of a red (green, etc.) H in G, and so on.

c-colouring

$[X]^k$

*mono-
chromatic*

In the above terminology, Ramsey's theorem can be expressed as follows: for every r there exists an n such that, given any n-set X, every 2-colouring of $[X]^2$ yields a monochromatic r-set $Y \subseteq X$. Interestingly, this assertion remains true for c-colourings of $[X]^k$ with arbitrary c and k—with almost exactly the same proof!

To avoid repetition, we shall use this opportunity to demonstrate a common alternative proof technique: we first prove an infinite version of the general Ramsey theorem (which is easier, because we need not worry about numbers), and then deduce the finite version by a so-called *compactness argument*.

Theorem 9.1.2. *Let k, c be positive integers, and X an infinite set. If $[X]^k$ is coloured with c colours, then X has an infinite monochromatic subset.*

[12.1.1]

---

[1] Note that Y is called monochromatic, but it is the elements of $[Y]^k$, not of Y, that are (equally) coloured.

Proof. We prove the theorem by induction on k, with c fixed. For $k = 1$ the assertion holds, so let $k > 1$ and assume the assertion for smaller values of k.

Let $[X]^k$ be coloured with c colours. We shall construct an infinite sequence $X_0, X_1, \ldots$ of infinite subsets of X and choose elements $x_i \in X_i$ with the following properties (for all i):

(i) $X_{i+1} \subseteq X_i \setminus \{x_i\}$;

(ii) all k-sets $\{x_i\} \cup Z$ with $Z \in [X_{i+1}]^{k-1}$ have the same colour, which we *associate* with x_i.

We start with $X_0 := X$ and pick $x_0 \in X_0$ arbitrarily. By assumption, X_0 is infinite. Having chosen an infinite set X_i and $x_i \in X_i$ for some i, we c-colour $[X_i \setminus \{x_i\}]^{k-1}$ by giving each set Z the colour of $\{x_i\} \cup Z$ from our c-colouring of $[X]^k$. By the induction hypothesis, $X_i \setminus \{x_i\}$ has an infinite monochromatic subset, which we choose as X_{i+1}. Clearly, this choice satisfies (i) and (ii). Finally, we pick $x_{i+1} \in X_{i+1}$ arbitrarily.

Since c is finite, one of the c colours is associated with infinitely many x_i. These x_i form an infinite monochromatic subset of X. $\qquad \square$

To deduce the finite version of Theorem 9.1.2, we make use of a standard graph-theoretical tool in combinatorics:

Lemma 9.1.3. (König's Infinity Lemma)
Let $V_0, V_1, \ldots$ *be an infinite sequence of disjoint non-empty finite sets, and let G be a graph on their union. Assume that every vertex v in a set V_n with $n \geqslant 1$ has a neighbour $f(v)$ in V_{n-1}. Then G contains an infinite path $v_0 v_1 \ldots$ with $v_n \in V_n$ for all n.*

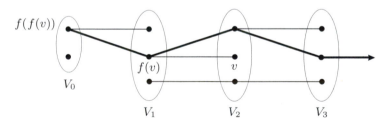

Fig. 9.1.2. König's infinity lemma

Proof. Let $\mathcal{P}$ be the set of all paths of the form $v\, f(v)\, f(f(v)) \ldots$ ending in V_0. Since V_0 is finite but $\mathcal{P}$ is infinite, infinitely many of the paths in $\mathcal{P}$ end at the same vertex $v_0 \in V_0$. Of these paths, infinitely many also agree on their penultimate vertex $v_1 \in V_1$, because V_1 is finite. Of those paths, infinitely many agree even on their vertex v_2 in V_2—and so on. Although the set of paths considered decreases from step to step, it is still infinite after any finite number of steps, so v_n gets defined for every $n \in \mathbb{N}$. By definition, each vertex v_n is adjacent to v_{n-1} on one of those paths, so $v_0 v_1 \ldots$ is indeed an infinite path. $\qquad \square$

Theorem 9.1.4. *For all $k, c, r \geqslant 1$ there exists an $n \geqslant k$ such that every*   [9.3.3]
n-set X has a monochromatic r-subset with respect to any c-colouring
of $[X]^k$.

Proof. As is customary in set theory, we denote by $n \in \mathbb{N}$ (also) the
set $\{0, \ldots, n-1\}$. Suppose the assertion fails for some k, c, r. Then for k, c, r
every $n \geqslant k$ there exist an n-set, without loss of generality the set n, and
a c-colouring $[n]^k \rightarrow c$ such that n contains no monochromatic r-set. Let
us call such colourings *bad*; we are thus assuming that for every $n \geqslant k$ *bad*
there exists a bad colouring of $[n]^k$. Our aim is to combine these into a   *colouring*
bad colouring of $[\mathbb{N}]^k$, which will contradict Theorem 9.1.2.

    For every $n \geqslant k$ let $V_n \neq \emptyset$ be the set of bad colourings of $[n]^k$. For
$n > k$, the restriction $f(g)$ of any $g \in V_n$ to $[n-1]^k$ is still bad, and
hence lies in V_{n-1}. By the infinity lemma, there is an infinite sequence
$g_k, g_{k+1}, \ldots$ of bad colourings $g_n \in V_n$ such that $f(g_n) = g_{n-1}$ for all
$n > k$. For every $m \geqslant k$, all colourings g_n with $n \geqslant m$ agree on $[m]^k$, so
for each $Y \in [\mathbb{N}]^k$ the value of $g_n(Y)$ coincides for all $n > \max Y$. Let
us define $g(Y)$ as this common value $g_n(Y)$. Then g is a bad colouring
of $[\mathbb{N}]^k$: every r-set $S \subseteq \mathbb{N}$ is contained in some sufficiently large n,
so S cannot be monochromatic since g coincides on $[n]^k$ with the bad
colouring g_n. $\square$

The least integer n associated with k, c, r as in Theorem 9.1.4 is the   *Ramsey*
Ramsey number for these parameters; we denote it by $R(k, c, r)$.   *number*
                                                                        $R(k, c, r)$

9.2 Ramsey numbers

Ramsey's theorem may be rephrased as follows: if $H = K^r$ and G
is a graph with sufficiently many vertices, then either G itself or its
complement $\overline{G}$ contains a copy of H as a subgraph. Clearly, the same is
true for any graph H, simply because $H \subseteq K^h$ for $h := |H|$.

    However, if we ask for the *least* n such that every graph G with   *Ramsey*
n vertices has the above property—this is the *Ramsey number* $R(H)$ *number*
of H—then the above question makes sense: if H has only few edges, it   *$R(H)$*
should embed more easily in G or $\overline{G}$, and we would expect $R(H)$ to be
smaller than the Ramsey number $R(h) = R(K^h)$.

    A little more generally, let $R(H_1, H_2)$ denote the least $n \in \mathbb{N}$ such $R(H_1, H_2)$
that $H_1 \subseteq G$ or $H_2 \subseteq \overline{G}$ for every graph G of order n. For most graphs
H_1, H_2, only very rough estimates are known for $R(H_1, H_2)$. Interesting-
ly, lower bounds given by random graphs (as in Theorem 11.3.3) are often
sharper than even the best bounds provided by explicit constructions.

    The following proposition describes one of the few cases where exact
Ramsey numbers are known for a relatively large class of graphs:

Proposition 9.2.1. *Let s, t be positive integers, and let T be a tree of order t. Then $R(T, K^s) = (s-1)(t-1) + 1$.*

(5.2.3)
(1.5.4)

Proof. The disjoint union of $s - 1$ graphs K^{t-1} contains no copy of T, while the complement of this graph, the complete $(s-1)$-partite graph K_{t-1}^{s-1}, does not contain K^s. This proves $R(T, K^s) \geqslant (s-1)(t-1) + 1$.

Conversely, let G be any graph of order $n = (s-1)(t-1) + 1$ whose complement contains no K^s. Then $s > 1$, and in any vertex colouring of G (in the sense of Chapter 5) at most $s - 1$ vertices can have the same colour. Hence, $\chi(G) \geqslant \lceil n/(s-1) \rceil = t$. By Corollary 5.2.3, G has a subgraph H with $\delta(H) \geqslant t - 1$, which by Corollary 1.5.4 contains a copy of T. $\qquad\square$

As the main result of this section, we shall now prove one of those rare general theorems providing a relatively good upper bound for the Ramsey numbers of a large class of graphs, a class defined in terms of a standard graph invariant. The theorem deals with the Ramsey numbers of sparse graphs: it says that the Ramsey number of graphs H with bounded maximum degree grows only linearly in $|H|$—an enormous improvement on the exponential bound from the proof of Theorem 9.1.1.

Theorem 9.2.2. (Chvátal, Rödl, Szemerédi & Trotter 1983)
For every positive integer Δ there is a constant c such that

$$R(H) \leqslant c\,|H|$$

(7.1.1)
(7.2.1)
(7.3.2)
(9.1.1)

for all graphs H with $\Delta(H) \leqslant \Delta$.

Proof. The basic idea of the proof is as follows. We wish to show that $H \subseteq G$ or $H \subseteq \overline{G}$ if $|G|$ is large enough (though not too large). Consider an ϵ-regular partition of G, as provided by the regularity lemma. If enough of the ϵ-regular pairs in this partition have positive density, we may hope to find a copy of H in G. If most pairs have zero or low density, we try to find H in $\overline{G}$. Let R, R' and R'' be the 'regularity graphs'[2] of G whose edges correspond to the pairs of density $\geqslant 0$; $\geqslant 1/2$; $< 1/2$; respectively. Then R is the edge-disjoint union of R' and R''.

Now to obtain $H \subseteq G$ or $H \subseteq \overline{G}$, it suffices by Lemma 7.3.2 to ensure that H is contained in a suitable 'inflated regularity graph' R'_s or R''_s. Since $\chi(H) \leqslant \Delta(H) + 1 \leqslant \Delta + 1$, this will be the case if $s \geqslant \alpha(H)$ and we can find a $K^{\Delta+1}$ in R' or in R''. But that is easy to ensure: we just need that $K^r \subseteq R$, where r is the Ramsey number of $\Delta + 1$, which will follow from Turán's theorem because R is dense.

Δ, d

ϵ_0

For the formal proof let now $\Delta \geqslant 1$ be given. On input $d := 1/2$ and Δ, Lemma 7.3.2 returns an ϵ_0; since the lemma's assertion about ϵ_0

---

[2] Later, we shall define R'' a little differently, so that it complies with our formal definition of a regularity graph.

becomes weaker if ϵ_0 is made smaller, we may assume that $\epsilon_0 < 1$. Let $m := R(\Delta + 1)$ be the Ramsey number of $\Delta + 1$. Let $\epsilon \leqslant \epsilon_0$ be positive but small enough that, for $k = m$ (and hence for all $k \geqslant m$),
$$2\epsilon < \frac{1}{m-1} - \frac{1}{k}. \tag{1}$$

Finally, let M be the integer returned by the regularity lemma (7.2.1) on input ϵ and m.

All the quantities defined so far depend only on Δ. We shall prove the theorem with
$$c := \frac{M}{\epsilon_0 (1 - \epsilon)}.$$

So let H with $\Delta(H) \leqslant \Delta$ be given, and let $s := |H|$. Let G be an arbitrary graph of order $n \geqslant c|H|$; we show that $H \subseteq G$ or $H \subseteq \overline{G}$.

By Lemma 7.2.1, G has an ϵ-regular partition $\{V_0, V_1, \ldots, V_k\}$ with exceptional set V_0 and $|V_1| = \ldots = |V_k| =: \ell$, where $m \leqslant k \leqslant M$. Then
$$\ell = \frac{n - |V_0|}{k} \geqslant \frac{n - \epsilon n}{M} = n\frac{1 - \epsilon}{M} \geqslant cs\frac{1 - \epsilon}{M} = \frac{s}{\epsilon_0}. \tag{2}$$

Let R be the regularity graph with parameters $\epsilon, \ell, 0$ corresponding to this partition. By definition, R has k vertices and
$$\|R\| \geqslant \binom{k}{2} - \epsilon k^2$$
$$= \tfrac{1}{2}k^2 \left(1 - \frac{1}{k} - 2\epsilon\right)$$
$$\underset{(1)}{>} \tfrac{1}{2}k^2 \left(1 - \frac{1}{k} - \frac{1}{m-1} + \frac{1}{k}\right)$$
$$= \tfrac{1}{2}k^2 \frac{m-2}{m-1}$$
$$\geqslant t_{m-1}(k)$$

edges. By Theorem 7.1.1, therefore, R has a subgraph $K = K^m$.

We now colour the edges of R with two colours: red if the edge corresponds to a pair (V_i, V_j) of density at least $1/2$, and green otherwise. Let R' be the spanning subgraph of R formed by the red edges, and R'' the spanning subgraph of R formed by the green edges and those whose corresponding pair has density exactly $1/2$. Then R' is a regularity graph of G with parameters ϵ, ℓ and $1/2$. And R'' is a regularity graph of $\overline{G}$, with the same parameters: as one easily checks, every pair (V_i, V_j) that is ϵ-regular for G is also ϵ-regular for $\overline{G}$.

r

By definition of m, our graph K contains a red or a green K^r, for $r := \chi(H) \leqslant \Delta + 1$. Correspondingly, $H \subseteq R'_s$ or $H \subseteq R''_s$. Since $\epsilon \leqslant \epsilon_0$ and $\ell \geqslant s/\epsilon_0$ by (2), both R' and R'' satisfy the requirements of Lemma 7.3.2, so $H \subseteq G$ or $H \subseteq \overline{G}$ as desired. $\qquad\square$

Ramsey-
minimal

So far in this section, we have been asking what is the least order of a graph G such that every 2-colouring of its edges yields a monochromatic copy of some given graph H. Rather than focusing on the order of G, we might alternatively try to minimize G itself, with respect to the subgraph relation. Given a graph H, let us call a graph G *Ramsey-minimal* for H if G is minimal with the property that every 2-colouring of its edges yields a monochromatic copy of H.

What do such Ramsey-minimal graphs look like? Are they unique? The following result, which we include for its pretty proof, answers the second question for some H:

Proposition 9.2.3. *If T is a tree but not a star, then infinitely many graphs are Ramsey-minimal for T.*

(1.5.4)
(5.2.3)
(11.2.2)

Proof. Let $|T| =: r$. We show that for every $n \in \mathbb{N}$ there is a graph of order at least n that is Ramsey-minimal for T.

Let us borrow the assertion of Theorem 11.2.2 from Chapter 11: by that theorem, there exists a graph G with chromatic number $\chi(G) > r^2$ and girth $g(G) > n$. If we colour the edges of G red and green, then the red and the green subgraph cannot both have an r-(vertex-)colouring in the sense of Chapter 5: otherwise we could colour the vertices of G with the pairs of colours from those colourings and obtain a contradiction to $\chi(G) > r^2$. So let $G' \subseteq G$ be monochromatic with $\chi(G') > r$. By Corollary 5.2.3, G' has a subgraph of minimum degree at least r, which contains a copy of T by Corollary 1.5.4.

Let $G^* \subseteq G$ be Ramsey-minimal for T. Clearly, G^* is not a forest: the edges of any forest can be 2-coloured (partitioned) so that no monochromatic subforest contains a path of length 3, let alone a copy of T. (Here we use that T is not a star, and hence contains a P^3.) So G^* contains a cycle, which has length $g(G) > n$ since $G^* \subseteq G$. In particular, $|G^*| > n$ as desired. $\qquad\square$

9.3 Induced Ramsey theorems

Ramsey's theorem can be rephrased as follows. For every graph $H = K^r$ there *exists* a graph G such that every 2-colouring of the edges of G yields a monochromatic $H \subseteq G$; as it turns out, this is witnessed by any large enough complete graph as G. Let us now change the problem slightly and ask for a graph G in which every 2-edge-colouring yields a mono-chromatic *induced* $H \subseteq G$, where H is now an arbitrary given graph.

This slight modification changes the character of the problem dramatically. What is needed now is no longer a simple proof that G is 'big enough' (as for Theorem 9.1.1), but a careful construction: the construction of a graph that, however we bipartition its edges, contains an induced copy of H with all edges in one partition class. We shall call such a graph a *Ramsey graph* for H.

Ramsey graph

The fact that such a Ramsey graph exists for every choice of H is one of the fundamental results of graph Ramsey theory. It was proved around 1973, independently by Deuber, by Erdős, Hajnal & Pósa, and by Rödl.

Theorem 9.3.1. *Every graph has a Ramsey graph. In other words, for every graph H there exists a graph G that, for every partition $\{E_1, E_2\}$ of $E(G)$, has an induced subgraph H with $E(H) \subseteq E_1$ or $E(H) \subseteq E_2$.*

We give two proofs. Each of these is highly individual, yet each offers a glimpse of true Ramsey theory: the graphs involved are used as hardly more than bricks in the construction, but the edifice is impressive.

First proof. In our construction of the desired Ramsey graph we shall repeatedly replace vertices of a graph $G = (V, E)$ already constructed by copies of another graph H. For a vertex set $U \subseteq V$ let $G[U \to H]$ denote the graph obtained from G by replacing the vertices $u \in U$ with copies $H(u)$ of H and joining each $H(u)$ completely to all $H(u')$ with $uu' \in E$ and to all vertices $v \in V \setminus U$ with $uv \in E$ (Fig. 9.3.1). Formally,

$G[U \to H]$

$H(u)$

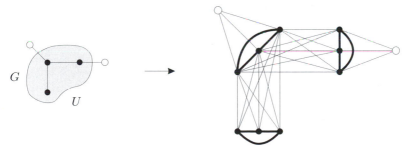

Fig. 9.3.1. A graph $G[U \to H]$ with $H = K^3$

$G[U \to H]$ is the graph on

$$(U \times V(H)) \cup ((V \setminus U) \times \{\emptyset\}),$$

in which two vertices (v, w) and (v', w') are adjacent if and only if either $vv' \in E$, or else $v = v' \in U$ and $ww' \in E(H)$.[3]

---

[3] The replacement of $V \setminus U$ by $(V \setminus U) \times \{\emptyset\}$ is just a formal device to ensure

We prove the following formal strengthening of Theorem 9.3.1:

$G(H_1, H_2)$

> For any two graphs H_1, H_2 there exists a graph $G =$
> $G(H_1, H_2)$ such that every edge colouring of G with the
> colours 1 and 2 yields either an induced $H_1 \subseteq G$ with all   (∗)
> its edges coloured 1 or an induced $H_2 \subseteq G$ with all its
> edges coloured 2.

This formal strengthening makes it possible to apply induction on
$|H_1| + |H_2|$, as follows.

x_i

H_i', H_i''

f^i

origin

G_1, G_2

   If either H_1 or H_2 has no edges (in particular, if $|H_1| + |H_2| \leqslant 1$),
then (∗) holds with $G = \overline{K^n}$ for large enough n. For the induction step,
we now assume that both H_1 and H_2 have at least one edge, and that
(∗) holds for all pairs (H_1', H_2') with smaller $|H_1'| + |H_2'|$.
   For each $i = 1, 2$, pick a vertex $x_i \in H_i$ that is incident with an
edge. Let $H_i' := H_i - x_i$, and let H_i'' be the subgraph of H_i' induced by
the neighbours of x_i.
   We shall construct a sequence $G^0, \ldots, G^n$ of disjoint graphs; G^n will
be the desired Ramsey graph $G(H_1, H_2)$. Along with the graphs G_i, we
shall define subsets $V^i \subseteq V(G^i)$ and a map

$$f \colon V^1 \cup \ldots \cup V^n \to V^0 \cup \ldots \cup V^{n-1}$$

such that

$$f(V^i) = V^{i-1} \tag{1}$$

for all $i \geqslant 1$. Writing $f^i := f \circ \ldots \circ f$ for the i-fold composition of f
whenever it is defined, and f^0 for the identity map on $V^0 = V(G^0)$, we
thus have $f^i(v) \in V^0$ for all $v \in V^i$. We call $f^i(v)$ the $origin$ of v.
   The subgraphs $G^i [V^i]$ will reflect the structure of G^0 as follows:

> Vertices in V^i with different origins are adjacent in G^i if
> and only if their origins are adjacent in G^0.   (2)

   Assertion (2) will not be used formally in the proof below. However,
it can help us to visualize the graphs G^i: every G^i (more precisely, every
$G^i [V^i]$—there will also be some vertices $x \in G^i - V^i$) is essentially an
inflated copy of G^0 in which every vertex $w \in G^0$ has been replaced by
the set of all vertices in V^i with origin w, and the map f links vertices
with the same origin across the various G^i.
   By the induction hypothesis, there are Ramsey graphs

$$G_1 := G(H_1, H_2') \quad \text{and} \quad G_2 := G(H_1', H_2) \, .$$

---

that all vertices of $G[U \to H]$ have the same form (v, w), and that $G[U \to H]$ is
formally disjoint from G.

Let G^0 be a copy of G_1, and set $V^0 := V(G^0)$. Let $W'_0, \ldots, W'_{n-1}$ be the subsets of V^0 spanning an H'_2 in G^0. Thus, n is defined as the number of induced copies of H'_2 in G^0, and we shall construct a graph G^i for every set W'_{i-1}, $i = 1, \ldots, n$. Since H_1 has an edge, $n \geqslant 1$: otherwise G^0 could not be a $G(H_1, H'_2)$. For $i = 0, \ldots, n-1$, let W''_i be the image of $V(H''_2)$ under some isomorphism $H'_2 \to G^0[W'_i]$. $\quad\quad G^0, V^0$ $\quad\quad W'_i$ $\quad\quad n$ $\quad\quad W''_i$

Assume now that $G^0, \ldots, G^{i-1}$ and $V^0, \ldots, V^{i-1}$ have been defined for some $i \geqslant 1$, and that f has been defined on $V^1 \cup \ldots \cup V^{i-1}$ and satisfies (1) for all $j \leqslant i$. We expand G^{i-1} to G^i in two steps. For the first step, consider the set U^{i-1} of all the vertices $v \in V^{i-1}$ whose origin $f^{i-1}(v)$ lies in W''_{i-1}. (For $i = 1$, this gives $U^0 = W''_0$.) Expand G^{i-1} to a graph $\tilde{G}^{i-1}$ by replacing every vertex $u \in U^{i-1}$ with a copy $G_2(u)$ of G_2, i.e. let $\quad\quad U^{i-1}$ $\quad\quad G_2(u)$

$$\tilde{G}^{i-1} := G^{i-1}[\,U^{i-1} \to G_2\,] \qquad\qquad \tilde{G}^{i-1}$$

(see Figures 9.3.2 and 9.3.3). Set $f(u') := u$ for all $u \in U^{i-1}$ and $u' \in G_2(u)$, and $f(v') := v$ for all $v' = (v, \emptyset)$ with $v \in V^{i-1} \smallsetminus U^{i-1}$. (Recall that $(v, \emptyset)$ is simply the unexpanded copy of a vertex $v \in G^{i-1}$ in $\tilde{G}^{i-1}$.) Let V^i be the set of those vertices v' or u' of $\tilde{G}^{i-1}$ for which f has thus been defined, i.e. the vertices that either correspond directly to a vertex v in V^{i-1} or else belong to an expansion $G_2(u)$ of such a vertex u. Then (1) holds for i. Also, if we assume (2) inductively for $i-1$, then (2) holds again for i (in $\tilde{G}^{i-1}$). The graph $\tilde{G}^{i-1}$ is already the 'essential part' of G^i: the part that looks like an inflated copy of G^0. $\quad\quad V^i$

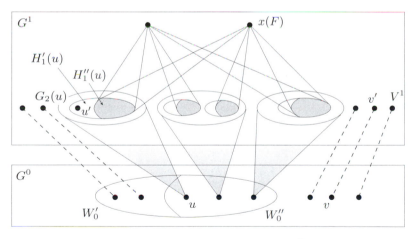

Fig. 9.3.2. The construction of G^1

In the second step we now extend $\tilde{G}^{i-1}$ to the desired graph G^i by adding some further vertices $x \notin V^i$. Let $\mathcal{F}$ denote the set of all families F of the form $\quad\quad \mathcal{F}$

$$F = \left(H'_1(u) \mid u \in U^{i-1} \right),$$

$H'_1(u)$

$x(F)$

$H''_1(u)$
G^i

x

y_u

$\hat{U}^{i-1}$

$\hat{G}^{i-1}$

H'

where each $H'_1(u)$ is an induced subgraph of $G_2(u)$ isomorphic to H'_1. (Less formally: $\mathcal{F}$ is the collection of ways to select from each $G_2(u)$ exactly one induced copy of H'_1.) For each $F \in \mathcal{F}$, add a vertex $x(F)$ to $\tilde{G}^{i-1}$ and join it to all the vertices of $H''_1(u)$ for every $u \in U^{i-1}$, where $H''_1(u)$ is the image of H''_1 under some isomorphism $H'_1 \to H'_1(u)$ (Fig. 9.3.2). Denote the resulting graph by G^i. This completes the inductive definition of the graphs $G^0, \dots, G^n$.

Let us now show that $G := G^n$ satisfies (*). To this end, we prove the following assertion (**) about G^i for $i = 0, \dots, n$:

> For every edge colouring with the colours 1 and 2, G^i contains either an induced H_1 coloured 1, or an induced H_2 coloured 2, or an induced subgraph H coloured 2 such that $V(H) \subseteq V^i$ and the restriction of f^i to $V(H)$ is an isomorphism between H and $G^0[W'_k]$ for some $k \in \{i, \dots, n-1\}$.   (**)

Note that the third of the above cases cannot arise for $i = n$, so (**) for n is equivalent to (*) with $G := G^n$.

For $i = 0$, (**) follows from the choice of G^0 as a copy of $G_1 = G(H_1, H'_2)$ and the definition of the sets W'_k. Now let $1 \leqslant i \leqslant n$, and assume (**) for smaller values of i.

Let an edge colouring of G^i be given. For each $u \in U^{i-1}$ there is a copy of G_2 in G^i:

$$G^i \supseteq G_2(u) \simeq G(H'_1, H_2).$$

If $G_2(u)$ contains an induced H_2 coloured 2 for some $u \in U^{i-1}$, we are done. If not, then every $G_2(u)$ has an induced subgraph $H'_1(u) \simeq H'_1$ coloured 1. Let F be the family of these graphs $H'_1(u)$, one for each $u \in U^{i-1}$, and let $x := x(F)$. If, for some $u \in U^{i-1}$, all the x–$H''_1(u)$ edges in G^i are also coloured 1, we have an induced copy of H_1 in G^i and are again done. We may therefore assume that each $H''_1(u)$ has a vertex y_u for which the edge xy_u is coloured 2. Let

$$\hat{U}^{i-1} := \{ y_u \mid u \in U^{i-1} \} \subseteq V^i.$$

Then f defines an isomorphism from

$$\hat{G}^{i-1} := G^i\left[\hat{U}^{i-1} \cup \{(v, \emptyset) \mid v \in V(G^{i-1}) \setminus U^{i-1}\}\right]$$

to G^{i-1}: just map every y_u to u and every $(v, \emptyset)$ to v. Our edge colouring of G^i thus induces an edge colouring of G^{i-1}. If this colouring yields an induced $H_1 \subseteq G^{i-1}$ coloured 1 or an induced $H_2 \subseteq G^{i-1}$ coloured 2, we have these also in $\hat{G}^{i-1} \subseteq G^i$ and are again home.

By (**) for $i-1$ we may therefore assume that G^{i-1} has an induced subgraph H' coloured 2, with $V(H') \subseteq V^{i-1}$, and such that the restriction of f^{i-1} to $V(H')$ is an isomorphism from H' to $G^0[W'_k] \simeq H'_2$

for some $k \in \{i-1, \ldots, n-1\}$. Let $\hat{H}'$ be the corresponding induced $\hat{H}'$
subgraph of $\hat{G}^{i-1} \subseteq G^i$ (also coloured 2); then $V(\hat{H}') \subseteq V^i$,

$$f^i(V(\hat{H}')) = f^{i-1}(V(H')) = W'_k \,,$$

and $f^i \colon \hat{H}' \to G^0[W'_k]$ is an isomorphism.

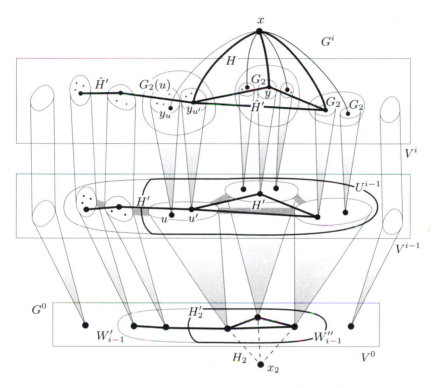

Fig. 9.3.3. A monochromatic copy of H_2 in G^i

If $k \geqslant i$, this completes the proof of $(**)$ with $H := \hat{H}'$; we therefore
assume that $k < i$, and hence $k = i-1$ (Fig. 9.3.3). By definition
of U^{i-1} and $\hat{G}^{i-1}$, the inverse image of W''_{i-1} under the isomorphism
$f^i \colon \hat{H}' \to G^0[W'_{i-1}]$ is a subset of $\hat{U}^{i-1}$. Since x is joined to precisely
those vertices of $\hat{H}'$ that lie in $\hat{U}^{i-1}$, and all these edges xy_u have colour 2,
the graph $\hat{H}'$ and x together induce in G^i a copy of H_2 coloured 2, and
the proof of $(**)$ is complete.   □

Let us return once more to the reformulation of Ramsey's theorem
considered at the beginning of this section: for every graph H there
exists a graph G such that every 2-colouring of the edges of G yields
a monochromatic $H \subseteq G$. The graph G for which this follows at once
from Ramsey's theorem is a sufficiently large complete graph. If we

ask, however, that G shall not contain any complete subgraphs larger than those in H, i.e. that $\omega(G) = \omega(H)$, the problem again becomes difficult—even if we do not require H to be induced in G.

Our second proof of Theorem 9.3.1 solves both problems at once: given H, we shall construct a Ramsey graph for H with the same clique number as H.

bipartite

For this proof, i.e. for the remainder of this section, let us view bipartite graphs P as triples (V_1, V_2, E), where V_1 and V_2 are the two vertex classes and $E \subseteq V_1 \times V_2$ is the set of edges. The reason for this more explicit notation is that we want embeddings between bipartite graphs to respect their bipartitions: given another bipartite graph $P' = (V_1', V_2', E')$, an injective map $\varphi \colon V_1 \cup V_2 \to V_1' \cup V_2'$ will be called an

embedding
$P \to P'$

embedding of P in P' if $\varphi(V_i) \subseteq V_i'$ for $i = 1, 2$ and $\varphi(v_1)\varphi(v_2)$ is an edge of P' if and only if $v_1 v_2$ is an edge of P. (Note that such embeddings are 'induced'.) Instead of $\varphi \colon V_1 \cup V_2 \to V_1' \cup V_2'$ we may simply write $\varphi \colon P \to P'$.

We need two lemmas.

Lemma 9.3.2. *Every bipartite graph can be embedded in a bipartite*

E

graph of the form $(X, [X]^k, E)$ *with* $E = \{\, xY \mid x \in Y \,\}$.

Proof. Let P be any bipartite graph, with vertex classes $\{\, a_1, \ldots, a_n \,\}$ and $\{\, b_1, \ldots, b_m \,\}$, say. Let X be a set with $2n + m$ elements, say

$$X = \{\, x_1, \ldots, x_n,\, y_1, \ldots, y_n,\, z_1, \ldots, z_m \,\};$$

we shall define an embedding $\varphi \colon P \to (X, [X]^{n+1}, E)$.

Let us start by setting $\varphi(a_i) := x_i$ for all $i = 1, \ldots, n$. Which $(n+1)$-sets $Y \subseteq X$ are suitable candidates for the choice of $\varphi(b_i)$ for a given vertex b_i? Clearly those adjacent exactly to the images of the neighbours of b_i, i.e. those satisfying

$$Y \cap \{\, x_1, \ldots, x_n \,\} = \varphi(N_P(b_i)). \tag{1}$$

Since $d(b_i) \leqslant n$, the requirement of (1) leaves at least one of the $n+1$ elements of Y unspecified. In addition to $\varphi(N_P(b_i))$, we may therefore include in each $Y = \varphi(b_i)$ the vertex z_i as an 'index'; this ensures that $\varphi(b_i) \neq \varphi(b_j)$ for $i \neq j$, even when b_i and b_j have the same neighbours in P. To specify the sets $Y = \varphi(b_i)$ completely, we finally fill them up with 'dummy' elements y_j until $|Y| = n + 1$. $\qquad\square$

Our second lemma already covers the bipartite case of the theorem: it says that every bipartite graph has a Ramsey graph—even a bipartite one.

Lemma 9.3.3. *For every bipartite graph P there exists a bipartite graph P' such that for every 2-colouring of the edges of P' there is an embedding $\varphi: P \to P'$ for which all the edges of $\varphi(P)$ have the same colour.*

Proof. We may assume by Lemma 9.3.2 that P has the form $(X, [X]^k, E)$ (9.1.4)
with $E = \{\, xY \mid x \in Y \,\}$. We show the assertion for the graph $P' :=$ P, X, k, E
$(X', [X']^{k'}, E')$, where $k' := 2k - 1$, X' is any set of cardinality P', X', k'

$$|X'| = R\left(k',\, 2\tbinom{k'}{k},\, k\,|X| + k - 1\right),$$

(this is the Ramsey number defined after Theorem 9.1.4), and

$$E' := \{\, x'Y' \mid x' \in Y' \,\}. \qquad\qquad\qquad E'$$

Let us then colour the edges of P' with two colours α and β. Of the α, β
$|Y'| = 2k - 1$ edges incident with a vertex $Y' \in [X']^{k'}$, at least k must
have the same colour. For each Y' we may therefore choose a fixed k-set
$Z' \subseteq Y'$ such that all the edges $x'Y'$ with $x' \in Z'$ have the same colour; Z'
we shall call this colour *associated* with Y'.   *associated*

The sets Z' can lie within their supersets Y' in $\binom{k'}{k}$ ways, as follows.
Let X' be linearly ordered. Then for every $Y' \in [X']^{k'}$ there is a unique
order-preserving bijection $\sigma_{Y'}: Y' \to \{\, 1, \dots, k' \,\}$, which maps Z' to one $\sigma_{Y'}$
of $\binom{k'}{k}$ possible images.
We now colour $[X']^{k'}$ with the $2\binom{k'}{k}$ elements of the set

$$[\{\, 1, \dots, k' \,\}]^k \times \{\, \alpha, \beta \,\}$$

as colours, giving each $Y' \in [X']^{k'}$ as its colour the pair $(\sigma_{Y'}(Z'), \gamma)$,
where γ is the colour α or β associated with Y'. Since $|X'|$ was chosen
as the Ramsey number with parameters k', $2\binom{k'}{k}$ and $k\,|X| + k - 1$, we
know that X' has a monochromatic subset W of cardinality $k\,|X| + k - 1$. W
All Z' with $Y' \subseteq W$ thus lie within their Y' in the same way, i.e. there
exists an $S \in [\{\, 1, \dots, k' \,\}]^k$ such that $\sigma_{Y'}(Z') = S$ for all $Y' \in [W]^{k'}$,
and all $Y' \in [W]^{k'}$ are associated with the same colour, say with α. α

We now construct the desired embedding φ of P in P'. We first $\varphi|_X$
define φ on $X =: \{\, x_1, \dots, x_n \,\}$, choosing images $\varphi(x_i) =: w_i \in W$ so x_i, w_i, n
that $w_i < w_j$ in our ordering of X' whenever $i < j$. Moreover, we choose
the w_i so that exactly $k - 1$ elements of W are smaller than w_1, exactly
$k - 1$ lie between w_i and w_{i+1} for $i = 1, \dots, n - 1$, and exactly $k - 1$
are bigger than w_n. Since $|W| = kn + k - 1$, this can indeed be done
(Fig. 9.3.4).

We now define φ on $[X]^k$. Given $Y \in [X]^k$, we wish to choose $\varphi|_{[X]^k}$
$\varphi(Y) =: Y' \in [X']^{k'}$ so that the neighbours of Y' among the vertices
in $\varphi(X)$ are precisely the images of the neighbours of Y in P, i.e. the

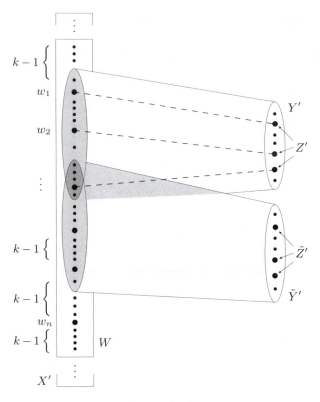

Fig. 9.3.4. The graph of Lemma 9.3.3

vertices $\varphi(x)$ with $x \in Y$, and so that all these edges at Y' are coloured α. To find such a set Y', we first fix its subset Z' as $\{\, \varphi(x) \mid x \in Y \,\}$ (these are k vertices of type w_i) and then extend Z' by $k' - k$ further vertices $u \in W \smallsetminus \varphi(X)$ to a set $Y' \in [W]^{k'}$, in such a way that Z' lies correctly within Y', i.e. so that $\sigma_{Y'}(Z') = S$. This can be done, because $k - 1 = k' - k$ other vertices of W lie between any two w_i. Then

$$Y' \cap \varphi(X) = Z' = \{\, \varphi(x) \mid x \in Y \,\},$$

so Y' has the correct neighbours in $\varphi(X)$, and all the edges between Y' and these neighbours are coloured α (because those neighbours lie in Z' and Y' is associated with α). Finally, φ is injective on $[X]^k$: the images Y' of different vertices Y are distinct, because their intersections with $\varphi(X)$ differ. Hence, our map φ is indeed an embedding of P in P'. $\square$

Second proof of Theorem 9.3.1. Let H be given as in the theorem, and let $n := R(r)$ be the Ramsey number of $r := |H|$. Then, for every 2-colouring of its edges, the graph $K = K^n$ contains a monochromatic copy of H—although not necessarily induced.

We start by constructing a graph G^0, as follows. Imagine the vertices of K to be arranged in a column, and replace every vertex by a row of $\binom{n}{r}$ vertices. Then each of the $\binom{n}{r}$ columns arising can be associated with one of the $\binom{n}{r}$ ways of embedding $V(H)$ in $V(K)$; let us furnish this column with the edges of such a copy of H. The graph G^0 thus arising consists of $\binom{n}{r}$ disjoint copies of H and $(n-r)\binom{n}{r}$ isolated vertices (Fig. 9.3.5).

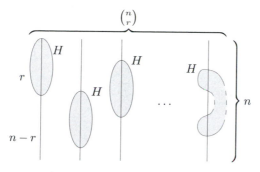

Fig. 9.3.5. The graph G^0

In order to define G^0 formally, we assume that $V(K) = \{1,\ldots,n\}$ and choose copies $H_1,\ldots,H_{\binom{n}{r}}$ of H in K with pairwise distinct vertex sets. (Thus, on each r-set in $V(K)$ we have one fixed copy H_j of H.) We then define

$$V(G^0) := \{(i,j) \mid i = 1,\ldots,n;\ j = 1,\ldots,\tbinom{n}{r}\}$$

$$E(G^0) := \bigcup_{j=1}^{\binom{n}{r}} \{(i,j)(i',j) \mid ii' \in E(H_j)\}.$$

G^0

The idea of the proof now is as follows. Our aim is to reduce the general case of the theorem to the bipartite case dealt with in Lemma 9.3.3. Applying the lemma iteratively to all the pairs of rows of G^0, we construct a very large graph G such that for every edge colouring of G there is an induced copy of G^0 in G that is monochromatic on all the bipartite subgraphs induced by its pairs of rows, i.e. in which edges between the same two rows always have the same colour. The projection of this $G^0 \subseteq G$ to $\{1,\ldots,n\}$ (by contracting its rows) then defines an edge colouring of K. By the choice of $|K|$, one of the $H_j \subseteq K$ will be monochromatic. But this H_j occurs with the same colouring in the jth column of our G^0, where it is an induced subgraph of G^0, and hence of G.

Formally, we shall define a sequence $G^0,\ldots,G^m$ of n-partite graphs G^k, with n-partition $\{V_1^k,\ldots,V_n^k\}$ say, and then let $G := G^m$. The graph G^0 has been defined above; let $V_1^0, \ldots, V_n^0$ be its rows:

V_i^0

$$V_i^0 := \{(i,j) \mid j = 1, \ldots, \binom{n}{r}\}.$$

e_k, m

i_1, i_2

P

P'

W_1, W_2

φ_p, q

Now let $e_1, \ldots, e_m$ be an enumeration of the edges of K. For $k = 0, \ldots, m-1$, construct G^{k+1} from G^k as follows. If $e_{k+1} = i_1 i_2$, say, let $P = (V_{i_1}^k, V_{i_2}^k, E)$ be the bipartite subgraph of G^k induced by its i_1th and i_2th row. By Lemma 9.3.3, P has a bipartite Ramsey graph $P' = (W_1, W_2, E')$. We wish to define $G^{k+1} \supseteq P'$ in such a way that every (monochromatic) embedding $P \to P'$ can be extended to an embedding $G^k \to G^{k+1}$. Let $\{\varphi_1, \ldots, \varphi_q\}$ be the set of all embeddings of P in P', and let

$$V(G^{k+1}) := V_1^{k+1} \cup \ldots \cup V_n^{k+1},$$

where

$$V_i^{k+1} := \begin{cases} W_1 & \text{for } i = i_1 \\ W_2 & \text{for } i = i_2 \\ \bigcup_{p=1}^{q}(V_i^k \times \{p\}) & \text{for } i \notin \{i_1, i_2\}. \end{cases}$$

(Thus for $i \neq i_1, i_2$, we take as V_i^{k+1} just q disjoint copies of V_i^k.) We now define the edge set of G^{k+1} so that the obvious extensions of φ_p to all of $V(G^k)$ become embeddings of G^k in G^{k+1}: for $p = 1, \ldots, q$, let $\psi_p \colon V(G^k) \to V(G^{k+1})$ be defined by

$$\psi_p(v) := \begin{cases} \varphi_p(v) & \text{for } v \in P \\ (v,p) & \text{for } v \notin P \end{cases}$$

and let

$$E(G^{k+1}) := \bigcup_{p=1}^{q} \{\psi_p(v)\psi_p(v') \mid vv' \in E(G^k)\}.$$

Now for every 2-colouring of its edges, G^{k+1} contains an induced copy $\psi_p(G^k)$ of G^k whose edges in P, i.e. those between its i_1th and i_2th row, have the same colour: just choose p so that $\varphi_p(P)$ is the monochromatic induced copy of P in P' that exists by Lemma 9.3.3.

We claim that $G := G^m$ satisfies the assertion of the theorem. So let a 2-colouring of the edges of G be given. By the construction of G^m from G^{m-1}, we can find in G^m an induced copy of G^{m-1} such that for $e_m = ii'$ all edges between the ith and the i'th row have the same colour. In the same way, we find inside this copy of G^{m-1} an induced copy of G^{m-2} whose edges between the ith and the i'th row have the same colour also for $ii' = e_{m-1}$. Continuing in this way, we finally arrive at an induced copy of G^0 in G such that, for each pair (i,i'), all the edges between V_i^0 and $V_{i'}^0$ have the same colour. As shown earlier, this G^0 contains a monochromatic induced copy H_j of H. $\qquad\square$

9.4 Ramsey properties and connectivity

According to Ramsey's theorem, every large enough graph G has a very dense or a very sparse induced subgraph of given order, a K^r or $\overline{K^r}$. If we assume that G is connected, we can say a little more:

Proposition 9.4.1. *For every $r \in \mathbb{N}$ there is an $n \in \mathbb{N}$ such that every connected graph of order at least n contains K^r, $K_{1,r}$ or P^r as an induced subgraph.*

Proof. Let $d+1$ be the Ramsey number of r, let $n > 1 + rd^r$, and let G be a graph of order at least n. If G has a vertex v of degree at least $d+1$ then, by Theorem 9.1.1 and the choice of d, either $N(v)$ induces a K^r in G or $\{v\} \cup N(v)$ induces a $K_{1,r}$. On the other hand, if $\Delta(G) \leqslant d$, then by Proposition 1.3.3 G has radius $> r$, and hence contains two vertices at a distance $\geqslant r$. Any shortest path in G between these two vertices contains a P^r. □

The collection of sparse graphs in Proposition 9.4.1, comprising just stars and paths, is smallest possible in the following sense. If $\mathcal{G}$ is *any* set of connected graphs with the same property, i.e. such that, given $r \in \mathbb{N}$, every large enough connected graph contains either a K^r or an induced copy of a graph of order $\geqslant r$ from $\mathcal{G}$, then $\mathcal{G}$ contains arbitrarily large stars and arbitrarily long paths. Indeed, since stars and paths are themselves connected but have no connected subgraphs other than stars and paths, $\mathcal{G}$ must contain arbitrarily large stars and paths—and Proposition 9.4.1 tells us that we need no more than these.

In principle, we could look for a set like $\mathcal{G}$ for any assumed connectivity k. We could try to find a 'minimal' set (in the above sense) of 'typical' k-connected graphs, one such that every large k-connected graph has a large subgraph in this set. Unfortunately, $\mathcal{G}$ seems to grow very quickly with k: already for $k = 2$ it becomes thoroughly messy if (as for $k = 1$) we insist that those subgraphs be induced. By relaxing our specification of containment from 'induced subgraph' to 'topological minor' and on to 'minor', however, we can give some neat characterizations up to $k = 4$.

Proposition 9.4.2. *For every $r \in \mathbb{N}$ there is an $n \in \mathbb{N}$ such that every 2-connected graph of order at least n contains C^r or $K_{2,r}$ as a topological minor.*

Proof. Let d be the n associated with r in Proposition 9.4.1, and let G be a 2-connected graph with more than $1 + rd^r$ vertices. By Proposition 1.3.3, either G has a vertex of degree $> d$ or $\mathrm{diam}(G) \geqslant \mathrm{rad}(G) > r$.

In the latter case let $a, b \in G$ be two vertices at distance $> r$. By Menger's theorem (3.3.5), G contains two independent a–b paths. These form a cycle of length $> r$.

(1.3.3)
(3.3.5)

Assume now that G has a vertex v of degree $> d$. Since G is 2-connected, $G - v$ is connected and thus has a spanning tree; let T be a minimal tree in $G - v$ that contains all the neighbours of v. Then every leaf of T is a neighbour of v. By the choice of d, either T has a vertex of degree $\geqslant r$ or T contains a path of length $\geqslant r$, without loss of generality linking two leaves. Together with v, such a path forms a cycle of length $\geqslant r$. A vertex u of degree $\geqslant r$ in T can be joined to v by r independent paths through T, to form a $TK_{2,r}$. $\square$

Theorem 9.4.3. (Oporowski, Oxley & Thomas 1993)
For every $r \in \mathbb{N}$ there is an $n \in \mathbb{N}$ such that every 3-connected graph of order at least n contains a wheel of order r or a $K_{3,r}$ as a minor.

Let us call a graph of the form $C^n * \overline{K^2}$ ($n \geqslant 4$) a *double wheel*, the 1-skeleton of a triangulation of the cylinder as in Fig. 9.4.1 a *crown*, and the 1-skeleton of a triangulation of the Möbius strip a *Möbius crown*.

Fig. 9.4.1. A crown and a Möbius crown

Theorem 9.4.4. (Oporowski, Oxley & Thomas 1993)
For every $r \in \mathbb{N}$ there is an $n \in \mathbb{N}$ such that every 4-connected graph with at least n vertices has a minor of order $\geqslant r$ that is a double wheel, a crown, a Möbius crown, or a $K_{4,s}$.

Note that the minors occurring in Theorems 9.4.3 and 9.4.4 are themselves 3- and 4-connected, respectively, and are not minors of one another. Thus in each case, the collection of minors is minimal in the sense discussed earlier.

Exercises

1.$^-$ Determine the Ramsey number $R(3)$.

2.   Deduce the case $k = 2$ (but c arbitrary) of Theorem 9.1.4 directly from Theorem 9.1.1.

      (Hint. Induction on c.)

3.$^+$ Construct a graph on $\mathbb{R}$ that has neither a complete nor an edgeless induced subgraph on $|\mathbb{R}| = 2^{\aleph_0}$ vertices. (So Ramsey's theorem does not extend to uncountable sets.)

      (Hint. Choose a well-ordering of $\mathbb{R}$, and compare it with the natural ordering. Use the fact that any countable union of countable sets is countable.)

4. Sketch a proof of the following theorem of Erdős and Szekeres: for every
    $k \in \mathbb{N}$ there is an $n \in \mathbb{N}$ such that among any n points in the plane,
    no three of them collinear, there are k points spanning a convex k-gon,
    i.e. such that none of them lies in the convex hull of the others.

    (Hint. Use the fact that $n \geqslant 4$ points span a convex polygon if and
    only if every four of them do.)

5. Prove the following result of Schur: for every $k \in \mathbb{N}$ there is an $n \in \mathbb{N}$
    such that, for every partition of $\{1, \ldots, n\}$ into k sets, at least one of
    the subsets contains numbers x, y, z such that $x + y = z$.

6. Let $(X, \leqslant)$ be a totally ordered set, and let $G = (V, E)$ be the graph
    on $V := [X]^2$ with $E := \{(x, y)(x', y') \mid x < y = x' < y'\}$.
    - (i) Show that G contains no triangle.
    - (ii) Show that $\chi(G)$ will get arbitrarily large if $|X|$ is chosen large
      enough.

    (Hint for (ii). $R(2, k, 3)$)

7. A family of sets is called a Δ-system if every two of the sets have the
    same intersection. Show that every infinite family of sets of the same
    finite cardinality contains an infinite Δ-system.

8. Prove the following weakening of Scott's Theorem 8.1.5: for every $r \in \mathbb{N}$
    and every tree T there exists a $k \in \mathbb{N}$ such that every graph G with
    $\chi(G) \geqslant k$ and $\omega(G) < r$ contains a subdivision of T in which no two
    branch vertices are adjacent in G (unless they are adjacent in T).

9. Use the infinity lemma to show that, given $k \in \mathbb{N}$, a countably infi-
    nite graph is k-colourable (in the sense of Chapter 5) if all its finite
    subgraphs are k-colourable.

10.[+] Let $m, n \in \mathbb{N}$, and assume that $m - 1$ divides $n - 1$. Show that every
    tree T of order m satisfies $R(T, K_{1,n}) = m + n - 1$.

    (Hint. Imitate the proof of Proposition 9.2.1.)

11.[+] Prove that $2^c < R(2, c, 3) \leqslant 3c!$ for every $c \in \mathbb{N}$.

    (Hint. Induction on c.)

12.[−] Derive the statement (∗) in the first proof of Theorem 9.3.1 from the
    theorem itself, i.e. show that (∗) is only formally stronger than the
    theorem.

13. Show that the Ramsey graph G for H constructed in the second proof
    of Theorem 9.3.1 does indeed satisfy $\omega(G) = \omega(H)$.

14. Show that, given any two graphs H_1 and H_2, there exists a graph
    $G = G(H_1, H_2)$ such that, for every vertex-colouring of G with colours
    1 and 2, there is either an induced copy of H_1 coloured 1 or an induced
    copy of H_2 coloured 2 in G.

    (Hint. Apply induction as in the first proof of Theorem 9.3.1. For
    the induction step, construct $G(H_1, H_2)$ from the disjoint union of
    $G(H_1, H_2')$ and $G(H_1', H_2)$ by joining some new vertices in a suitable
    way.)

15. Show that every infinite connected graph contains an infinite path or an infinite star.

    (Hint. Infinity lemma.)

16.⁻ The K^r from Ramsey's theorem, last sighted in Proposition 9.4.1, conspicuously fails to make an appearance from Proposition 9.4.2 onwards. Can it be excused?

Notes

Due to increased interaction with research on random and pseudo-random[4] structures (the latter being provided, for example, by the regularity lemma), the Ramsey theory of graphs has recently seen a period of major activity and advance. Theorem 9.2.2 is an early example of this development.

For the more classical approach, the introductory text by R.L. Graham, B.L. Rothschild & J.H. Spencer, *Ramsey Theory* (2nd edn.), Wiley 1990, makes stimulating reading. This book includes a chapter on graph Ramsey theory, but is not confined to it. A more recent general survey is given by J. Nešetřil in the *Handbook of Combinatorics* (R.L. Graham, M. Grötschel & L. Lovász, eds.), North-Holland 1995. The Ramsey theory of infinite sets forms a substantial part of combinatorial set theory, and is treated in depth in P. Erdős, A. Hajnal, A. Máté & R. Rado, *Combinatorial Set Theory*, North-Holland 1984. An attractive collection of highlights from various branches of Ramsey theory, including applications in algebra, geometry and point-set topology, is offered in B. Bollobás, *Graph Theory*, Springer GTM 63, 1979.

König's infinity lemma, or *König's lemma* for short, is contained in the first-ever book on the subject of graph theory: D. König, *Theorie der endlichen und unendlichen Graphen*, Akademische Verlagsgesellschaft, Leipzig 1936. The *compactness* technique for deducing finite results from infinite (or vice versa), hinted at in Section 9.1, is less mysterious than it sounds. As long as 'infinite' means 'countably infinite', it is precisely the art of applying the infinity lemma (as in the proof of Theorem 9.1.4), no more no less. For larger infinite sets, the same argument becomes equivalent to the well-known theorem of Tychonov that arbitrary products of compact spaces are compact— which has earned the compactness argument its name. Details can be found in Ch. 6, Thm. 10 of Bollobás, and in Graham, Rothschild & Spencer, Ch. 1, Thm. 4. Another frequently used version of the general compactness argument is *Rado's selection lemma*; see A. Hajnal's chapter on infinite combinatorics in the Handbook cited above.

Theorem 9.2.2 is due to V. Chvátal, V. Rödl, E. Szemerédi & W.T. Trotter, The Ramsey number of a graph with bounded maximum degree, *J. Combin. Theory B* **34** (1983), 239–243. Our proof follows the sketch in J. Komlós & M. Simonovits, Szemerédi's Regularity Lemma and its applications in graph theory, in (D. Miklos, V.T. Sós & T. Szőnyi, eds.) *Paul Erdős is 80*, Vol. 2,

---

[4] Concrete graphs whose structure resembles the structure expected of a random graph are called *pseudo-random*. For example, the bipartite graphs spanned by an ε-regular pair of vertex sets in a graph are pseudo-random.

Proc. Colloq. Math. Soc. János Bolyai (1996). The theorem marks a break-through towards a conjecture of Burr and Erdős (1975), which asserts that the Ramsey numbers of graphs with bounded *average* degree are linear: for every $d \in \mathbb{N}$, the conjecture says, there exists a constant c such that $R(H) \leqslant c|H|$ for all graphs H with $d(H) \leqslant d$. This conjecture has been verified also for the class of planar graphs (Chen & Schelp 1993) and, more generally, for the class of graphs not containing K^r (for any fixed r) as a topological minor (Rödl & Thomas 1996). See Nešetřil's Handbook chapter for references.

Our first proof of Theorem 9.3.1 is based on W. Deuber, A generalization of Ramsey's theorem, in (A. Hajnal, R. Rado & V.T. Sós, eds.) *Infinite and finite sets*, North-Holland 1975. The same volume contains the alternative proof of this theorem by Erdős, Hajnal and Pósa. Rödl proved the same result in his MSc thesis at the Charles University, Prague, in 1973. Our second proof of Theorem 9.3.1, which preserves the clique number of H for G, is due to J. Nešetřil & V. Rödl, A short proof of the existence of restricted Ramsey graphs by means of a partite construction, *Combinatorica* **1** (1981), 199–202.

The two theorems in Section 9.4 are due to B. Oporowski, J. Oxley & R. Thomas, Typical subgraphs of 3- and 4-connected graphs, *J. Combin. Theory B* **57** (1993), 239–257.

10   Hamilton Cycles

In Chapter 1.8 we briefly discussed the problem of when a graph contains an Euler tour, a closed walk traversing every edge exactly once. The simple Theorem 1.8.1 solved that problem quite satisfactorily. Let us now ask the analogous question for vertices: when does a graph G contain a closed walk that contains every vertex of G exactly once? If $|G| \geqslant 3$, then any such walk is a cycle: a *Hamilton cycle* of G. If G has a Hamilton cycle, it is called *hamiltonian*. Similarly, a path in G containing every vertex of G is a *Hamilton path*.

Hamilton cycle

Hamilton path

To determine whether or not a given graph has a Hamilton cycle is much harder than deciding whether it is Eulerian, and no good characterization[1] is known of the graphs that do. We shall begin this chapter by presenting the standard sufficient conditions for the existence of a Hamilton cycle (Sections 10.1 and 10.2). The rest of the chapter is then devoted to the beautiful theorem of Fleischner that the 'square' of every 2-connected graph has a Hamilton cycle. This is one of the main results in the field of Hamilton cycles. The simple proof we present (due to Říha) is still a little longer than other proofs in this book, but not difficult.

10.1 Simple sufficient conditions

What kind of condition might be sufficient for the existence of a Hamilton cycle in a graph G? Purely global assumptions, like high edge density, will not be enough: we cannot do without the local property that every vertex has at least two neighbours. But neither is any large (but constant) minimum degree sufficient: it is easy to find graphs without a Hamilton cycle whose minimum degree exceeds any given constant bound.

---

[1] The notion of a 'good characterization' can be made precise; see the introduction to Chapter 12.5 and the notes for Chapter 12.

The following classic result derives its significance from this background:

Theorem 10.1.1. (Dirac 1952)
Every graph with $n \geqslant 3$ vertices and minimum degree at least $n/2$ has a Hamilton cycle.

Proof. Let $G = (V, E)$ be a graph with $|G| = n \geqslant 3$ and $\delta(G) \geqslant n/2$. Then G is connected: otherwise, the degree of any vertex in the smallest component C of G would be less than $|C| \leqslant n/2$.

Let $P = x_0 \ldots x_k$ be a longest path in G. By the maximality of P, all the neighbours of x_0 and all the neighbours of x_k lie on P. Hence at least $n/2$ of the vertices $x_0, \ldots, x_{k-1}$ are adjacent to x_k, and at least $n/2$ of these same $k < n$ vertices x_i are such that $x_0 x_{i+1} \in E$. By the pigeon hole principle, there is a vertex x_i that has both properties, so we have $x_0 x_{i+1} \in E$ and $x_i x_k \in E$ for some $i < k$ (Fig. 10.1.1).

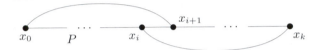

Fig. 10.1.1. Finding a Hamilton cycle in the proof Theorem 10.1.1

We claim that the cycle $C := x_0 x_{i+1} P x_k x_i P x_0$ is a Hamilton cycle of G. Indeed, since G is connected, C would otherwise have a neighbour in $G - C$, which could be combined with a spanning path of C into a path longer than P. $\square$

Theorem 10.1.1 is best possible in that we cannot replace the bound of $n/2$ with $\lfloor n/2 \rfloor$: if n is odd and G is the union of two copies of $K^{\lceil n/2 \rceil}$ meeting in one vertex, then $\delta(G) = \lfloor n/2 \rfloor$ but $\kappa(G) = 1$, so G cannot have a Hamilton cycle. In other words, the high level of the bound of $\delta \geqslant n/2$ is needed to ensure, if nothing else, that G is 2-connected: a condition just as trivially necessary for hamiltonicity as a minimum degree of at least 2. It would seem, therefore, that prescribing some high (constant) value for κ rather than for δ stands a better chance of implying hamiltonicity. However, this is not so; indeed it is easy for any k to modify the above set of examples into arbitrarily large graphs that are k-connected (and hence contain long cycles in terms of k; cf. Ex. 14, Ch. 3), but whose circumference is bounded above by a function of k.

There is another invariant with a similar property: a low independence number $\alpha(G)$ ensures that G has long cycles (Ex. 13, Ch. 5), though not necessarily a Hamilton cycle. Put together, however, the two assumptions of high connectivity and low independence number surprisingly complement each other to produce a sufficient condition for hamiltonicity:

Proposition 10.1.2. *Every graph G with $|G| \geqslant 3$ and $\kappa(G) \geqslant \alpha(G)$ has a Hamilton cycle.*

Proof. Put $\kappa(G) =: k$, and let C be a longest cycle in G. Enumerate the vertices of C cyclically, say as $V(C) = \{ v_i \mid i \in \mathbb{Z}_n \}$ with $v_i v_{i+1} \in E(C)$ for all $i \in \mathbb{Z}_n$. If C is not a Hamilton cycle, pick a vertex $v \in G - C$ and a v–C fan $\mathcal{F} = \{ P_i \mid i \in I \}$ in G, where $I \subseteq \mathbb{Z}_n$ and each P_i ends in v_i. Let $\mathcal{F}$ be chosen with maximum cardinality; then $vv_j \notin E(G)$ for any $j \notin I$, and

$$|\mathcal{F}| \geqslant \min \{ k, |C| \} \tag{1}$$

(3.3.3)

k

by Menger's theorem (3.3.3).

For every $i \in I$, we have $i+1 \notin I$: otherwise, $(C \cup P_i \cup P_{i+1}) - v_i v_{i+1}$ would be a cycle longer than C (Fig. 10.1.2, left). Thus $|\mathcal{F}| < |C|$, and hence $|I| = |\mathcal{F}| \geqslant k$ by (1). Furthermore, $v_{i+1} v_{j+1} \notin E(G)$ for all $i, j \in I$, as otherwise $(C \cup P_i \cup P_j) + v_{i+1} v_{j+1} - v_i v_{i+1} - v_j v_{j+1}$ would be a cycle longer than C (Fig. 10.1.2, right). Hence $\{ v_{i+1} \mid i \in I \} \cup \{ v \}$ is a set of $k+1$ or more independent vertices in G, contradicting $\alpha(G) \leqslant k$. $\qquad\square$

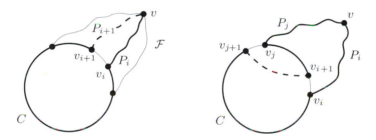

Fig. 10.1.2. Two cycles longer than C

It may come as a surprise to learn that hamiltonicity for planar graphs is related to the four colour problem. As we noted in Chapter 6.6, the four colour theorem is equivalent to the non-existence of a planar snark, i.e. to the assertion that every bridgeless planar cubic graph has a 4-flow. It is easily checked that 'bridgeless' can be replaced with '3-connected' in this assertion, and that every hamiltonian graph has a 4-flow (Ex. 12, Ch. 6). For a proof of the four colour theorem, therefore, it would suffice to show that every 3-connected planar cubic graph has a Hamilton cycle!

Unfortunately, this is not the case: the first counterexample was found by Tutte in 1946. Ten years later, Tutte proved the following deep theorem as a best possible weakening:

Theorem 10.1.3. (Tutte 1956)
Every 4-connected planar graph has a Hamilton cycle.

10.2 Hamilton cycles and degree sequences

Historically, Dirac's theorem formed the point of departure for the discovery of a series of weaker and weaker degree conditions, all sufficient for hamiltonicity. The development culminated in a single theorem that encompasses all the earlier results: the theorem we shall prove in this section.

degree-sequence   If G is a graph with n vertices and degrees $d_1 \leqslant \ldots \leqslant d_n$, then the n-tuple $(d_1, \ldots, d_n)$ is called the *degree-sequence* of G. Note that this sequence is unique, even though G has several vertex enumerations giving rise to its degree sequence. Let us call an arbitrary integer sequence *hamiltonian sequence* $(a_1, \ldots, a_n)$ *hamiltonian* if every graph with n vertices and a degree sequence pointwise greater than $(a_1, \ldots, a_n)$ is hamiltonian. (A sequence *pointwise greater* $(d_1, \ldots, d_n)$ is *pointwise greater* than $(a_1, \ldots, a_n)$ if $d_i \geqslant a_i$ for all i.)

The following theorem characterizes all hamiltonian sequences:

Theorem 10.2.1. (Chvátal 1972)
An integer sequence $(a_1, \ldots, a_n)$ such that $0 \leqslant a_1 \leqslant \ldots \leqslant a_n < n$ and $n \geqslant 3$ is hamiltonian if and only if the following holds for every $i < n/2$:

$$a_i \leqslant i \Rightarrow a_{n-i} \geqslant n - i.$$

$(a_1, \ldots, a_n)$ *Proof.* Let $(a_1, \ldots, a_n)$ be an arbitrary integer sequence such that $0 \leqslant a_1 \leqslant \ldots \leqslant a_n < n$ and $n \geqslant 3$. We first assume that this sequence satisfies the condition of the theorem and prove that it is hamiltonian. Suppose not; then there exists a graph $G = (V, E)$ with a degree sequence $(d_1, \ldots, d_n)$ $(d_1, \ldots, d_n)$ such that

$$d_i \geqslant a_i \qquad \text{for all } i \qquad (1)$$

$G = (V, E)$ but G has no Hamilton cycle. Let G be chosen with the maximum number of edges, and let $(v_1, \ldots, v_n)$ be an enumeration of V with $d(v_i) = d_i$ for all i. By (1), our assumptions for $(a_1, \ldots, a_n)$ transfer to $(d_1, \ldots, d_n)$, i.e.,

$$d_i \leqslant i \Rightarrow d_{n-i} \geqslant n - i \qquad \text{for all } i < n/2. \qquad (2)$$

x, y Let x, y be distinct and non-adjacent vertices in G, such that $d(x) \leqslant d(y)$ and $d(x) + d(y)$ is maximum. One easily checks that the degree sequence of $G + xy$ is pointwise greater than $(d_1, \ldots, d_n)$, and hence than $(a_1, \ldots, a_n)$. Hence, by the maximality of G, the new edge xy lies on a Hamilton cycle H of $G + xy$. Then $H - xy$ is a Hamilton path $x_1, \ldots, x_n$ $x_1, \ldots, x_n$ in G, with $x_1 = x$ and $x_n = y$ say.

As in the proof of Dirac's theorem, we now consider the index sets

$$I := \{\, i \mid xx_{i+1} \in E \,\} \quad \text{and} \quad J := \{\, j \mid x_j y \in E \,\}.$$

Then $I \cup J \subseteq \{1, \ldots, n-1\}$, and $I \cap J = \emptyset$ because G has no Hamilton cycle. Hence

$$d(x) + d(y) = |I| + |J| < n, \tag{3}$$

so $h := d(x) < n/2$ by the choice of x. h

Since $x_i y \notin E$ for all $i \in I$, all these x_i were candidates for the choice of x (together with y). Our choice of $\{x, y\}$ with $d(x) + d(y)$ maximum thus implies that $d(x_i) \leqslant d(x)$ for all $i \in I$. Hence G has at least $|I| = h$ vertices of degree at most h, so $d_h \leqslant h$. By (2), this implies that $d_{n-h} \geqslant n - h$, i.e. the $h+1$ vertices $v_{n-h}, \ldots, v_n$ all have degree at least $n - h$. Since $d(x) = h$, one of these vertices, z say, is not adjacent z
to x. Since

$$d(x) + d(z) \geqslant h + (n - h) = n,$$

this contradicts the choice of x and y by (3).

For the converse implication we now assume that our sequence $(a_1, \ldots, a_n)$ fails to satisfy the condition of the theorem, so there exists an $h < n/2$ with h

$$a_h \leqslant h \quad \text{and} \quad a_{n-h} < n - h. \tag{4}$$

We shall exhibit a graph G with vertices $v_1, \ldots, v_n$ and degree sequence $v_1, \ldots, v_n$
$(d_1, \ldots, d_n)$ such that

$$d(v_i) = d_i \geqslant a_i \qquad \text{for all } i \tag{5}$$

but G has no Hamilton cycle. Set

$$E(G) := \{ v_i v_j \mid i, j > h \} \cup \{ v_i v_j \mid i \leqslant h; \; j > n - h \}.   G$$

In other words, G is the union of a K^{n-h} on the vertices $v_{h+1}, \ldots, v_n$ and a $K_{h,h}$ with partition sets $\{v_1, \ldots, v_h\}$ and $\{v_{n-h+1}, \ldots, v_n\}$ (Fig. 10.2.1).

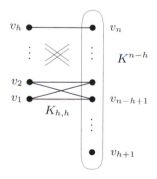

Fig. 10.2.1. A graph G as in (5)

As one easily checks, every cycle in G that contains $v_1, \ldots, v_h$ lies inside the $K_{h,h}$. But $h < n/2$ and hence $2h < n$, so G has some further vertices outside the $K_{h,h}$. Therefore G has no Hamilton cycle.

Let us now verify (5). Clearly, G has the degree sequence

$$\Big(\underbrace{h, \ldots, h}_{h \text{ times}}, \underbrace{n-h-1, \ldots, n-h-1}_{n-2h \text{ times}}, \underbrace{n-1, \ldots, n-1}_{h \text{ times}} \Big).$$

Hence by (4),

$$a_i \leqslant a_h \leqslant h = d_i \qquad\qquad (i = 1, \ldots, h)$$
$$a_i \leqslant a_{n-h} \leqslant n-h-1 = d_i \qquad (i = h+1, \ldots, n-h)$$
$$a_i \leqslant n-1 = d_i \qquad\qquad (i = n-h+1, \ldots, n)$$

as desired.   □

By applying Theorem 10.2.1 to $G * K^1$, one can easily prove the following adaptation of the theorem to Hamilton paths. Let an integer sequence be called *path-hamiltonian* if every graph with a pointwise greater degree sequence has a Hamilton path.

Corollary 10.2.2. *An integer sequence $(a_1, \ldots, a_n)$ such that $n \geqslant 2$ and $0 \leqslant a_1 \leqslant \ldots \leqslant a_n < n$ is path-hamiltonian if and only if every $i \leqslant n/2$ is such that $a_i < i \Rightarrow a_{n+1-i} \geqslant n - i$.*   □

10.3 Hamilton cycles in the square of a graph

G^d

Given a graph G and a positive integer d, we denote by G^d the graph on $V(G)$ in which two vertices are adjacent if and only if they have distance at most d in G. Clearly, $G = G^1 \subseteq G^2 \subseteq \ldots$ Our goal in this section is to prove the following fundamental result:

Theorem 10.3.1. (Fleischner 1974)
If G is a 2-connected graph, then G^2 has a Hamilton cycle.

We begin with three simple lemmas. Let us say that an edge $e \in G^2$
bridges   *bridges* a vertex $v \in G$ if its ends are neighbours of v in G.

Lemma 10.3.2. *Let $P = v_0 \ldots v_k$ be a path ($k \geqslant 1$), and let G be the graph obtained from P by adding two vertices u, w, together with the edges uv_1 and wv_k (Fig. 10.3.1).*

(i) *P^2 contains a path Q from v_0 to v_1 with $V(Q) = V(P)$ and $v_{k-1}v_k \in E(Q)$, such that each of the vertices $v_1, \ldots, v_{k-1}$ is bridged by an edge of Q.*

(ii) *G^2 contains disjoint paths Q from v_0 to v_k and Q' from u to w, such that $V(Q) \cup V(Q') = V(G)$ and each of the vertices $v_1, \ldots, v_k$ is bridged by an edge of Q or Q'.*

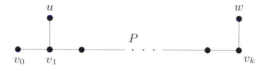

Fig. 10.3.1. The graph G in Lemma 10.3.2

Proof. (i) If k is even, let $Q := v_0v_2 \ldots v_{k-2}v_kv_{k-1}v_{k-3} \ldots v_3v_1$. If k is odd, let $Q := v_0v_2 \ldots v_{k-1}v_kv_{k-2} \ldots v_3v_1$.

(ii) If k is even, let $Q := v_0v_2 \ldots v_{k-2}v_k$; if k is odd, let $Q := v_0v_1v_3 \ldots v_{k-2}v_k$. In both cases, let Q' be the u–w path on the remaining vertices of G^2. □

Lemma 10.3.3. *Let $G = (V, E)$ be a cubic multigraph with a Hamilton cycle C. Let $e \in E(C)$ and $f \in E \smallsetminus E(C)$ be edges with a common end v (Fig. 10.3.2). Then there exists a closed walk in G that traverses e once, every other edge of C once or twice, and every edge in $E \smallsetminus E(C)$ once. This walk can be chosen to contain the triple (e, v, f), that is, it traverses e in the direction of v and then leaves v by the edge f.*

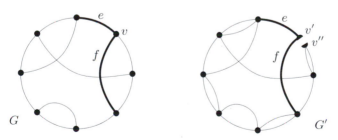

Fig. 10.3.2. The multigraphs G and G' in Lemma 10.3.3

Proof. By Proposition 1.2.1, C has even length. Replace every other edge of C by a double edge, in such a way that e does not get replaced. In the arising 4-regular multigraph G', split v into two vertices v', v'', making v' incident with e and f, and v'' incident with the other two edges at v. By Theorem 1.8.1 this multigraph has an Euler tour, which induces the desired walk in G. □

(1.2.1)
(1.8.1)

Lemma 10.3.4. *For every 2-connected graph G and $x \in V(G)$, there is a cycle $C \subseteq G$ that contains x as well as a vertex $y \neq x$ with $N_G(y) \subseteq V(C)$.*

Proof. If G has a Hamilton cycle, there is nothing more to show. If not, let $C' \subseteq G$ be any cycle containing x; such a cycle exists, since G is 2-connected. Let D be a component of $G - C'$. Assume that C' and D are chosen so that $|D|$ is minimum. Since G is 2-connected, D has at least two neighbours on C'. Then C' contains a path P between two such neighbours u and v, whose interior $\mathring{P}$ does not contain x and has no neighbour in D (Fig. 10.3.3). Replacing P in C' by a u–v path through D, we obtain a cycle C that contains x and a vertex $y \in D$. If y had a neighbour z in $G - C$, then z would lie in a component $D' \subsetneq D$ of $G - C$, contradicting the choice of C' and D. Hence all the neighbours of y lie on C, and C satisfies the assertion of the lemma. $\square$

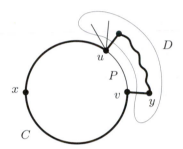

Fig. 10.3.3. The proof of Lemma 10.3.4

Proof of Theorem 10.3.1. We show by induction on $|G|$ that, given any vertex $x^* \in G$, there is a Hamilton cycle H in G^2 with the following property:

$$\text{Both edges of } H \text{ at } x^* \text{ lie in } G. \qquad (*)$$

x^*
C
y^*

For $|G| = 3$, we have $G = K^3$ and the assertion is trivial. So let $|G| \geqslant 4$, assume the assertion for graphs of smaller order, and let $x^* \in V(G)$ be given. By Lemma 10.3.4, there is a cycle $C \subseteq G$ that contains both x^* and a vertex $y^* \neq x^*$ whose neighbours in G all lie on C.

If C is a Hamilton cycle of G, there is nothing to show; so assume that $G - C \neq \emptyset$. Consider a component D of $G - C$. Let $\tilde{D}$ denote the graph $G/(G-D)$ obtained from G by contracting $G - D$ into a new vertex $\tilde{x}$. If $|D| = 1$, set $\mathcal{P}(D) := \{D\}$. If $|D| > 1$, then $\tilde{D}$ is again 2-connected. Hence, by the induction hypothesis, $\tilde{D}^2$ has a Hamilton cycle $\tilde{C}$ whose edges at $\tilde{x}$ both lie in $\tilde{D}$. Note that the path $\tilde{C} - \tilde{x}$ may have some edges that do not lie in G^2: edges joining two neighbours of $\tilde{x}$ that have no common neighbour in G (and are themselves non-adjacent in G). Let $\tilde{E}$ denote the set of these edges, and let $\mathcal{P}(D)$ denote the set

$\mathcal{P}(D)$

$\tilde{C}$

$\mathcal{P}(D)$

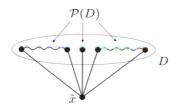

Fig. 10.3.4. $\mathcal{P}(D)$ consists of three paths, one of which is trivial

of components of $(\tilde{C} - \tilde{x}) - \tilde{E}$; this is a set of paths in G^2 whose ends are adjacent to $\tilde{x}$ in $\tilde{D}$ (Fig. 10.3.4).

Let $\mathcal{P}$ denote the union of the sets $\mathcal{P}(D)$ over all components D of $G - C$. Clearly, $\mathcal{P}$ has the following properties: $\qquad \mathcal{P}$

> The elements of $\mathcal{P}$ are pairwise disjoint paths in G^2 avoiding C, and $V(G) = V(C) \cup \bigcup_{P \in \mathcal{P}} V(P)$. Every end y of a path $P \in \mathcal{P}$ has a neighbour on C in G; we choose such a neighbour and call it the *foot* of P at y. $\qquad (1)$

$\qquad\qquad\qquad\qquad\qquad\qquad\qquad\qquad\qquad\qquad\qquad\qquad$ *foot*

If $P \in \mathcal{P}$ is trivial, then P has exactly one foot. If P is non-trivial, then P has a foot at each of its ends. These two feet need not be distinct, however; so any non-trivial P has either one or two feet.

We shall now modify $\mathcal{P}$ a little, preserving the properties summarized under (1); no properties of $\mathcal{P}$ other than those will be used later in the proof. If a vertex of C is a foot of two distinct paths $P, P' \in \mathcal{P}$, say at $y \in P$ and at $y' \in P'$, then yy' is an edge and $Pyy'P'$ is a path in G^2; we replace P and P' in $\mathcal{P}$ by this path. We repeat this modification of $\mathcal{P}$ until the following holds:

$$\text{No vertex of } C \text{ is a foot of two distinct paths in } \mathcal{P}. \qquad (2)$$

For $i = 1, 2$ let $\mathcal{P}_i \subseteq \mathcal{P}$ denote the set of all paths in $\mathcal{P}$ with exactly i $\qquad \mathcal{P}_1, \mathcal{P}_2$ feet, and let $X_i \subseteq V(C)$ denote the set of all feet of paths in $\mathcal{P}_i$. Then $\qquad X_1, X_2$ $X_1 \cap X_2 = \emptyset$ by (2), and $y^* \notin X_1 \cup X_2$.

Let us also simplify G a little; again, these changes will affect neither the paths in $\mathcal{P}$ nor the validity of (1) and (2). First, we shall assume from now on that all elements of $\mathcal{P}$ are paths in G itself, not just in G^2. This assumption may give us some additional edges for G^2, but we shall not use these in our construction of the desired Hamilton cycle H. (In particular, H will contain all the paths from $\mathcal{P}$ whole, as subpaths.) Thus if H lies in G^2 and satisfies $(*)$ for the modified version of G, it will do so also for the original. For every $P \in \mathcal{P}$, we further delete all P–C edges in G except those between the ends of P and its corresponding feet. Finally, we delete all chords of C in G. We are thus assuming without loss of generality:

> The only edges of G between C and a path $P \in \mathcal{P}$ are
> the two edges between the ends of P and its corresponding
> feet. (If $|P| = 1$, these two edges coincide.) The only edges   (3)
> of G with both ends on C are the edges of C itself.

Our goal is to construct the desired Hamilton cycle H of G^2 from the paths in $\mathcal{P}$ and suitable paths in C^2. As a first approximation, we shall construct a closed walk W in the graph

$\tilde{G}$
$$\tilde{G} := G - \bigcup \mathcal{P}_1 \,,$$

a walk that will already satisfy a $(*)$-type condition and traverse every path in $\mathcal{P}_2$ exactly once. Later, we shall modify W so that it passes through every vertex of C exactly once and, finally, so as to include the paths from $\mathcal{P}_1$. For the construction of W we assume that $\mathcal{P}_2 \neq \emptyset$; the case of $\mathcal{P}_2 = \emptyset$ is much simpler and will be treated later.

We start by choosing a fixed cyclic orientation of C, a bijection $i \mapsto v_i$ from $\mathbb{Z}_{|C|}$ to $V(C)$ with $v_i v_{i+1} \in E(C)$ for all $i \in \mathbb{Z}_{|C|}$. Let us think of this orientation as clockwise; then every vertex $v_i \in C$ has a *right*

v^+, *right*   neighbour $v_i^+ := v_{i+1}$ and a *left* neighbour $v_i^- := v_{i-1}$. Accordingly, the
v^-, *left*   edge $v^- v$ lies to the *left* of v, the edge $v v^+$ lies on its *right*, and so on.

A non-trivial path $P = v_i v_{i+1} \dots v_{j-1} v_j$ in C such that $V(P) \cap X_2 =$
interval $\{ v_i, v_j \}$ will be called an *interval*, with *left end* v_i and *right end* v_j. Thus, C is the union of $|X_2| = 2 |\mathcal{P}_2|$ intervals. As usual, we write $P =:$
$[v, w]$ *etc.* $[v_i, v_j]$ and set $(v_i, v_j) := \mathring{P}$ as well as $[v_i, v_j) := P\mathring{v}_j$ and $(v_i, v_j] := \mathring{v}_i P$. For intervals $[u, v]$ and $[v, w]$ with a common end v we say that $[u, v]$ lies to the *left* of $[v, w]$, and $[v, w]$ lies to the *right* of $[u, v]$. We denote
I^*, P^* the unique interval $[v, w]$ with $x^* \in (v, w]$ as I^*, the path in $\mathcal{P}_2$ with
Q^* foot w as P^*, and the path $I^* w P^*$ as Q^*.

For the construction of W, we may think of $\tilde{G}$ as a multigraph M on X_2 whose edges are the intervals on C and the paths in $\mathcal{P}_2$ (with their feet as ends). By (2), M is cubic, so we may apply Lemma 10.3.3 with
W $e := I^*$ and $f := P^*$. The lemma provides us with a closed walk W in $\tilde{G}$ which traverses I^* once, every other interval of C once or twice, and every path in $\mathcal{P}_2$ once. Moreover, W contains Q^* as a subpath. The two edges at x^* of this path lie in G; in this sense, W already satisfies $(*)$.

Let us now modify W so that W passes through every vertex of C exactly once. Simultaneously, we shall prepare for the later inclusion of
$e(v)$ the paths from $\mathcal{P}_1$ by defining a map $v \mapsto e(v)$ that is injective on X_1 and assigns to every $v \in X_1$ an edge $e(v)$ of the modified W with the following property:

> The edge $e(v)$ either bridges v or is incident with it. In the
> latter case, $e(v) \in C$ and $e(v) \neq v x^*$. $(**)$

For simplicity, we shall define the map $v \mapsto e(v)$ on all of $V(C) \smallsetminus X_2$, a set that includes X_1 by (2). To ensure injectivity on X_1, we only have to make sure that no edge $vw \in C$ is chosen both as $e(v)$ and as $e(w)$. Indeed, since $|X_1| \geqslant 2$ if injectivity is a problem, and $\mathcal{P}_2 \neq \emptyset$ by assumption, we have $|C - y^*| \geqslant |X_1| + 2\,|\mathcal{P}_2| \geqslant 4$ and hence $|C| \geqslant 5$; thus, no edge of G^2 can bridge more than one vertex of C, or bridge a vertex of C and lie on C at the same time.

For our intended adjustments of W at the vertices of C, we consider the intervals of C one at a time. By definition of W, every interval is of one of the following three types:

Type 1: W traverses I once;

Type 2: W traverses I twice, in one direction and back immediately afterwards (formally: W contains a triple (e, x, e) with $x \in X_2$ and $e \in E(I)$);

Type 3: W traverses I twice, on separate occasions (i.e., there is no triple as above).

By definition of W, the interval I^* is of type 1. The vertex x in the definition of a type 2 interval will be called the *dead end* of that interval. Finally, since Q^* is a subpath of W and W traverses both I^* and P^* only once, we have:

The interval to the right of I^* is of type 2 and has its dead end on the left. (4)

Consider a fixed interval $I = [x_1, x_2]$. Let y_1 be the neighbour of x_1, and y_2 the neighbour of x_2 on a path in $\mathcal{P}_2$. Let I^- denote the interval to the left of I.

Suppose first that I is of type 1. We then leave W unchanged on I. If $I \neq I^*$ we choose as $e(v)$, for each $v \in \mathring{I}$, the edge to the left of v. As $I^- \neq I^*$ by (4), and hence $x_1 \neq x^*$, these choices of $e(v)$ satisfy (∗∗). If $I = I^*$, we define $e(v)$ as the edge left of v if $v \in (x_1, x^*] \cap \mathring{I}$, and as the edge right of v if $v \in (x^*, x_2)$. These choices of $e(v)$ are again compatible with (∗∗).

Suppose now that I is of type 2. Assume first that x_2 is the dead end of I. Then W contains the walk $y_1 x_1 I x_2 I x_1 I^-$ (possibly in reverse order). We now apply Lemma 10.3.2 (i) with $P := y_1 x_1 I \mathring{x}_2$, and replace in W the subwalk $y_1 x_1 I x_2 I x_1$ by the y_1–x_1 path $Q \subseteq G^2$ of the lemma (Fig. 10.3.5). Then $V(Q) = V(P) \smallsetminus \{y_1, x_1\} = V(\mathring{I})$. The vertices

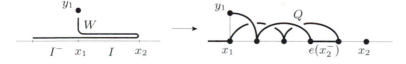

Fig. 10.3.5. How to modify W on an interval of type 2

$v \in (x_1, x_2^-)$ are each bridged by an edge of Q, which we choose as $e(v)$.
As $e(x_2^-)$ we choose the edge to the left of x_2^- (unless $x_2^- = x_1$). This
edge, too, lies on Q, by the lemma. Moreover, by (4) it is not incident with x^* (since x_2 is the dead end of I, by assumption) and hence
satisfies (**). The case that x_1 is the dead end of I can be treated in
the same way: using Lemma 10.3.2 (i), we replace in W the subwalk
$y_2 x_2 I x_1 I x_2$ by a y_2–x_2 path $Q \subseteq G^2$ with $V(\mathring{Q}) = V(\mathring{I})$, choose as $e(v)$
for $v \in (x_1^+, x_2)$ an edge of Q bridging v, and define $e(x_1^+)$ as the edge
to the right of x_1^+ (unless $x_1^+ = x_2$).

Suppose finally that I is of type 3. Since W traverses the edge $y_1 x_1$
only once and the interval I^- no more than twice, W contains $y_1 x_1 I$
and $I^- \cup I$ as subpaths, and I^- is of type 1. By (4), however, $I^- \neq I^*$.
Hence, when $e(v)$ was defined for the vertices $v \in I^-$, the rightmost edge
$x_1^- x_1$ of I^- was not chosen as $e(v)$ for any v, so we may now replace this
edge. Since W traverses I^+ no more than twice, it must traverse the
edge $x_2 y_2$ immediately after one of its two subpaths $y_1 x_1 I$ and $x_1^- x_1 I$.
Take the starting vertex of this subpath (y_1 or x_1^-) as the vertex u in
Lemma 10.3.2 (ii), and the other vertex in $\{y_1, x_1^-\}$ as v_0; moreover, set
$v_k := x_2$ and $w := y_2$. Then the lemma enables us to replace these two
subpaths of W between $\{y_1, x_1^-\}$ and $\{x_2, y_2\}$ by disjoint paths in G^2
(Fig. 10.3.6), and furthermore assigns to every vertex $v \in \mathring{I}$ an edge $e(v)$
of one of those paths, bridging v.

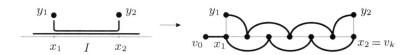

Fig. 10.3.6. A type 3 modification for the case $u = y_1$ and k odd

Following the above modifications, W is now a closed walk in $\tilde{G}^2$.
Let us check that, moreover, W contains every vertex of $\tilde{G}$ exactly once.
For vertices of the paths in $\mathcal{P}_2$ this is clear, because W still traverses every
such path once and avoids it otherwise. For the vertices of $C - X_2$, it
follows from the above modifications by Lemma 10.3.2. So how about
the vertices in X_2?

Let $x \in X_2$ be given, and let y be its neighbour on a path in $\mathcal{P}_2$. Let
I_1 denote the interval I that satisfied $yxI \subseteq W$ before the modification
of W, and let I_2 denote the other interval ending in x. If I_1 is of type 1,
then I_2 is of type 2 with dead end x. In this case, x was retained in W
when W was modified on I_1 but skipped when W was modified on I_2,
and is thus contained exactly once in W now. If I_1 is of type 2, then x
is not its dead end, and I_2 is of type 1. The subwalk of W that started
with yx and then went along I_1 and back, was replaced with a y–x path.
This path is now followed on W by the unchanged interval I_2, so in this
case too the vertex x is now contained in W exactly once. Finally, if I_1

is of type 3, then x was contained in one of the replacement paths Q, Q' from Lemma 10.3.2 (ii); as these paths were disjoint by the assertion of the lemma, x is once more left on W exactly once.

We have thus shown that W, after the modifications, is a closed walk in $\tilde{G}^2$ containing every vertex of $\tilde{G}$ exactly once, so W defines a Hamilton cycle $\tilde{H}$ of $\tilde{G}^2$. Since W still contains the path Q^*, $\tilde{H}$ satisfies $(*)$. $\tilde{H}$

Up until now, we have assumed that $\mathcal{P}_2$ is non-empty. If $\mathcal{P}_2 = \emptyset$, let us set $\tilde{H} := \tilde{G} = C$; then, again, $\tilde{H}$ satisfies $(*)$. It remains to turn $\tilde{H}$ into a Hamilton cycle H of G^2 by incorporating the paths from $\mathcal{P}_1$. In order to be able to treat the case of $\mathcal{P}_2 = \emptyset$ along with the case of $\mathcal{P}_2 \neq \emptyset$, we define a map $v \mapsto e(v)$ also when $\mathcal{P}_2 = \emptyset$, as follows: for every $v \in C - y^*$, set $\tilde{H}$

$$e(v) := \begin{cases} vv^+ & \text{if } v \in [\,x^*, y^*) \\ vv^- & \text{if } v \in (y^*, x^*). \end{cases}$$

(Here, $[\,x^*, y^*)$ and (y^*, x^*) denote the obvious paths in C defined analogously to intervals.) As before, this map $v \mapsto e(v)$ is defined on a superset of X_1 and is injective on X_1; recall that y^* cannot lie in X_1 by definition. Moreover, this map satisfies the following sharpening of $(**)$ for every v in its domain:

> The edge $e(v)$ is incident with v, lies on C, and is distinct $(***)$
> from vx^*.

Let $P \in \mathcal{P}_1$ be a path to be incorporated into $\tilde{H}$, say with foot P, v
$v \in X_1$ and ends y_1, y_2. (If $|P| = 1$, then $y_1 = y_2$.) Our aim is to replace y_1, y_2
the edge $e := e(v)$ in $\tilde{H}$ by P; we thus have to show that the ends of P e
are joined to those of e by suitable edges of G^2.

By (2) and (3), v has only two neighbours in $\tilde{G}$, its neighbours x_1, x_2 on C. If v is incident with e, i.e. if $e = vx_i$ with $i \in \{1, 2\}$, we replace e by the path $vy_1 Py_2 x_i \subseteq G^2$ (Fig. 10.3.7). If v is not incident with e,

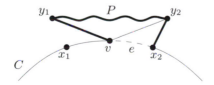

Fig. 10.3.7. Replacing the edge e in $\tilde{H}$

then $(***)$ fails, so $\mathcal{P}_2 \neq \emptyset$ and e bridges v by $(**)$. Then $e = x_1 x_2$, and we replace e by the path $x_1 y_1 Py_2 x_2 \subseteq G^2$ (Fig. 10.3.8). Since $v \mapsto e(v)$ is injective on X_1, assertion (2) implies that all these modifications

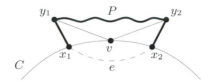

Fig. 10.3.8. Replacing the edge e in $\tilde{H}$

of $\tilde{H}$ (one for every $P \in \mathcal{P}_1$) can be performed independently, and hence produce a Hamilton cycle H of G^2.

Let us finally check that H satisfies $(*)$, i.e. that both edges of H at x^* lie in G. Since $(*)$ holds for $\tilde{H}$, it suffices to show that any edge $e = x^* z$ of $\tilde{H}$ that is not in H (and hence has the form $e = e(v)$ for some $v \in X_1$) was replaced by an x^*–z path whose first edge lies in G.

Where can the vertex v lie? Let us show that v must be incident with e. If not, then $(***)$ fails for e, so $\mathcal{P}_2 \neq \emptyset$ and e bridges v (as above). Now $\mathcal{P}_2 \neq \emptyset$ and $v \in X_1$ together imply that $|C - y^*| \geqslant |X_1| + 2 |\mathcal{P}_2| \geqslant 3$, so $|C| \geqslant 4$. The fact that e bridges v thus contradicts $e \in \tilde{G}$, by (3).

So v is indeed incident with e. Hence $v \in \{ x^*, z \}$ by definition of e, while $e \neq vx^*$ by $(**)$ or $(***)$. Thus $v = x^*$, and e was replaced by a path of the form $x^* y_1 P y_2 z$. Since $x^* y_1$ is an edge of G, this replacement again preserves $(*)$. Therefore H does indeed satisfy $(*)$, and our induction is complete. $\qquad\qquad\qquad\qquad\qquad\qquad\qquad\qquad\qquad\qquad\square$

We close the chapter with a far-reaching conjecture generalizing Dirac's theorem:

Conjecture. (Seymour 1974)
Let G be a graph of order $n \geqslant 3$, and let k be a positive integer. If G has minimum degree

$$\delta(G) \geqslant \frac{k}{k+1} \, n \,,$$

then G has a Hamilton cycle H such that $H^k \subseteq G$.

For $k = 1$, this is precisely Dirac's theorem. The case $k = 2$ had already been conjectured by Pósa in 1963 and was proved for large n by Komlós, Sárközy & Szemerédi (1996).

Exercises

1. Show that every uniquely 3-edge-colourable cubic graph is hamiltonian. ('Unique' means that all 3-edge-colourings induce the same edge partition.)

2. Prove or disprove the following strengthening of Proposition 10.1.2: 'Every k-connected graph G with $\chi(G) \geqslant |G|/k$ has a Hamilton cycle.'

3. Given a graph G, consider a maximal sequence $G_0, \ldots, G_k$ such that $G_0 = G$ and $G_{i+1} = G_i + x_i y_i$ for $i = 0, \ldots, k-1$, where x_i, y_i are two non-adjacent vertices of G_i satisfying $d_{G_i}(x_i) + d_{G_i}(y_i) \geqslant |G|$. The last graph of the sequence, G_k, is called the *Hamilton closure* of G. Show that this graph depends only on G, not on the choice of the sequence $G_0, \ldots, G_k$.

4. Let x, y be two nonadjacent vertices of a connected graph G, with $d(x) + d(y) \geqslant |G|$. Show that G has a Hamilton cycle if and only if $G + xy$ has one. Using the previous exercise, deduce the following strengthening of Dirac's theorem: if $d(x) + d(y) \geqslant |G|$ for every two non-adjacent vertices $x, y \in G$, then G has a Hamilton cycle.

5. Given an even positive integer k, construct for every $n \geqslant k$ a k-regular graph of order $2n + 1$.

    (Hint. Induction on k with n fixed; for the induction step consider $\overline{G}$.)

6.⁻ Find a hamiltonian graph whose degree sequence is not hamiltonian.

7.⁻ Let G be a graph with fewer than i vertices of degree at most i, for every $i < |G|/2$. Use Chvátal's theorem to show that G is hamiltonian. (Thus in particular, Chvátal's theorem implies Dirac's theorem.)

8. Find a connected graph G whose square G^2 has no Hamilton cycle.

9.⁺ Show by induction on $|G|$ that the third power G^3 of a connected graph G contains a Hamilton path between any two vertices. Deduce that G^3 is hamiltonian.

10. Show that the square of a 2-connected graph contains a Hamilton path between any two vertices.

    (Hint. Condition $(*)$ in the proof of Fleischner's theorem.)

11. An oriented complete graph is called a *tournament*. Show that every tournament contains a (directed) Hamilton path.

12.⁺ Let G be a graph in which every vertex has odd degree. Show that every edge of G lies on an even number of Hamilton cycles.

    (Hint. Let $xy \in E(G)$ be given. The Hamilton cycles through xy correspond to the Hamilton paths in $G - xy$ from x to y. Consider the set $\mathcal{H}$ of all Hamilton paths in $G - xy$ starting at x, and show that an even number of these end in y. To show this, define a graph on $\mathcal{H}$ so that the desired assertion follows from Proposition 1.2.1. (How can a Hamilton path $P \in \mathcal{H}$ be modified into another? In how many ways? What has this to do with the degree in G of the last vertex of P?))

Notes

The problem of finding a Hamilton cycle in a graph has the same kind of origin as its Euler tour counterpart and the four colour problem: all three problems come from mathematical puzzles older than graph theory itself. What began as a game invented by W.R. Hamilton in 1857—in which 'Hamilton cycles' had to be found on the graph of the dodecahedron—reemerged over a hundred years later as a combinatorial optimization problem of prime importance: the *travelling salesman problem*. Here, a salesman has to visit a number of customers, and his problem is to arrange these in a suitable circular route. (For reasons not included in the mathematical brief, the route has to be such that after visiting a customer the salesman does not pass through that town again.) Much of the motivation for considering Hamilton cycles comes from variations of this algorithmic problem.

A detailed discussion of the various degree conditions for hamiltonicity referred to at the beginning of Section 10.2 can be found in R. Halin, *Graphentheorie*, Wissenschaftliche Buchgesellschaft 1980. All the relevant references for Sections 10.1 and 10.2 can be found there, or in B. Bollobás, *Extremal Graph Theory*, Academic Press 1978.

The 'proof' of the four colour theorem indicated at the end of Section 10.1, which is based on the (false) premise that every 3-connected cubic planar graph is hamiltonian, is usually attributed to the Scottish mathematician P.G. Tait. Following Kempe's flawed proof of 1879 (see the notes for Chapter 5), it seems that Tait believed to be in possession of at least one 'new proof of Kempe's theorem'. However, when he addressed the Edinburgh Mathematical Society on this subject in 1883, he seems to have been aware that he could not—really—prove the above statement about Hamilton cycles. His account in P.G. Tait, Listing's topologie, *Phil. Mag.* **17** (1884), 30–46, makes some entertaining reading.

A shorter proof of Tutte's theorem that 4-connected planar graphs are hamiltonian was given by C. Thomassen, A theorem on paths in planar graphs, *J. Graph Theory* **7** (1983), 169–176. Tutte's counterexample to Tait's assumption that even 3-connectedness suffices (at least for cubic graphs) is shown in Bollobás, and in J.A. Bondy & U.S.R. Murty, *Graph Theory with Applications*, Macmillan 1976 (where Tait's attempted proof is discussed in some detail).

Proposition 10.1.2 is due to Chvátal & Erdős (1972). Our proof of Fleischner's theorem is based on S. Říha, A new proof of the theorem by Fleischner, *J. Combin. Theory B* **52** (1991), 117–123. Seymour's conjecture is from P.D. Seymour, Problem 3, in (T.P. McDonough and V.C. Mavron, eds.) *Combinatorics*, Cambridge University Press 1974. Pósa's conjecture was proved for large n by J. Komlós, G.N. Sárközy & E. Szemerédi, On the square of a Hamiltonian cycle in dense graphs, *Random Structures and Algorithms* **9** (1996), 193–211.

11 Random Graphs

At various points in this book, we already encountered the following fundamental theorem of Erdős: *for every integer k there is a graph G with $g(G) > k$ and $\chi(G) > k$.* In plain English: there exist graphs combining arbitrarily large girth with arbitrarily high chromatic number.

How could one prove such a theorem? The standard approach would be to construct a graph with those two properties, possibly in steps by induction on k. However, this is anything but straightforward: the global nature of the second property forced by the first, namely, that the graph should have high chromatic number 'overall' but be acyclic (and hence 2-colourable) locally, flies in the face of any attempt to build it up, constructively, from smaller pieces that have the same or similar properties.

In his pioneering paper of 1959, Erdős took a radically different approach: for each n he defined a probability space on the set of graphs with n vertices, and showed that, for some carefully chosen probability measures, the probability that an n-vertex graph has both of the above properties is positive for all large enough n.

This approach, now called the *probabilistic method*, has since unfolded into a sophisticated and versatile proof technique, in graph theory as much as in other branches of discrete mathematics. The theory of *random graphs* is now a subject in its own right. The aim of this chapter is to offer an elementary but rigorous introduction to random graphs: no more than is necessary to understand its basic concepts, ideas and techniques, but enough to give an inkling of the power and elegance hidden behind the calculations.

Erdős's theorem asserts the existence of a graph with certain properties: it is a perfectly ordinary assertion showing no trace of the randomness employed in its proof. There are also results in random graphs that are generically random even in their statement: these are theorems about *almost all* graphs, a notion we shall meet in Section 11.3. In the

last section, we give a detailed proof of a theorem of Erdős and Rényi
that illustrates a proof technique frequently used in random graphs, the
so-called *second moment method*.

11.1 The notion of a random graph

V

$\mathcal{G}$

Let V be a fixed set of n elements, say $V = \{0, \ldots, n-1\}$. Our aim is
to turn the set $\mathcal{G}$ of all graphs on V into a probability space, and then
to consider the kind of questions typically asked about random objects:
What is the probability that a graph $G \in \mathcal{G}$ has this or that property?
What is the expected value of a given invariant on G, say its expected
girth or chromatic number?

Intuitively, we should be able to generate G randomly as follows.
For each $e \in [V]^2$ we decide by some random experiment whether or not
e shall be an edge of G; these experiments are performed independently,
and for each the probability of success—i.e. of accepting e as an edge

p

for G—is equal to some fixed[1] number $p \in [0, 1]$. Then if G_0 is some
fixed graph on V, with m edges say, the elementary event $\{G_0\}$ has a

q

probability of $p^m q^{\binom{n}{2} - m}$ (where $q := 1 - p$): with this probability, our
randomly generated graph G is this particular graph G_0. (The probability that G is *isomorphic* to G_0 will usually be greater.) But if the
probabilities of all the elementary events are thus determined, then so
is the entire probability measure of our desired space $\mathcal{G}$. Hence all that
remains to be checked is that such a probability measure on $\mathcal{G}$, one for
which all individual edges occur independently with probability p, does
indeed exist.[2]

In order to construct such a measure on $\mathcal{G}$ formally, we start by
defining for every potential edge $e \in [V]^2$ its own little probability space

Ω_e

P_e

$\mathcal{G}(n, p)$

$\Omega_e := \{0_e, 1_e\}$, choosing $P_e(\{1_e\}) := p$ and $P_e(\{0_e\}) := q$ as the
probabilities of its two elementary events. As our desired probability
space $\mathcal{G} = \mathcal{G}(n, p)$ we then take the product space

Ω

$$\Omega := \prod_{e \in [V]^2} \Omega_e \, .$$

---

[1] Often, the value of p will depend on the cardinality n of the set V on which our
random graphs are generated; thus, p will be the value $p = p(n)$ of some function
$n \mapsto p(n)$. Note, however, that V (and hence n) is fixed for the definition of $\mathcal{G}$:
for each n separately, we are constructing a probability space of the graphs G on
$V = \{0, \ldots, n-1\}$, and within each space the probability that $e \in [V]^2$ is an edge
of G has the same value for all e.

[2] Any reader ready to believe this may skip ahead now to the end of Proposition 11.1.1, without missing anything.

Thus, formally, an element of Ω is a map ω assigning to every $e \in [V]^2$ either 0_e or 1_e, and the probability measure P on Ω is the product measure of all the measures P_e. In practice, of course, we identify ω with the graph G on V whose edge set is

$$E(G) = \{ e \mid \omega(e) = 1_e \},$$

and call G a *random graph* on V with edge probability p.

Following standard probabilistic terminology, we may now call any set of graphs on V an *event* in $\mathcal{G}(n, p)$. In particular, for every $e \in [V]^2$ the set

$$A_e := \{ \omega \mid \omega(e) = 1_e \}$$

of all graphs G on V with $e \in E(G)$ is an event: the event that e is an edge of G. For these events, we can now prove formally what had been our guiding intuition all along:

Proposition 11.1.1. *The events A_e are independent and occur with probability p.*

Proof. By definition,

$$A_e = \{ 1_e \} \times \prod_{e' \neq e} \Omega_{e'}.$$

Since P is the product measure of all the measures P_e, this implies

$$P(A_e) = p \cdot \prod_{e' \neq e} 1 = p.$$

Similarly, if $\{ e_1, \ldots, e_k \}$ is any subset of $[V]^2$, then

$$P(A_{e_1} \cap \ldots \cap A_{e_k}) = P\Big(\{ 1_{e_1} \} \times \ldots \times \{ 1_{e_k} \} \times \prod_{e \notin \{ e_1, \ldots, e_k \}} \Omega_e \Big)$$
$$= p^k$$
$$= P(A_{e_1}) \cdots P(A_{e_k}).$$

$\square$

As noted before, P is determined uniquely by the value of p and our assumption that the events A_e are independent. In order to calculate probabilities in $\mathcal{G}(n, p)$, it therefore generally suffices to work with these two assumptions: our concrete model for $\mathcal{G}(n, p)$ has served its purpose and will not be needed again.

As a simple example of such a calculation, consider the event that G contains some fixed graph H on a subset of V as a subgraph; let $|H| =: k$ and $\|H\| =: \ell$. The probability of this event $H \subseteq G$ is the product of the probabilities A_e over all the edges $e \in H$, so $P[H \subseteq G] = p^\ell$. In

contrast, the probability that H is an *induced* subgraph of G is $p^\ell q^{\binom{k}{2}-\ell}$:
now the edges missing from H are required to be missing from G too,
and they do so independently with probability q.

The probability P_H that G has an induced subgraph *isomorphic*
to H is usually more difficult to compute: since the possible instances
of H on subsets of V overlap, the events that they occur in G are not
independent. However, the sum (over all k-sets $U \subseteq V$) of the probabilities $P[H \simeq G[U]]$ is always an upper bound for P_H, since P_H is the
measure of the union of all those events. For example, if $H = \overline{K^k}$, we
have the following trivial upper bound on the probability that G contains
an induced copy of H:

[11.2.1]
[11.3.4]

Lemma 11.1.2. *For all integers n, k with $n \geqslant k \geqslant 2$, the probability
that $G \in \mathcal{G}(n, p)$ has a set of k independent vertices is at most*

$$P[\alpha(G) \geqslant k] \leqslant \binom{n}{k} q^{\binom{k}{2}}.$$

Proof. The probability that a fixed k-set $U \subseteq V$ is independent in
G is $q^{\binom{k}{2}}$. The assertion thus follows from the fact that there are only
$\binom{n}{k}$ such sets U. □

Analogously, the probability that $G \in \mathcal{G}(n, p)$ contains a K^k is at
most

$$P[\omega(G) \geqslant k] \leqslant \binom{n}{k} p^{\binom{k}{2}}.$$

Now if k is fixed, and n is small enough that these bounds for the prob-
abilities $P[\alpha(G) \geqslant k]$ and $P[\omega(G) \geqslant k]$ sum to less than 1, then $\mathcal{G}$
contains graphs that have neither property: graphs which contain nei-
ther a K^k nor a $\overline{K^k}$ induced. But then any such n is a lower bound for
the Ramsey number of k!

As the following theorem shows, this lower bound is quite close to
the upper bound of 2^{2k-3} implied by the proof of Theorem 9.1.1:

Theorem 11.1.3. (Erdős 1947)
For every integer $k \geqslant 3$, the Ramsey number of k satisfies

$$R(k) > 2^{k/2}.$$

Proof. For $k = 3$ we have $R(3) = 6 > 2^{3/2}$ (exercise), so let $k \geqslant 4$.
We show that, for all $n \leqslant 2^{k/2}$ and $G \in \mathcal{G}(n, \frac{1}{2})$, the probabilities
$P[\alpha(G) \geqslant k]$ and $P[\omega(G) \geqslant k]$ are both less than $\frac{1}{2}$.

Since $p = q = \frac{1}{2}$, Lemma 11.1.2 and the analogous assertion for $\omega(G)$
imply the following for all $n \leqslant 2^{k/2}$ (use that $k! > 2^k$ for $k \geqslant 4$):

$$P\left[\,\alpha(G)\geqslant k\,\right],\ P\left[\,\omega(G)\geqslant k\,\right]\ \leqslant\ \binom{n}{k}\left(\tfrac{1}{2}\right)^{\binom{k}{2}}$$
$$<\ \left(n^{k}/2^{k}\right)2^{-\frac{1}{2}k(k-1)}$$
$$\leqslant\ \left(2^{k^{2}/2}/2^{k}\right)2^{-\frac{1}{2}k(k-1)}$$
$$=\ 2^{-k/2}$$
$$<\ \tfrac{1}{2}\,.$$

$\square$

In the context of random graphs, each of the familiar graph invariants (like average degree, connectivity, girth, chromatic number, and so on) may be interpreted as a non-negative *random variable* on $\mathcal{G}(n,p)$, a function

random variable

$$X\colon\mathcal{G}(n,p)\to[\,0,\infty)\,.$$

The *mean* or *expected* value of X is the number

mean

expectation

$$E(X):=\sum_{G\in\mathcal{G}(n,p)}P(\{\,G\,\})\cdot X(G)\,.$$

$E(X)$

Note that the operator E, the *expectation*, is linear: we have $E(X+Y)=E(X)+E(Y)$ and $E(\lambda X)=\lambda E(X)$ for any two random variables X,Y on $\mathcal{G}(n,p)$ and $\lambda\in\mathbb{R}$.

Computing the mean of a random variable X can be a simple and effective way to establish the existence of a graph G such that $X(G)<a$ for some fixed $a>0$ and, moreover, G has some desired property $\mathcal{P}$. Indeed, if the expected value of X is small, then $X(G)$ cannot be large for more than a few graphs in $\mathcal{G}(n,p)$, because $X(G)\geqslant 0$ for all $G\in\mathcal{G}(n,p)$. Hence X must be small for many graphs in $\mathcal{G}(n,p)$, and it is reasonable to expect that among these we may find one with the desired property $\mathcal{P}$.

This simple idea lies at the heart of countless non-constructive existence proofs using random graphs, including the proof of Erdős's theorem presented in the next section. Quantified, it takes the form of the following lemma, whose proof follows at once from the definition of the expectation and the additivity of P:

Lemma 11.1.4. (Markov's Inequality)
Let $X\geqslant 0$ be a random variable on $\mathcal{G}(n,p)$ and $a>0$. Then

[11.2.2]
[11.4.1]
[11.4.3]

$$P\left[\,X\geqslant a\,\right]\leqslant E(X)/a\,.$$

Proof.

$$E(X)=\sum_{G\in\mathcal{G}(n,p)}P(\{\,G\,\})\cdot X(G)$$

$$\geq \sum_{\substack{G \in \mathcal{G}(n,p) \\ X(G) \geq a}} P(\{G\}) \cdot X(G)$$

$$\geq \sum_{\substack{G \in \mathcal{G}(n,p) \\ X(G) \geq a}} P(\{G\}) \cdot a$$

$$= P[X \geq a] \cdot a.$$

$\square$

Since our probability spaces are finite, the expectation can often be computed by a simple application of *double counting*, a standard combinatorial technique we met before in the proofs of Corollary 4.2.8 and Theorem 5.5.3. For example, if X is a random variable on $\mathcal{G}(n,p)$ that counts the number of subgraphs of G in some fixed set $\mathcal{H}$ of graphs on V, then $E(X)$, by definition, counts the number of pairs (G, H) such that $H \subseteq G$, each weighted with the probability of $\{G\}$. Algorithmically, we compute $E(X)$ by going through the graphs $G \in \mathcal{G}(n,p)$ in an 'outer loop' and performing, for each G, an 'inner loop' that runs through the graphs $H \in \mathcal{H}$ and counts '$P(\{G\})$' whenever $H \subseteq G$. Alternatively, we may count the same set of weighted pairs with H in the outer and G in the inner loop: this amounts to adding up, over all $H \subseteq \mathcal{H}$, the probabilities $P[H \subseteq G]$.

To illustrate this once in detail, let us compute the expected number of cycles of some given length $k \geq 3$ in a random graph $G \in \mathcal{G}(n,p)$. So let $X: \mathcal{G}(n,p) \to \mathbb{N}$ be the random variable that assigns to every random graph G its number of k-cycles, the number of subgraphs isomorphic to C^k. Let us write

X

$(n)_k$
$$(n)_k := n(n-1)(n-2)\cdots(n-k+1)$$

for the number of sequences of k distinct elements of a given n-set.

[11.2.2]
[11.4.3]
Lemma 11.1.5. *The expected number of k-cycles in $G \in \mathcal{G}(n,p)$ is*

$$E(X) = \frac{(n)_k}{2k} p^k.$$

Proof. For every k-cycle C with vertices in $V = \{0, \ldots, n-1\}$, the vertex set of the graphs in $\mathcal{G}(n,p)$, let $X_C: \mathcal{G}(n,p) \to \{0,1\}$ denote the *indicator random variable* of C:

$$X_C: G \mapsto \begin{cases} 1 & \text{if } C \subseteq G; \\ 0 & \text{otherwise.} \end{cases}$$

Since X_C takes only 1 as a positive value, its expectation $E(X_C)$ equals the measure $P[X_C = 1]$ of the set of all graphs in $\mathcal{G}(n,p)$ that contain C. But this is just the probability that $C \subseteq G$:

$$E(X_C) = P[C \subseteq G] = p^k. \tag{1}$$

How many such cycles $C = v_0 \ldots v_{k-1}v_0$ are there? There are $(n)_k$ sequences $v_0 \ldots v_{k-1}$ of distinct vertices in V, and each cycle is identified by $2k$ of those sequences—so there are exactly $(n)_k/2k$ such cycles.

Our random variable X assigns to every graph G its number of k-cycles. Clearly, this is the sum of all the values $X_C(G)$, where C varies over the $(n)_k/2k$ cycles of length k with vertices in V:

$$X = \sum_C X_C.$$

Since the expectation is linear, (1) thus implies

$$E(X) = E\left(\sum_C X_C\right) = \sum_C E(X_C) = \frac{(n)_k}{2k} p^k$$

as claimed. $\square$

11.2 The probabilistic method

Very roughly, the *probabilistic method* in discrete mathematics has developed from the following idea. In order to prove the existence of an object with some desired property, one defines a probability space on some larger—and certainly non-empty—class of objects, and then shows that an element of this space has the desired property with positive probability. The 'objects' inhabiting this probability space may be of any kind: partitions or orderings of the vertices of some fixed graph arise as naturally as mappings, embeddings or, of course, graphs themselves. In this section, we illustrate the probabilistic method by giving a detailed account of one of its earliest results: of Erdős's classic theorem on large girth and chromatic number.

Erdős's theorem says that, given any positive integer k, there is a graph G with girth $g(G) > k$ and chromatic number $\chi(G) > k$. Let us call cycles of length at most k *short*, and sets of $|G|/k$ or more vertices *big*. For a proof of Erdős's theorem, it suffices to find a graph G without short cycles and without big independent sets of vertices: then the colour classes in any vertex colouring of G are *small* (not big), so we need more than k colours to colour G. [short, big/small]

How can we find such a graph G? If we choose p small enough, then a random graph in $\mathcal{G}(n,p)$ is unlikely to contain any (short) cycles. If we choose p large enough, then G is unlikely to have big independent vertex sets. So the question is: do these two ranges of p overlap, that is, can we choose p so that, for some n, it is both small enough to give $P[g \leqslant k] < \frac{1}{2}$ and large enough for $P[\alpha \geqslant n/k] < \frac{1}{2}$? If so, then

$\mathcal{G}(n,p)$ will contain at least one graph without either short cycles or big independent sets.

Unfortunately, such a choice of p is impossible: the two ranges of p do not overlap! As we shall see in Section 11.4, we must keep p below n^{-1} to make the occurrence of short cycles in G unlikely—but for any such p there will most likely be no cycles in G at all (Exercise 18), so G will be bipartite and hence have at least $n/2$ independent vertices.

But all is not lost. In order to make big independent sets unlikely, we shall fix p above n^{-1}, at $n^{\epsilon-1}$ for some $\epsilon > 0$. Fortunately, though, if ϵ is small enough then this will produce only *few* short cycles in G, even compared with n (rather than, more typically, with n^k). If we then delete a vertex in each of those cycles, the graph H obtained will have no short cycles, and its independence number $\alpha(H)$ will be at most that of G. Since H is not much smaller than G, its chromatic number will thus still be large, so we have found a graph with both large girth and large chromatic number.

To prepare for the formal proof of Erdős's theorem, we first show that an edge probability of $p = n^{\epsilon-1}$ is indeed always large enough to ensure that $G \in \mathcal{G}(n,p)$ 'almost surely' has no big independent set of vertices. More precisely, we prove the following slightly stronger assertion:

Lemma 11.2.1. *Let $k > 0$ be an integer, and let $p = p(n)$ be a function of n such that $p \geqslant (6k \ln n)n^{-1}$ for n large. Then*

$$\lim_{n\to\infty} P\left[\alpha \geqslant \tfrac{1}{2}n/k\right] = 0.$$

(11.1.2) *Proof.* For all integers n, r with $n \geqslant r \geqslant 2$, and all $G \in \mathcal{G}(n,p)$, Lemma 11.1.2 implies

$$P[\alpha \geqslant r] \leqslant \binom{n}{r} q^{\binom{r}{2}}$$
$$\leqslant n^r q^{\binom{r}{2}}$$
$$= \left(n q^{(r-1)/2}\right)^r$$
$$\leqslant \left(n e^{-p(r-1)/2}\right)^r;$$

here, the last inequality follows from the fact that $1 - p \leqslant e^{-p}$ for all p. (Compare the functions $x \mapsto e^x$ and $x \mapsto x + 1$ for $x = -p$.) Now if $p \geqslant (6k \ln n)n^{-1}$ and $r \geqslant \tfrac{1}{2}n/k$, then the term under the exponent satisfies

$$ne^{-p(r-1)/2} = ne^{-pr/2 + p/2}$$
$$\leqslant ne^{-(3/2)\ln n + p/2}$$
$$\leqslant n n^{-3/2} e^{1/2}$$
$$= \sqrt{e}/\sqrt{n} \xrightarrow[n\to\infty]{} 0.$$

Since $p \geqslant (6k \ln n)n^{-1}$ for n large, we thus obtain for $r := \lceil \frac{1}{2}n/k \rceil$

$$\lim_{n \to \infty} P[\alpha \geqslant \tfrac{1}{2}n/k] = \lim_{n \to \infty} P[\alpha \geqslant r] = 0,$$

as claimed. $\square$

Theorem 11.2.2. (Erdős 1959)
For every integer k there exists a graph H with girth $g(H) > k$ and   [9.2.3]
chromatic number $\chi(H) > k$.

(11.1.4)
Proof. Assume that $k \geqslant 3$, fix ϵ with $0 < \epsilon < 1/k$, and let $p := n^{\epsilon-1}$. Let   (11.1.5)
$X(G)$ denote the number of *short* cycles in a random graph $G \in \mathcal{G}(n,p)$, p, ϵ, X
i.e. its number of cycles of length at most k.

By Lemma 11.1.5, we have

$$E(X) = \sum_{i=3}^{k} \frac{(n)_i}{2i} p^i \leqslant \tfrac{1}{2} \sum_{i=3}^{k} n^i p^i \leqslant \tfrac{1}{2}(k-2)\, n^k p^k \,;$$

note that $(np)^i \leqslant (np)^k$, because $np = n^\epsilon \geqslant 1$. By Lemma 11.1.4,

$$\begin{aligned}
P[X \geqslant n/2] &\leqslant E(X)/(n/2) \\
&\leqslant (k-2)\, n^{k-1} p^k \\
&= (k-2)\, n^{k-1} n^{(\epsilon-1)k} \\
&= (k-2)\, n^{k\epsilon-1}.
\end{aligned}$$

As $k\epsilon - 1 < 0$ by our choice of ϵ, this implies that

$$\lim_{n \to \infty} P[X \geqslant n/2] = 0.$$

Let n be large enough that $P[X \geqslant n/2] < \frac{1}{2}$ and $P[\alpha \geqslant \frac{1}{2}n/k] < \frac{1}{2}$; n
the latter is possible by our choice of p and Lemma 11.2.1. Then there
is a graph $G \in \mathcal{G}(n,p)$ with fewer than $n/2$ short cycles and $\alpha(G) <
\frac{1}{2}n/k$. From each of those cycles delete a vertex, and let H be the graph
obtained. Then $|H| \geqslant n/2$ and H has no short cycles, so $g(H) > k$. By
definition of G,

$$\chi(H) \geqslant \frac{|H|}{\alpha(H)} \geqslant \frac{n/2}{\alpha(G)} > k\,.$$

$\square$

Corollary 11.2.3. *There are graphs with arbitrarily large girth and
arbitrarily large values of the invariants κ, ε and δ.*

Proof. Apply Corollary 5.2.3 and Theorem 1.4.2. $\square$ (1.4.2)
(5.2.3)

11.3 Properties of almost all graphs

property

A *graph property* is a class of graphs that is closed under isomorphism, one that contains with every graph G also the graphs isomorphic to G. If $p = p(n)$ is a fixed function (possibly constant), and $\mathcal{P}$ is a graph property, we may ask how the probability $P[G \in \mathcal{P}]$ behaves for $G \in \mathcal{G}(n, p)$ as $n \to \infty$. If this probability tends to 1, we say that $G \in \mathcal{P}$ for *almost*

almost all
etc.

all (or *almost every*) $G \in \mathcal{G}(n, p)$, or that $G \in \mathcal{P}$ *almost surely*; if it tends to 0, we say that *almost no* $G \in \mathcal{G}(n, p)$ has the property $\mathcal{P}$. (For example, in Lemma 11.2.1 we proved that, for a certain p, almost no $G \in \mathcal{G}(n, p)$ has a set of more than $\frac{1}{2}n/k$ independent vertices.)

To illustrate the new concept let us show that, for constant p, every fixed abstract[3] graph H is an induced subgraph of almost all graphs:

Proposition 11.3.1. *For every constant $p \in (0, 1)$ and every graph H, almost every $G \in \mathcal{G}(n, p)$ contains an induced copy of H.*

Proof. Let H be given, and $k := |H|$. If $n \geqslant k$ and $U \subseteq \{0, \ldots, n-1\}$ is a fixed set of k vertices of G, then $G[U]$ is isomorphic to H with a certain probability $r > 0$. This probability r depends on p, but not on n (why not?). Now G contains a collection of $\lfloor n/k \rfloor$ disjoint such sets U. The probability that none of the corresponding graphs $G[U]$ is isomorphic to H is $(1 - r)^{\lfloor n/k \rfloor}$, since these events are independent by the disjointness of the edges sets $[U]^2$. Thus

$$P[H \not\subseteq G \text{ induced}] \leqslant (1 - r)^{\lfloor n/k \rfloor} \xrightarrow[n \to \infty]{} 0,$$

which implies the assertion.   □

The following lemma is a simple device enabling us to deduce that quite a number of natural graph properties (including that of Proposition 11.3.1) are shared by almost all graphs. Given $i, j \in \mathbb{N}$, let $\mathcal{P}_{i,j}$

$\mathcal{P}_{i,j}$

denote the property that the graph considered contains, for any disjoint vertex sets U, W with $|U| \leqslant i$ and $|W| \leqslant j$, a vertex $v \notin U \cup W$ that is adjacent to all the vertices in U but to none in W.

Lemma 11.3.2. *For every constant $p \in (0, 1)$ and $i, j \in \mathbb{N}$, almost every graph $G \in \mathcal{G}(n, p)$ has the property $\mathcal{P}_{i,j}$.*

---

[3] The word 'abstract' is used to indicate that only the isomorphism type of H is known or relevant, not its actual vertex and edge sets. In our context, it indicates that the word 'subgraph' is used in the usual sense of 'isomorphic to a subgraph'.

Proof. For fixed U, W and $v \in G - (U \cup W)$, the probability that v is adjacent to all the vertices in U but to none in W, is

$$p^{|U|} q^{|W|} \geqslant p^i q^j.$$

Hence, the probability that no suitable v exists for these U and W, is

$$(1 - p^{|U|} q^{|W|})^{n - |U| - |W|} \leqslant (1 - p^i q^j)^{n - i - j}$$

(for $n \geqslant i + j$), since the corresponding events are independent for different v. As there are no more than n^{i+j} pairs of such sets U, W in $V(G)$, the probability that some such pair has no suitable v is at most

$$n^{i+j} (1 - p^i q^j)^{n - i - j},$$

which tends to zero as $n \to \infty$ since $1 - p^i q^j < 1$. $\qquad\square$

Corollary 11.3.3. *For every constant $p \in (0, 1)$ and $k \in \mathbb{N}$, almost every graph in $\mathcal{G}(n, p)$ is k-connected.*

Proof. By Lemma 11.3.2, it is enough to show that every graph in $\mathcal{P}_{2, k-1}$ is k-connected. But this is easy: any graph in $\mathcal{P}_{2, k-1}$ has order at least $k + 2$, and if W is a set of fewer than k vertices, then by definition of $\mathcal{P}_{2, k-1}$ any other two vertices x, y have a common neighbour $v \notin W$; in particular, W does not separate x from y. $\qquad\square$

In the proof of Corollary 11.3.3, we showed substantially more than was asked for: rather than finding, for any two vertices $x, y \notin W$, some x–y path avoiding W, we showed that x and y have a common neighbour outside W; thus, all the paths needed to establish the desired connectivity could in fact be chosen of length 2. What seemed like a clever trick in this particular proof is in fact indicative of a more fundamental phenomenon for constant edge probabilities: by an easy result in logic, any statement about graphs expressed by quantifying over vertices only (rather than over sets or sequences of vertices)[4] is either almost surely true or almost surely false. All such statements, or their negations, are in fact immediate consequences of an assertion that the graph has property $\mathcal{P}_{i,j}$, for some suitable i, j.

As a last example of an 'almost all' result we now show that almost every graph has a surprisingly high chromatic number:

---

[4] In the terminology of logic: any first order sentence in the language of graph theory

Proposition 11.3.4. *For every constant $p \in (0, 1)$ and every $\epsilon > 0$, almost every graph $G \in \mathcal{G}(n, p)$ has chromatic number*

$$\chi(G) > \frac{\log(1/q)}{2 + \epsilon} \cdot \frac{n}{\log n} .$$

(11.1.2)   *Proof.* For every n and $k \geqslant 2$, Lemma 11.1.2 implies

$$P[\alpha \geqslant k] \leqslant \binom{n}{k} q^{\binom{k}{2}}$$
$$\leqslant n^k q^{\binom{k}{2}}$$
$$= q^{k \frac{\log n}{\log q} + \frac{1}{2}k(k-1)}$$
$$= q^{\frac{k}{2}\left(-\frac{2 \log n}{\log(1/q)} + k - 1\right)}.$$

For

$$k := (2 + \epsilon) \frac{\log n}{\log(1/q)}$$

the exponent of this expression tends to infinity with n, so the expression itself tends to zero. Hence, almost every $G \in \mathcal{G}(n, p)$ is such that in any vertex colouring of G no k vertices can have the same colour, so every colouring uses more than

$$\frac{n}{k} = \frac{\log(1/q)}{2 + \epsilon} \cdot \frac{n}{\log n}$$

colours. $\square$

By a result of Bollobás (1988), Proposition 11.3.4 is sharp in the following sense: if we replace ϵ by $-\epsilon$, then the lower bound given for χ turns into an upper bound.

Most of the results of this section have the interesting common feature that the values of p played no role whatsoever: if almost every graph in $\mathcal{G}(n, \frac{1}{2})$ had the property considered, then the same was true for almost every graph in $\mathcal{G}(n, 1/1000)$. How could this happen?

Such insensitivity of our random model to changes of p was certainly not intended: after all, among all the graphs with a certain property $\mathcal{P}$ it is often those having $\mathcal{P}$ 'only just' that are the most interesting—for those graphs are most likely to have different properties too, properties to which $\mathcal{P}$ might thus be set in relation. (The proof of Erdős's theorem is a good example.) For most properties, however—and this explains the above phenomenon—the critical order of magnitude of p around which the property will 'just' occur or not occur lies far below any constant value of p: it is typically a function of n tending to zero as $n \to \infty$.

Let us then see what happens if p is allowed to vary with n. Almost immediately, a fascinating picture unfolds. For edge probabilities p whose order of magnitude lies below n^{-2}, a random graph $G \in \mathcal{G}(n, p)$ almost surely has no edges at all. As p grows, G acquires more and more structure: from about $p = \sqrt{n}\, n^{-2}$ onwards, it almost surely has a component with more than two vertices, these components grow into trees, and around $p = n^{-1}$ the first cycles are born. Soon, some of these will have several crossing chords, making the graph non-planar. At the same time, one component outgrows the others, until it devours them around $p = (\log n)n^{-1}$, making the graph connected. Hardly later, at $p = (1 + \epsilon)(\log n)n^{-1}$, our graph almost surely has a Hamilton cycle!

It has become customary to compare this development of random graphs as p grows to the evolution of an organism: for each $p = p(n)$, one thinks of the properties shared by almost all graphs in $\mathcal{G}(n, p)$ as properties of 'the' typical random graph $G \in \mathcal{G}(n, p)$, and studies how G changes its features with the growth rate of p. As with other species, the evolution of random graphs happens in relatively sudden jumps: the critical edge probabilities mentioned above are thresholds below which almost no graph and above which almost every graph has the property considered. More precisely, we call a real function $t = t(n)$ with $t(n) \neq 0$ for all n a *threshold function* for a graph property $\mathcal{P}$ if the following holds for all $p = p(n)$, and $G \in \mathcal{G}(n, p)$:

threshold function

$$\lim_{n \to \infty} P[G \in \mathcal{P}] = \begin{cases} 0 & \text{if } p/t \to 0 \text{ as } n \to \infty \\ 1 & \text{if } p/t \to \infty \text{ as } n \to \infty. \end{cases}$$

If $\mathcal{P}$ has a threshold function t, then clearly any positive multiple ct of t is also a threshold function for $\mathcal{P}$; thus, threshold functions in the above sense are only ever unique up to a multiplicative constant.[5]

Which graph properties have threshold functions? Natural candidates for such properties are *increasing* ones, properties closed under the addition of edges. (Graph properties of the form $\{ G \mid G \supseteq H \}$, with H fixed, are common increasing properties; connectedness is another.) And indeed, Bollobás & Thomason (1987) have shown that all increasing properties, trivial exceptions aside, have threshold functions.

In the next section we shall study a general method to compute threshold functions.

---

[5] Our notion of threshold reflects only the crudest interesting level of screening: for some properties, such as connectedness, one can define sharper thresholds where the constant factor is crucial. Note also the role of the constant factor in our comparison of connectedness with hamiltonicity in the previous paragraph.

11.4 Threshold functions and second moments

Consider a graph property of the form

$$\mathcal{P} = \{\, G \mid X(G) > 0 \,\},$$

$X \geqslant 0$ where $X \geqslant 0$ is a random variable on $\mathcal{G}(n,p)$. Countless properties can be expressed naturally in this way; if X denotes the number of spanning trees, for example, then $\mathcal{P}$ corresponds to connectedness.

How could we prove that $\mathcal{P}$ has a threshold function t? Any such proof will consist of two parts: a proof that almost no $G \in \mathcal{G}(n,p)$ has $\mathcal{P}$ when p is small compared with t, and one showing that almost every G has $\mathcal{P}$ when p is large.

If X is integral, we may use Markov's inequality for the first part of the proof and find an upper bound for $E(X)$ instead of $P[X > 0]$: if $E(X)$ is small then $X(G)$ can be positive—and hence at least 1—only for few $G \in \mathcal{G}(n,p)$. Besides, the expectation is much easier to calculate than probabilities: without worrying about such things as independence or incompatibility of events, we may compute the expectation of a sum of random variables—for example, of indicator random variables—simply by adding up their individual expected values.

For the second part of the proof, things are more complicated. In order to show that $P[X > 0]$ is large, it is not enough to bound $E(X)$ from below: since X is not bounded above, $E(X)$ may be large simply because X is very large on just a few graphs G—so X may still be zero for most $G \in \mathcal{G}(n,p)$.[6] In order to prove that $P[X > 0] \to 1$, we thus have to show that this cannot happen, i.e. that X does not deviate a lot from its mean too often.

The following elementary tool from probability theory achieves just that. As is customary, we write

μ

$$\mu := E(X)$$

and define $\sigma \geqslant 0$ by setting

σ^2

$$\sigma^2 := E\big((X - \mu)^2\big)\,.$$

This quantity σ^2 is called the *variance* or *second moment* of X; by definition, it is a (quadratic) measure of how much X deviates from its mean. Since E is linear, the defining term for σ^2 expands to

$$\sigma^2 = E(X^2 - 2\mu X + \mu^2) = E(X^2) - \mu^2.$$

---

[6] For some p between n^{-1} and $(\log n)n^{-1}$, for example, almost every $G \in \mathcal{G}(n,p)$ has an isolated vertex (and hence no spanning tree), but its expected number of spanning trees tends to infinity with n! See the Exercise 13 for details.

Note that μ and σ^2 always refer to a random variable on some fixed probability space. In our setting, where we consider the spaces $\mathcal{G}(n, p)$, both quantities are functions of n.

The following lemma says exactly what we need: that X cannot deviate a lot from its mean too often.

Lemma 11.4.1. (Chebyshev's Inequality)
For all real $\lambda > 0$,

$$P\big[\,|X - \mu| \geqslant \lambda\,\big] \leqslant \sigma^2/\lambda^2.$$

Proof. By Lemma 11.1.4 and definition of σ^2, (11.1.4)

$$P\big[\,|X - \mu| \geqslant \lambda\,\big] = P\big[\,(X - \mu)^2 \geqslant \lambda^2\,\big] \leqslant \sigma^2/\lambda^2.$$

$\square$

For a proof that $X(G) > 0$ for almost all $G \in \mathcal{G}(n, p)$, Chebyshev's inequality can be used as follows:

Lemma 11.4.2. *If $\mu > 0$ for n large, and $\sigma^2/\mu^2 \to 0$ as $n \to \infty$, then $X(G) > 0$ for almost all $G \in \mathcal{G}(n, p)$.*

Proof. Any graph G with $X(G) = 0$ satisfies $|X(G) - \mu| = \mu$. Hence Lemma 11.4.1 implies with $\lambda := \mu$ that

$$P\,[\,X = 0\,] \;\leqslant\; P\big[\,|X - \mu| \geqslant \mu\,\big] \;\leqslant\; \sigma^2/\mu^2 \;\xrightarrow[n \to \infty]{}\; 0\,.$$

Since $X \geqslant 0$, this means that $X > 0$ almost surely, i.e. that $X(G) > 0$ for almost all $G \in \mathcal{G}(n, p)$. $\square$

As the main result of this section, we now prove a theorem that will at once give us threshold functions for a number of natural properties. Given a graph H, we denote by $\mathcal{P}_H$ the graph property of containing a copy of H as a subgraph. We shall call H *balanced* if $\varepsilon(H') \leqslant \varepsilon(H)$ for all subgraphs H' of H.

$\mathcal{P}_H$

balanced

Theorem 11.4.3. (Erdős & Rényi 1960)
If H is a balanced graph with k vertices and $\ell \geqslant 1$ edges, then $t(n) := n^{-k/\ell}$ is a threshold function for $\mathcal{P}_H$.

k, ℓ

t

(11.1.4)
(11.1.5)
X

$\mathcal{H}$

h

p, γ

Proof. Let $X(G)$ denote the number of subgraphs of G isomorphic to H. Given $n \in \mathbb{N}$, let $\mathcal{H}$ denote the set of all graphs isomorphic to H whose vertices lie in $\{0, \ldots, n-1\}$, the vertex set of the graphs $G \in \mathcal{G}(n,p)$:

$$\mathcal{H} := \{H' \mid H' \simeq H, \ V(H') \subseteq \{0, \ldots, n-1\}\}.$$

Given $H' \in \mathcal{H}$ and $G \in \mathcal{G}(n,p)$, we shall write $H' \subseteq G$ to express that H' itself—not just an isomorphic copy of H'—is a subgraph of G.

By h we denote the number of isomorphic copies of H on a fixed k-set; clearly, $h \leqslant k!$. As there are $\binom{n}{k}$ possible vertex sets for the graphs in $\mathcal{H}$, we thus have

$$|\mathcal{H}| = \binom{n}{k} h \leqslant \binom{n}{k} k! \leqslant n^k. \tag{1}$$

Given $p = p(n)$, we set $\gamma := p/t$; then

$$p = \gamma n^{-k/\ell}. \tag{2}$$

We have to show that almost no $G \in \mathcal{G}(n,p)$ lies in $\mathcal{P}_H$ if $\gamma \to 0$ as $n \to \infty$, and that almost all $G \in \mathcal{G}(n,p)$ lie in $\mathcal{P}_H$ if $\gamma \to \infty$ as $n \to \infty$.

For the first part of the proof, we find an upper bound for $E(X)$, the expected number of subgraphs of G isomorphic to H. As in the proof of Lemma 11.1.5, double counting gives

$$E(X) = \sum_{H' \in \mathcal{H}} P[H' \subseteq G]. \tag{3}$$

For every fixed $H' \in \mathcal{H}$, we have

$$P[H' \subseteq G] = p^\ell, \tag{4}$$

because $\|H\| = \ell$. Hence,

$$E(X) \underset{(3,4)}{=} |\mathcal{H}| p^\ell \underset{(1,2)}{\leqslant} n^k (\gamma n^{-k/\ell})^\ell = \gamma^\ell. \tag{5}$$

Thus if $\gamma \to 0$ as $n \to \infty$, then

$$P[G \in \mathcal{P}_H] = P[X \geqslant 1] \leqslant E(X) \leqslant \gamma^\ell \underset{n \to \infty}{\longrightarrow} 0$$

by Markov's inequality (11.1.4), so almost no $G \in \mathcal{G}(n,p)$ lies in $\mathcal{P}_H$.

We now come to the second part of the proof: we show that almost all $G \in \mathcal{G}(n, p)$ lie in $\mathcal{P}_H$ if $\gamma \to \infty$ as $n \to \infty$. Note first that, for $n \geqslant k$,

$$
\begin{aligned}
\binom{n}{k} n^{-k} &= \frac{1}{k!} \left(\frac{n}{n} \cdots \frac{n-k+1}{n} \right) \\
&\geqslant \frac{1}{k!} \left(\frac{n-k+1}{n} \right)^k \\
&\geqslant \frac{1}{k!} \left(1 - \frac{k-1}{k} \right)^k ;
\end{aligned}
\tag{6}
$$

thus, n^k exceeds $\binom{n}{k}$ by no more than a factor independent of n.

Our goal is to apply Lemma 11.4.2, and hence to bound $\sigma^2/\mu^2 = \left(E(X^2) - \mu^2 \right)/\mu^2$ from above. As in (3) we have

$$
E(X^2) = \sum_{(H', H'') \in \mathcal{H}^2} P\left[H' \cup H'' \subseteq G \right].
\tag{7}
$$

Let us then calculate these probabilities $P\left[H' \cup H'' \subseteq G \right]$. Given $H', H'' \in \mathcal{H}$, we have

$$
P\left[H' \cup H'' \subseteq G \right] = p^{2\ell - \|H' \cap H''\|}.
$$

Since H is balanced, $\varepsilon(H' \cap H'') \leqslant \varepsilon(H) = \ell/k$. With $|H' \cap H''| =: i$ this yields $\|H' \cap H''\| \leqslant i\ell/k$, so by $0 \leqslant p \leqslant 1$,

         i

$$
P\left[H' \cup H'' \subseteq G \right] \leqslant p^{2\ell - i\ell/k}.
\tag{8}
$$

We have now estimated the individual summands in (7); what does this imply for the sum as a whole? Since (8) depends on the parameter $i = |H' \cap H''|$, we partition the range $\mathcal{H}^2$ of the sum in (7) into the subsets

$$
\mathcal{H}_i^2 := \left\{ (H', H'') \in \mathcal{H}^2 : |H' \cap H''| = i \right\}, \qquad i = 0, \dots, k,
$$

         $\mathcal{H}_i^2$

and calculate for each $\mathcal{H}_i^2$ the corresponding sum

$$
A_i := \sum_i P\left[H' \cup H'' \subseteq G \right]
$$

         A_i

by itself. (Here, as below, we use $\sum_i$ to denote sums over all pairs $(H', H'') \in \mathcal{H}_i^2$.)

         $\sum_i$

If $i = 0$ then H' and H'' are disjoint, so the events $H' \subseteq G$ and $H'' \subseteq G$ are independent. Hence,

$$
A_0 = \sum_0 P\left[H' \cup H'' \subseteq G \right]
$$

$$\underset{(3)}{=} \sum_0 P[H' \subseteq G] \cdot P[H'' \subseteq G]$$

$$\leqslant \sum_{(H',H'') \in \mathcal{H}^2} P[H' \subseteq G] \cdot P[H'' \subseteq G]$$

$$= \left(\sum_{H' \in \mathcal{H}} P[H' \subseteq G] \right) \cdot \left(\sum_{H'' \in \mathcal{H}} P[H'' \subseteq G] \right)$$

$$\underset{(3)}{=} \mu^2. \tag{9}$$

Let us now estimate A_i for $i \geqslant 1$. Writing $\sum'$ for $\sum_{H' \in \mathcal{H}}$ and $\sum''$ for $\sum_{H'' \in \mathcal{H}}$, we note that $\sum_i$ can be written as $\sum' \sum''_{|H' \cap H''|=i}$. For fixed H' (corresponding to the first sum $\sum'$), the second sum ranges over

$$\binom{k}{i}\binom{n-k}{k-i} h$$

summands: the number of graphs $H'' \in \mathcal{H}$ with $|H'' \cap H'| = i$. Hence, for all $i \geqslant 1$ and suitable constants c_1, c_2 independent of n,

$$A_i = \sum_i P[H' \cup H'' \subseteq G]$$

$$\underset{(8)}{\leqslant} \sum' \binom{k}{i}\binom{n-k}{k-i} h\, p^{2\ell} p^{-i\ell/k}$$

$$\underset{(2)}{=} |\mathcal{H}| \binom{k}{i}\binom{n-k}{k-i} h\, p^{2\ell} \left(\gamma\, n^{-k/\ell} \right)^{-i\ell/k}$$

$$\leqslant |\mathcal{H}|\, p^{\ell} c_1\, n^{k-i}\, h\, p^{\ell} \gamma^{-i\ell/k}\, n^i$$

$$\underset{(5)}{=} \mu\, c_1 n^k h\, p^{\ell} \gamma^{-i\ell/k}$$

$$\underset{(6)}{\leqslant} \mu\, c_2 \binom{n}{k} h\, p^{\ell} \gamma^{-i\ell/k}$$

$$\underset{(1,5)}{=} \mu^2 c_2 \gamma^{-i\ell/k}$$

$$\leqslant \mu^2 c_2 \gamma^{-\ell/k}$$

if $\gamma \geqslant 1$. By definition of the A_i, this implies with $c_3 := kc_2$ that

$$E(X^2)/\mu^2 \underset{(7)}{=} \left(A_0/\mu^2 + \sum_{i=1}^{k} A_i/\mu^2 \right) \underset{(9)}{\leqslant} 1 + c_3 \gamma^{-\ell/k}$$

and hence

$$\frac{\sigma^2}{\mu^2} = \frac{E(X^2) - \mu^2}{\mu^2} \leqslant c_3 \gamma^{-\ell/k} \underset{\gamma \to \infty}{\longrightarrow} 0.$$

By Lemma 11.4.2, therefore, $X > 0$ almost surely, i.e. almost all $G \in \mathcal{G}(n,p)$ have a subgraph isomorphic to H and hence lie in $\mathcal{P}_H$. □

Theorem 11.4.3 allows us to read off threshold functions for a number of natural graph properties.

Corollary 11.4.4. *If $k \geqslant 3$, then $t(n) = n^{-1}$ is a threshold function for the property of containing a k-cycle.* □

Interestingly, the threshold function in Corollary 11.4.4 is independent of the cycle length k considered: in the evolution of random graphs, cycles of all (constant) lengths appear at about the same time!

There is a similar phenomenon for trees. Here, the threshold function does depend on the order of the tree considered, but not on its shape:

Corollary 11.4.5. *If T is a tree of order $k \geqslant 2$, then $t(n) = n^{-k/(k-1)}$ is a threshold function for the property of containing a copy of T.*

We finally have the following result for complete subgraphs:

Corollary 11.4.6. *If $k \geqslant 2$, then $t(n) = n^{-2/(k-1)}$ is a threshold function for the property of containing a K^k.*

Proof. K^k is balanced, because $\varepsilon(K^i) = \frac{1}{2}(i-1) < \frac{1}{2}(k-1) = \varepsilon(K^k)$ for $i < k$. With $\ell := \|K^k\| = \frac{1}{2}k(k-1)$, we obtain $n^{-k/\ell} = n^{-2/(k-1)}$. □

It is not difficult to adapt the proof of Theorem 11.4.3 to the case that H is unbalanced. The threshold then becomes $t(n) = n^{-1/\varepsilon'(H)}$, where $\varepsilon'(H) := \max\{\varepsilon(F) \mid F \subseteq H\}$; see Exercise 21.

Exercises

1.⁻ What is the probability that a random graph in $\mathcal{G}(n,p)$ has exactly m edges, for $0 \leqslant m \leqslant \binom{n}{2}$ fixed?

2.⁻ What is the expected number of edges in $G \in \mathcal{G}(n,p)$?

3.⁻ What is the expected number of K^r-subgraphs in $G \in \mathcal{G}(n,p)$?

4. Characterize the graphs that occur as a subgraph in every graph of sufficiently large average degree.

   (Hint. Erdős's theorem.)

5. In the usual terminology of measure spaces (and in particular, of probability spaces), the phrase 'almost all' is used to refer to a set of points whose complement has measure zero. Rather than considering a limit of probabilities in $\mathcal{G}(n,p)$ as $n \to \infty$, would it not be more natural to define a probability space on the set of *all* finite graphs (one copy of each) and to investigate properties of 'almost all' graphs in this space, in the sense above?

6. Show that if almost all $G \in \mathcal{G}(n,p)$ have a graph property $\mathcal{P}_1$ and almost all $G \in \mathcal{G}(n,p)$ have a graph property $\mathcal{P}_2$, then almost all $G \in \mathcal{G}(n,p)$ have both properties, i.e. have the property $\mathcal{P}_1 \cap \mathcal{P}_2$.

7.⁻ Show that, for p constant, almost every graph in $\mathcal{G}(n,p)$ has diameter 2.

8. Show that almost no graph in $\mathcal{G}(n,p)$ has a separating complete subgraph.

9. Derive Proposition 11.3.1 from Lemma 11.3.2.

10.⁺ (i) Show that with probability 1 an infinite random graph $G \in \mathcal{G}(\aleph_0, p)$ has all the properties $\mathcal{P}_{i,j}$ $(i, j \in \mathbb{N})$.

    (ii) Show that any two (infinite) graphs having all the properties $\mathcal{P}_{i,j}$ are isomorphic. (Hence, by (i), there is essentially only one countably infinite random graph.)

    (Hint for (ii). Enumerate the vertices of G and G' jointly, and construct an isomorphism $G \to G'$ inductively.)

11. Let $\epsilon > 0$ and $p = p(n) > 0$, and let $r \geqslant (1+\epsilon)(2\ln n)/p$ be an integer-valued function of n. Show that almost no graph in $\mathcal{G}(n,p)$ contains r independent vertices.

    (Hint. Imitate the proof of Lemma 11.2.1.)

12. Show that for every graph H there exists a function $p = p(n)$ such that $\lim_{n\to\infty} p(n) = 0$ but almost every $G \in \mathcal{G}(n,p)$ contains an induced copy of H.

    (Hint. Imitate the proof of Proposition 11.3.1. To bound the probabilities involved, use the inequality $1 - x \leqslant e^{-x}$ as in the proof of Lemma 11.2.1.)

13.⁺ (i) Show that, for every $\epsilon > 0$ and $p = (1-\epsilon)(\ln n)n^{-1}$, almost every $G \in \mathcal{G}(n,p)$ has an isolated vertex.

    (ii) Find a probability $p = p(n)$ such that almost every $G \in \mathcal{G}(n,p)$ is disconnected but the expected number of spanning trees of G tends to infinity as $n \to \infty$.

    (Hint for (ii): A theorem of Cayley states that K^n has exactly n^{n-2} spanning trees.)

14. Find an increasing graph property without a threshold function, and a property that is not increasing but has a threshold function.

15.⁻ Let H be a graph of order k, and let h denote the number of graphs isomorphic to H on some fixed set of k elements. Show that $h \leqslant k!$. For which graphs H does equality hold?

16.⁻ For every $k \geqslant 1$, find a threshold function for $\{\, G \mid \Delta(G) \geqslant k \,\}$.

(Hint. This is a result from the text in disguise.)

17.⁻ Given $d \in \mathbb{N}$, is there a threshold function for the property of containing a d-dimensional cube (see Ex. 2, Ch. 1)? If so, which; if not, why not?

18. Show that $t(n) = n^{-1}$ is also a threshold function for the property of containing *any* cycle.

19. Does the property of containing any tree of order k (for $k \geqslant 2$ fixed) have a threshold function? If so, which?

20.⁺ Given a graph H, let $\mathcal{P}$ be the property of containing an induced copy of H. If H is complete then, by Corollary 11.4.6, $\mathcal{P}$ has a threshold function. Show that $\mathcal{P}$ has no threshold function if H is not complete.

(Hint. Show first that no such threshold function $t = t(n)$ can tend to zero as $n \to \infty$. Then use Exercise 12.)

21.⁺ Prove the following version of Theorem 11.4.3 for unbalanced subgraphs. Let H be any graph with at least one edge, and put $\varepsilon'(H) :=$ $\max \{\, \varepsilon(F) \mid \emptyset \neq F \subseteq H \,\}$. Then the threshold function for $\mathcal{P}_H$ is $t(n) = n^{-1/\varepsilon'(H)}$.

(Hint. Imitate the proof of Theorem 11.4.3. Instead of the sets $\mathcal{H}_i$, consider the sets $\mathcal{H}_F^2 := \{\, (H, H'') \in \mathcal{H}^2 \mid H' \cap H'' = F \,\}$. Replace the distinction between the cases of $i = 0$ and $i > 0$ by the distinction between the cases of $\|F\| = 0$ and $\|F\| > 0$.)

Notes

There are a number of monographs and texts on the subject of random graphs. The most comprehensive of these is B. Bollobás, *Random Graphs*, Academic Press 1985. Another advanced but very readable monograph is S. Janson, T. Łuczak & A. Ruciński, *Topics in Random Graphs*, in preparation; this concentrates on areas developed since *Random Graphs* was published. E.M. Palmer, *Graphical Evolution*, Wiley 1985, covers material similar to parts of *Random Graphs* but is written in a more elementary way. Compact introductions going beyond what is covered in this chapter are given by B. Bollobás, *Graph Theory*, Springer GTM 63, 1979, and by M. Karoński, *Handbook of Combinatorics* (R.L. Graham, M. Grötschel & L. Lovász, eds.), North-Holland 1995.

A stimulating advanced introduction to the use of random techniques in discrete mathematics more generally is given by N. Alon & J.H. Spencer, *The Probabilistic Method*, Wiley 1992. One of the attractions of this book lies in the way it shows probabilistic methods to be relevant in proofs of entirely deterministic theorems, where nobody would suspect it. Another example for this phenomenon is Alon's proof of Theorem 5.4.1; see the notes for Chapter 5.

The probabilistic method had its first origins in the 1940s, one of its earliest results being Erdős's probabilistic lower bound for Ramsey numbers (Theorem 11.1.3). Lemma 11.3.2 about the properties $\mathcal{P}_{i,j}$ is taken from Bollobás's Springer text cited above. A very readable rendering of the proof that, for constant p, every first order sentence about graphs is either almost surely true or almost surely false, is given by P. Winkler, Random structures and zero-one laws, in (N.W. Sauer et al., eds.) *Finite and Infinite Combinatorics in Sets and Logic* (NATO ASI Series C **411**), Kluwer 1993.

The seminal paper on graph evolution is P. Erdős & A. Rényi, On the evolution of random graphs, *Publ. Math. Inst. Hungar. Acad. Sci.* **5** (1960), 17–61. This paper also includes Theorem 11.4.3 and its proof. The generalization of this theorem to unbalanced subgraphs was first proved by Bollobás in 1981, using advanced methods; a simple adaptation of the original Erdős-Renyi proof was found by Ruciński & Vince (1986), and is presented in Karoński's Handbook chapter.

There is another way of defining a random graph G, which is just as natural and common as the model we considered. Rather than choosing the edges of G independently, we choose the entire graph G uniformly at random from among all the graphs on $\{0, \ldots, n-1\}$ that have exactly $M = M(n)$ edges: then each of these graphs occurs with the same probability of $\binom{N}{M}$, where $N := \binom{n}{2}$. Just as we studied the likely properties of the graphs in $\mathcal{G}(n,p)$ for different functions $p = p(n)$, we may investigate how the properties of G in the other model depend on the function $M(n)$. If M is close to pN, the expected number of edges of a graph in $\mathcal{G}(n,p)$, then the two models behave very similarly. It is then largely a matter of convenience which of them to consider; see Bollobás for details.

In order to study threshold phenomena in more detail, one often considers the following *random graph process*: starting with a $\overline{K^n}$ as stage zero, one chooses additional edges one by one (uniformly at random) until the graph is complete. This is a simple example of a Markov chain, whose Mth stage corresponds to the 'uniform' random graph model described above. A survey about threshold phenomena in this setting is given by T. Łuczak, The phase transition in a random graph, in (D. Miklos, V.T. Sós & T. Szőnyi, eds.) *Paul Erdős is 80*, Vol. 2, Proc. Colloq. Math. Soc. János Bolyai (1996).

12 Minors, Trees, and WQO

Our goal in this last chapter is a single theorem, one which dwarfs any other result in graph theory and may doubtless be counted among the deepest theorems that mathematics has to offer: *in every infinite set of graphs there are two such that one is a minor of the other.* This *minor theorem*, inconspicuous though it may look at first glance, has made a fundamental impact both outside graph theory and within. Its proof, due to Neil Robertson and Paul Seymour, takes well over 500 pages.

So we have to be modest: of the actual proof of the minor theorem, this chapter will convey only a very rough impression. However, as with most truly fundamental results, the proof has sparked off the development of methods of quite independent interest and potential. This is true particularly for the use of *tree-decompositions*, a technique we shall meet in Sections 12.3 and 12.4. Section 12.1 gives an introduction to *well-quasi-ordering*, a concept central to the minor theorem. In Section 12.2 we apply this concept to prove the minor theorem for trees. The chapter finishes with an overview in Section 12.5 of the proof of the general minor theorem, and of some of its immediate consequences.

12.1 Well-quasi-ordering

A reflexive and transitive relation is called a *quasi-ordering*. A quasi-ordering $\leqslant$ on X is a *well-quasi-ordering*, and the elements of X are *well-quasi-ordered* by $\leqslant$, if for every infinite sequence $x_0, x_1, \ldots$ in X

well-quasi-ordering

good pair

there are indices $i < j$ such that $x_i \leqslant x_j$. Then (x_i, x_j) is a *good pair* of this sequence. A sequence containing a good pair is a *good sequence*; thus, a quasi-ordering on X is a well-quasi-ordering if and only if every infinite sequence in X is good. An infinite sequence is *bad* if it is not good.

good sequence

bad sequence

Proposition 12.1.1. *A quasi-ordering $\leqslant$ on X is a well-quasi-ordering if and only if X contains neither an infinite antichain nor an infinite strictly decreasing sequence $x_0 > x_1 > \ldots$.*

(9.1.2)

Proof. The forward implication is trivial. Conversely, let $x_0, x_1, \ldots$ be any infinite sequence in X. Let K be the complete graph on $\mathbb{N} = \{0, 1, \ldots\}$. Colour the edges ij ($i < j$) of K with three colours: green if $x_i \leqslant x_j$, red if $x_i > x_j$, and amber if x_i, x_j are incomparable. By Ramsey's theorem (9.1.2), K has an infinite induced subgraph whose edges all have the same colour. If there is neither an infinite antichain nor an infinite strictly decreasing sequence in X, then this colour must be green, i.e. $x_0, x_1, \ldots$ has an infinite subsequence in which every pair is good. In particular, the sequence $x_0, x_1, \ldots$ is good. $\square$

In the proof of Proposition 12.1.1, we showed more than was needed: rather than finding a single good pair in $x_0, x_1, \ldots$, we found an infinite increasing subsequence. We have thus shown the following:

Corollary 12.1.2. *If X is well-quasi-ordered, then every infinite sequence in X has an infinite increasing subsequence.* $\square$

The following lemma, and the idea of its proof, are fundamental to the theory of well-quasi-ordering. Let $\leqslant$ be a quasi-ordering on a set X.

$\leqslant$

For finite subsets $A, B \subseteq X$, write $A \leqslant B$ if there is an injective mapping $f \colon A \to B$ such that $a \leqslant f(a)$ for all $a \in A$. This naturally extends $\leqslant$ to

$[X]^{<\omega}$

a quasi-ordering on $[X]^{<\omega}$, the set of all finite subsets of X.

[12.2.1]

Lemma 12.1.3. *If X is well-quasi-ordered by $\leqslant$, then so is $[X]^{<\omega}$.*

Proof. Suppose that $\leqslant$ is a well-quasi-ordering on X but not on $[X]^{<\omega}$. We start by constructing a bad sequence $(A_n)_{n \in \mathbb{N}}$ in $[X]^{<\omega}$, as follows. Given $n \in \mathbb{N}$, assume inductively that A_i has been defined for every $i < n$, and that there exists a bad sequence in $[X]^{<\omega}$ starting with $A_0, \ldots, A_{n-1}$. (This is clearly true for $n = 0$: by assumption, $[X]^{<\omega}$ contains a bad sequence, and this has the empty sequence as an initial segment.) Choose $A_n \in [X]^{<\omega}$ so that some bad sequence in $[X]^{<\omega}$ starts with $A_0, \ldots, A_n$ and $|A_n|$ is minimum.

Clearly, $(A_n)_{n \in \mathbb{N}}$ is a bad sequence in $[X]^{<\omega}$; in particular, $A_n \neq \emptyset$ for all n. For each n pick an element $a_n \in A_n$ and set $B_n := A_n \setminus \{a_n\}$.

By Corollary 12.1.2, the sequence $(a_n)_{n \in \mathbb{N}}$ has an infinite increasing subsequence $(a_{n_i})_{i \in \mathbb{N}}$. By the minimal choice of A_{n_0}, the sequence

$$A_0, \ldots, A_{n_0-1}, B_{n_0}, B_{n_1}, B_{n_2}, \ldots$$

is good; consider a good pair. Since $(A_n)_{n \in \mathbb{N}}$ is bad, this pair cannot have the form (A_i, A_j) or (A_i, B_j), as $B_j \leqslant A_j$. So it has the form (B_i, B_j). Extending the injection $B_i \to B_j$ by $a_i \mapsto a_j$, we deduce again that (A_i, A_j) is good, a contradiction. $\qquad\qquad\qquad\qquad\qquad\square$

12.2 The minor theorem for trees

The minor theorem can be expressed by saying that the finite graphs are well-quasi-ordered by the minor relation $\preccurlyeq$: by Proposition 12.1.1 and the obvious fact that no strictly descending sequence of minors can be infinite, this is equivalent to the non-existence of an infinite antichain, the formulation chosen earlier.

In this section, we prove a strong version of the minor theorem for trees:

Theorem 12.2.1. (Kruskal 1960)
The finite trees are well-quasi-ordered by the topological minor relation.

We shall base the proof of Theorem 12.2.1 on the following strengthening of the topological minor relation for trees. If T is a tree and $r \in T$ is any fixed vertex, we call the pair (T, r) a *rooted tree*, and r its *root*.   *rooted tree*
Often, we simply write T for (T, r). Given two rooted trees (T, r) and (T', r'), we write $(T, r) \leqslant (T', r')$ if there exists an isomorphism φ be- $\leqslant$
tween some subdivision of T and a subtree T'' of T' such that the path $r'T'\varphi(r)$ contains no vertex of T'' other than $\varphi(r)$ (Fig. 12.2.1). As one easily checks, this is a quasi-ordering on the class of all rooted trees.

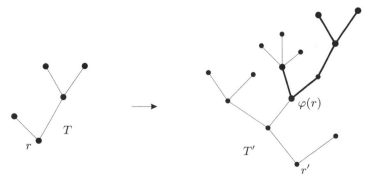

Fig. 12.2.1. $(T, r) \leqslant (T', r')$

(12.1.3)

Proof of Theorem 12.2.1. Since $(T, r) \leqslant (T', r')$ implies that T is a topological minor of T', it suffices to show that the rooted trees are well-quasi-ordered by $\leqslant$.

We proceed as in the proof of Lemma 12.1.3. Suppose the rooted trees are not well-quasi-ordered by $\leqslant$. Given $n \in \mathbb{N}$, assume inductively that we have chosen a sequence $(T_0, r_0), \ldots, (T_{n-1}, r_{n-1})$ of n rooted trees such that some bad sequence of rooted trees starts with this sequence. Choose as (T_n, r_n) a minimal-order rooted tree such that some bad sequence starts with $(T_0, r_0), \ldots, (T_n, r_n)$.

Clearly, $((T_n, r_n))_{n \in \mathbb{N}}$ is a bad sequence; in particular, none of the trees T_n is trivial. For each n, let A_n denote the set of all rooted trees (T, r) such that T is a component of $T_n - r_n$ and r is adjacent to r_n in T_n. Let us prove that the set $A := \bigcup_{n \in \mathbb{N}} A_n$ of all these rooted trees is well-quasi-ordered.

Let $(T^k)_{k \in \mathbb{N}}$ be any sequence of rooted trees in A. For every $k \in \mathbb{N}$ choose an $n = n(k)$ such that $T^k \in A_n$. Pick a k with smallest $n(k)$. Then

$$T_0, \ldots, T_{n(k)-1}, T^k, T^{k+1}, \ldots$$

is a good sequence, by the minimal choice of $T_{n(k)}$ and $T^k \subsetneqq T_{n(k)}$. Let (T, T') be a good pair of this sequence. Since $(T_n)_{n \in \mathbb{N}}$ is bad, T cannot be among the first $n(k)$ members $T_0, \ldots, T_{n(k)-1}$ of our sequence: then T' would be some T^i with $i \geqslant k$, i.e.

$$T \leqslant T' = T^i \leqslant T_{n(i)} \,;$$

since $n(k) \leqslant n(i)$ by the choice of k, this would make $(T, T_{n(i)})$ a good pair in the bad sequence $(T_n)_{n \in \mathbb{N}}$. Hence (T, T') is a good pair also in $(T^k)_{k \in \mathbb{N}}$, completing the proof that A is well-quasi-ordered.

By Lemma 12.1.3, the sequence $(A_n)_{n \in \mathbb{N}}$ in $[A]^{<\omega}$ has a good pair (A_i, A_j); let $f \colon A_i \to A_j$ be an injection such that $T \leqslant f(T)$ for all $T \in A_i$. Thus for every $T \in A_i$, with root r say, some subdivision of T is isomorphic to a subtree T'' of $T' := f(T)$ such that the path in T' between the image r'' of r under the isomorphism and the root r' of T' contains no vertex of T'' other than r''. Since r' is a neighbour of r_j, this means that the path $P := r'' T_j r_j$, likewise, meets T'' only in r''. As f is injective, the union of all these graphs $T'' \cup P$ (one for every $T \in A_i$) is a subtree of T_j isomorphic to a subdivision of T_i, the paths P corresponding to the edges of T_i at r_i. Hence (T_i, T_j) is a good pair in our original bad sequence of rooted trees, a contradiction. $\qquad\square$

12.3 Tree-decompositions

Trees are graphs with some very distinctive and fundamental properties; consider Theorem 1.5.1 and Corollary 1.5.2, or the more sophisticated example of Kruskal's theorem. It is therefore legitimate to ask to what degree those properties can be transferred to more general graphs, graphs that are not themselves trees but tree-like in some sense.[1] In this section, we study a concept of tree-likeness that permits generalizations of all the tree properties referred to above (including Kruskal's theorem), and which plays a crucial role in the proof of the minor theorem.

Let G be a graph, T a tree, and let $\mathcal{V} = (V_t)_{t \in T}$ be a family of vertex sets $V_t \subseteq V(G)$ indexed by the vertices t of T. The pair $(T, \mathcal{V})$ is called a *tree-decomposition* of G if it satisfies the following three conditions:

tree-decomposition

(T1) $V(G) = \bigcup_{t \in T} V_t$;

(T2) for every edge $e \in G$ there exists a $t \in T$ such that both ends of e lie in V_t;

(T3) $V_{t_1} \cap V_{t_3} \subseteq V_{t_2}$ whenever $t_1, t_2, t_3 \in T$ satisfy $t_2 \in t_1 T t_3$.

Conditions (T1) and (T2) together say that G is the union of the subgraphs $G[V_t]$; we call these subgraphs the *parts* of $(T, \mathcal{V})$ and say that $(T, \mathcal{V})$ is a tree-decomposition of G *into* these parts. Condition (T3) implies that the parts of $(T, \mathcal{V})$ are organized roughly like a tree (Fig. 12.3.1).

parts

into

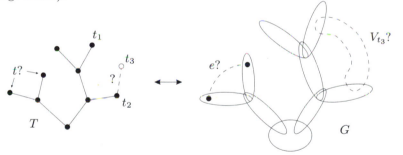

Fig. 12.3.1. Edges and parts ruled out by (T2) and (T3)

Before we discuss the role that tree-decompositions play in the proof of the minor theorem, let us note some of their basic properties. Consider a fixed tree-decomposition $(T, \mathcal{V})$ of G, with $\mathcal{V} = (V_t)_{t \in T}$ as above. As an immediate consequence of the definition of tree-decomposition, note that $(T, \mathcal{V})$ induces a decomposition on every subgraph of G:

$(T, \mathcal{V}), V_t$

Lemma 12.3.1. *For every $H \subseteq G$, the pair $\bigl(T, (V_t \cap V(H))_{t \in T}\bigr)$ is a tree-decomposition of H.*   □

[12.4.2]
[12.4.3]
[12.4.5]

---

[1] What exactly this 'sense' should be will depend both on the property considered and on its intended application.

One of the most basic features of a tree-decomposition is that it transfers the separation properties of its tree to the graph decomposed:

[12.4.5]   **Lemma 12.3.2.** *Let $t_1 t_2$ be any edge of T, let T_1, T_2 be the components of $T - t_1 t_2$, with $t_1 \in T_1$ and $t_2 \in T_2$, and put $U_1 := \bigcup_{t \in T_1} V_t$ and $U_2 := \bigcup_{t \in T_2} V_t$. Then $V_{t_1} \cap V_{t_2} = U_1 \cap U_2$, and $V_{t_1} \cap V_{t_2}$ separates U_1 from U_2 in G (Fig. 12.3.2).*

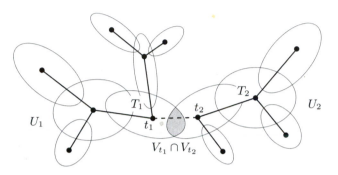

Fig. 12.3.2. $V_{t_1} \cap V_{t_2}$ separates U_1 from U_2

Proof. Both t_1 and t_2 lie on every t–t' path in T with $t \in T_1$ and $t' \in T_2$. Therefore $U_1 \cap U_2 \subseteq V_{t_1} \cap V_{t_2}$ by (T3), and hence $U_1 \cap U_2 = V_{t_1} \cap V_{t_2}$. It remains to show that G has no edge $u_1 u_2$ with $u_1 \in U_1 \smallsetminus U_2$ and $u_2 \in U_2 \smallsetminus U_1$. If $u_1 u_2$ is such an edge, then by (T2) there is a $t \in T$ with $u_1, u_2 \in V_t$. By the choice of u_1 and u_2 we have neither $t \in T_2$ nor $t \in T_1$, a contradiction.   □

Lemma 12.3.3. *Given a set $W \subseteq V(G)$, there is either a $t \in T$ such that $W \subseteq V_t$, or there are vertices $w_1, w_2 \in W$ and an edge $t_1 t_2 \in T$ such that w_1, w_2 lie outside the set $V_{t_1} \cap V_{t_2}$ and are separated by it in G.*

Proof. Let us orient the edges of T as follows. Given an edge $t_1 t_2 \in T$, define U_1, U_2 as in Lemma 12.3.2; then $V_{t_1} \cap V_{t_2} = U_1 \cap U_2$ separates U_1 from U_2. If $V_{t_1} \cap V_{t_2}$ does not separate any two vertices of W that lie outside it, we can find an $i \in \{1, 2\}$ such that $W \subseteq U_i$, and orient $t_1 t_2$ towards t_i. Now let t be the last vertex of a maximal directed path in T; then all the edges of T at t are directed towards t. Hence W lies in the intersection of all the corresponding sets U_i, which is precisely V_t.   □

The following special case of Lemma 12.3.3 is used particularly often:

[12.4.3]   **Lemma 12.3.4.** *Any complete subgraph of G is contained in some part of $(T, \mathcal{V})$.*   □

Lemma 12.3.4 can be used to show that the minimal complete separators of a graph H define a tree-decomposition of H into a unique set of parts (Exercise 16). Similarly, if H can be constructed recursively from a set $\mathcal{H}$ of graphs by pasting along complete subgraphs, then H has a tree-decomposition into elements of $\mathcal{H}$. For example, by Wagner's Theorem 8.3.4, any graph without a K^5 minor has a supergraph with a tree-decomposition into maximal planar graphs and copies of the Wagner graph W, and similarly for graphs without K^4 minors (see Proposition 12.4.3).

As indicated by Figure 12.3.1, the parts of $(T, \mathcal{V})$ reflect the structure of the tree T, so in this sense the graph G decomposed resembles a tree. However, this is valuable only inasmuch as the structure of G within each part is negligible: the smaller the parts, the closer the resemblance.

This observation motivates the following definition. The *width* of $(T, \mathcal{V})$ is the number

$$\max \left\{ |V_t| - 1 : t \in T \right\},$$

width

and the *tree-width* tw(G) of G is the least width of any tree-decomposition of G. As one easily checks,[2] trees themselves have tree-width 1.

tree-width
tw(G)

Graphs of bounded tree-width are sufficiently similar to trees that it becomes possible to adapt the proof of Kruskal's theorem to the class of these graphs; very roughly, one has to iterate the 'minimal bad subsequence' argument from the proof of Lemma 12.1.3 tw(G) times. This takes us a step further towards a proof of the minor theorem:

Theorem 12.3.5. (Robertson & Seymour 1990)
For every integer $k > 0$, the graphs of tree-width $< k$ are well-quasi-ordered by the minor relation.

In order to make use of Theorem 12.3.5 for a proof of the general minor theorem, we should be able to say something special about the graphs it does not cover, i.e. to deduce some information about a graph from the assumption that its tree-width is large. Our next theorem achieves just that: it identifies a canonical obstruction to small tree-width, a structural phenomenon that occurs in a graph if and only if its tree-width is large.[3]

Let $\mathcal{C}$ be a set of connected subsets of $V(G)$. Extending our terminology of Chapter 2.1, let us say that a set $U \subseteq V(G)$ *covers* $\mathcal{C}$ if it meets

covers

---

[2] Indeed the '-1' in the definition of width serves no other purpose than to make this statement true.

[3] The obstructions actually used in the proof of the minor theorem, so-called *tangles*, are similar in spirit to those of Theorem 12.3.6; they are more powerful, but also more complicated to describe.

touch   every element of $\mathcal{C}$. Moreover, we say that two elements of $\mathcal{C}$ *touch* if they meet or G contains an edge between them. Now if every two sets in $\mathcal{C}$ touch, we can find a $t \in T$ such that V_t covers $\mathcal{C}$ (Exercise 17). This implies the easy backward direction of the following characterization of the graphs of tree-width $\geqslant k$:

Theorem 12.3.6. (Seymour & Thomas 1993)
Let $k > 0$ be an integer. A graph G has tree-width $\geqslant k$ if and only if G contains a collection of mutually touching connected vertex sets that cannot be covered by $\leqslant k$ vertices.

Given any two vertices $t_1, t_2 \in T$, Lemma 12.3.2 implies that every V_t with $t \in t_1 T t_2$ separates V_{t_1} from V_{t_2} in G. Let us call our tree-*linked/lean*   decomposition $(T, \mathcal{V})$ of G *linked*, or *lean*,[4] if it satisfies the following condition:

(T4) given any $s \in \mathbb{N}$ and $t_1, t_2 \in T$, either G contains s disjoint V_{t_1}–V_{t_2} paths or there exists a $t \in t_1 T t_2$ such that $|V_t| < s$.

The 'branches' in a lean tree-decomposition are thus stripped of any bulk not necessary to maintain their connecting qualities: if a branch is thick (the parts along a path in T large), then G is highly connected along this branch.

In our quest for tree-decompositions into 'small' parts, we now have two criteria to choose between: the global 'worst case' criterion of width, which ensures that T is nontrivial (unless G is complete) but says nothing about the tree-likeness of G among parts other than the largest, and the more subtle local criterion of leanness, which ensures tree-likeness everywhere along T but might be difficult to achieve except with trivial or near-trivial T. Surprisingly, though, it is always possible to find a tree-decomposition that is optimal with respect to both criteria at once:

Theorem 12.3.7. (Thomas 1990)
Every graph G has a lean tree-decomposition of width $\mathrm{tw}(G)$.

The proof of Theorem 12.3.7 is not too long but technical, so we shall not present it. The fact that this theorem gives us a very useful property of minimum-width tree-decompositions 'for free' has made it a valuable tool wherever tree-decompositions are applied.

Tree-decompositions can lead to intuitive structural characterizations of graph properties. The reformulation of Wagner's theorem (see above) is the classical example; another is the following characterization of chordal graphs:

---

[4] depending on which of the two dual aspects of (T4) we wish to emphasize

Proposition 12.3.8. *G is chordal if and only if G has a tree-decompo-* [12.4.3]
sition into complete parts.

Proof. We apply induction on $|G|$. We first assume that G has a tree- (5.5.1)
decomposition $(T, \mathcal{V})$ such that $G[V_t]$ is complete for every $t \in T$; let
us choose $(T, \mathcal{V})$ with $|T|$ minimum. If $|T| \leqslant 1$, then G is complete and
hence chordal. If $|T| \geqslant 2$, then $T = (T_1 \cup T_2) + t_1 t_2$ as in Lemma 12.3.2.
By (T1) and (T2), $G = G_1 \cup G_2$ with $G_i := G[\bigcup_{t \in T_i} V_t]$ $(i = 1, 2)$, and
$V(G_1 \cap G_2) = V_{t_1} \cap V_{t_2}$ by the lemma; thus, $G_1 \cap G_2$ is complete. Since
$(T_i, (V_t)_{t \in T_i})$ is a tree-decomposition of G_i into complete parts, both
G_i are chordal by the induction hypothesis. (By the choice of $(T, \mathcal{V})$,
neither G_i is a subgraph of $G[V_{t_1} \cap V_{t_2}] = G_1 \cap G_2$, so both G_i are
indeed smaller than G.) Since $G_1 \cap G_2$ is complete, any induced cycle in
G lies in G_1 or in G_2 and hence has a chord, so G too is chordal.

Conversely, assume that G is chordal. If G is complete, there is
nothing to show. If not then, by Proposition 5.5.1, G is the union of
smaller chordal graphs G_1, G_2 with $G_1 \cap G_2$ complete. By the induction
hypothesis, G_1 and G_2 have tree-decompositions $(T_1, \mathcal{V}_1)$ and $(T_2, \mathcal{V}_2)$
into complete parts. By Lemma 12.3.4, $G_1 \cap G_2$ lies inside one of those
parts in each case, say with indices $t_1 \in T_1$ and $t_2 \in T_2$. As one easily
checks, $((T_1 \cup T_2) + t_1 t_2, \mathcal{V}_1 \cup \mathcal{V}_2)$ is a tree-decomposition of G into complete parts. □

Corollary 12.3.9.
$$\mathrm{tw}(G) = \min \{\omega(H) - 1 \mid V(H) = V(G);\ E(H) \supseteq E(G);\ H \text{ chordal}\}$$

Proof. By Lemma 12.3.4 and Proposition 12.3.8, each of the graphs H
considered for the minimum has a tree-decomposition of width $\omega(H) - 1$.
Every such tree-decomposition is also one of G, so $\mathrm{tw}(G) \leqslant \omega(H) - 1$
for every H.

Conversely, we construct an H as above with $\omega(H) - 1 \leqslant \mathrm{tw}(G)$. Let
$(T, \mathcal{V})$ be a tree-decomposition of G of width $\mathrm{tw}(G)$. For every $t \in T$ let
K_t denote the complete graph on V_t, and put $H := \bigcup_{t \in T} K_t$. Clearly,
$(T, \mathcal{V})$ is also a tree-decomposition of H. By Proposition 12.3.8, H is
chordal, and by Lemma 12.3.4, $\omega(H) - 1$ is at most the width of $(T, \mathcal{V})$,
i.e. at most $\mathrm{tw}(G)$. □

12.4 Tree-width and forbidden minors

If $\mathcal{X}$ is any set or class of graphs, then the class

$$\mathrm{Forb}_{\preccurlyeq}(\mathcal{X}) := \{G \mid G \not\succcurlyeq X \text{ for all } X \in \mathcal{X}\}$$ $\mathrm{Forb}_{\preccurlyeq}(\mathcal{X})$

of all graphs without a minor in $\mathcal{X}$ is a graph property, i.e. is closed under

forbidden
minors
isomorphism.[5] When it is written as above, we say that this property
is expressed by specifying the graphs $X \in \mathcal{X}$ as *forbidden* (or *excluded*)
minors.

(1.7.3)
By Proposition 1.7.3, $\mathrm{Forb}_{\preccurlyeq}(\mathcal{X})$ is closed under taking minors: if
$H \preccurlyeq G \in \mathrm{Forb}_{\preccurlyeq}(\mathcal{X})$ then $H \in \mathrm{Forb}_{\preccurlyeq}(\mathcal{X})$. Graph properties that are
hereditary   closed under taking minors will be called *hereditary* in this chapter.
Every hereditary property can in turn be expressed by forbidden minors:

Proposition 12.4.1. *A graph property $\mathcal{P}$ can be expressed by forbidden
minors if and only if it is hereditary.*

$\overline{\mathcal{P}}$
Proof. For the 'if' part, note that $\mathcal{P} = \mathrm{Forb}_{\preccurlyeq}(\overline{\mathcal{P}})$, where $\overline{\mathcal{P}}$ is the
complement of $\mathcal{P}$. $\square$

In Section 12.5, we shall return to the general question of how a given
hereditary property is best represented by forbidden minors. In this
section, we are interested in one particular type of hereditary property:
bounded tree-width.

Proposition 12.4.2. *For every $k > 0$, the property of having tree-width
$< k$ is hereditary.*

(1.7.1)
(12.3.1)
Proof. By Lemma 12.3.1 and Proposition 1.7.1, it suffices to prove
that the tree-width of a graph cannot increase when we contract an
edge. So let $(T, (V_t)_{t \in T})$ be a tree-decomposition of a graph G, and
let $xy \in G$ be an edge. For every $t \in T$ with $V_t \cap \{x, y\} \neq \emptyset$ set
$V'_t := (V_t \smallsetminus \{x, y\}) \cup \{v_{xy}\}$; for all other $t \in T$ put $V'_t := V_t$. We
show that $(T, (V'_t)_{t \in T})$ is a tree-decomposition of $G' := G/xy$; then, in
particular, $\mathrm{tw}(G') \leqslant \mathrm{tw}(G)$.

The assertions (T1) and (T2) for G' follow immediately from the
corresponding assertions for G. To establish (T3) for G', we only have to
show that, given $x \in V_{t_1}$ and $y \in V_{t_3}$, every V_{t_2} with $t_2 \in t_1 T t_3$ contains
either x or y (so that $v_{xy} \in V'_{t_2}$). By (T2) for G, there exists a $t \in T$
with $x, y \in V_t$. Hence every V_{t_2} with $t_2 \in t_1 T t$ contains x, and every V_{t_2}
with $t_2 \in t T t_3$ contains y. The assertion thus follows from the fact that
$t_1 T t_3 \subseteq t_1 T t \cup t T t_3$. $\square$

By Propositions 12.4.1 and 12.4.2, the property of having tree-width
$< k$ can be expressed by forbidden minors. Choosing their set $\mathcal{X}$ as small
as possible, we find that $\mathcal{X} = \{ K^3 \}$ for $k = 2$: the graphs of tree-width
< 2 are precisely the forests. For $k = 3$, we have $\mathcal{X} = \{ K^4 \}$:

Proposition 12.4.3. *A graph has tree-width < 3 if and only if it has
no K^4 minor.*

---

[5] As usual, we abbreviate $\mathrm{Forb}_{\preccurlyeq}(\{ X \})$ to $\mathrm{Forb}_{\preccurlyeq}(X)$.

(8.3.1)
(12.3.1)
(12.3.4)
(12.3.8)

Proof. By Lemma 12.3.4, we have $\mathrm{tw}(K^4) \geqslant 3$. By Proposition 12.4.2, therefore, a graph of tree-width < 3 cannot contain K^4 as a minor.

Conversely, let G be a graph without a K^4 minor; we assume that $|G| \geqslant 3$. Add edges to G until the graph G' obtained is edge-maximal without a K^4 minor. By Proposition 8.3.1, G' can be constructed recursively from triangles by pasting along K^2s. By induction on the number of recursion steps and Lemma 12.3.4, every graph constructible in this way has a tree-decomposition into triangles (as in the proof of Proposition 12.3.8). Such a tree-decomposition of G' has width 2, and by Lemma 12.3.1 it is also a tree-decomposition of G. $\qquad\square$

A question converse to the above is to ask for which X (other than K^3 and K^4) the tree-width of the graphs in $\mathrm{Forb}_{\preccurlyeq}(X)$ is bounded. Interestingly, it is not difficult to show that any such X must be planar. Indeed, consider the graph on $\{1, \ldots, n\}^2$ with the edge set

$$\{\, (i,j)(i',j') : |i - i'| + |j - j'| = 1 \,\};$$

this graph is called the $n \times n$ *grid*. Clearly, the $n \times n$ grid is planar (for every n), and hence lies in every class $\mathrm{Forb}_{\preccurlyeq}(X)$ with non-planar X. On the other hand, it is not difficult to show that the tree-width of the $n \times n$ grid tends to infinity with n (Exercise 19). Therefore, the tree-width of the graphs in $\mathrm{Forb}_{\preccurlyeq}(X)$ cannot be bounded unless X is planar.

grid

The following deep and surprising theorem says that, conversely, the tree-width of the graphs in $\mathrm{Forb}_{\preccurlyeq}(X)$ is bounded for every planar X:

Theorem 12.4.4. (Robertson & Seymour 1986)
The tree-width of the graphs in $\mathrm{Forb}_{\preccurlyeq}(X)$ is bounded if and only if X is planar.

The proof of Theorem 12.4.4 is too involved to be presented here. However, there is a similar result on the related notion of 'path-width', which we shall prove instead: its proof is much simpler, but it gives an indication of some of the techniques used for the proof of Theorem 12.4.4.

A tree-decomposition whose tree is a path is called a *path-decomposition*. We usually denote a path-decomposition $(P, \mathcal{V})$ simply by listing the sets $V_1, \ldots, V_s \in \mathcal{V}$ in the order defined by P. The least width of a path-decomposition of G is the *path-width* $\mathrm{pw}(G)$ of G.

path-decomposition

path-width
$\mathrm{pw}(G)$

The analogue of Theorem 12.4.4 for path-width is obtained simply by replacing planarity with acyclicity:

Theorem 12.4.5. (Robertson & Seymour 1983)
The path-width of the graphs in $\mathrm{Forb}_{\preccurlyeq}(X)$ is bounded if and only if X is a forest.

The forward implication of Theorem 12.4.5 is again easy. All we have to show is that trees can have arbitrarily large path-width: since

$\mathrm{Forb}_{\preccurlyeq}(X)$ contains all trees if X has a cycle, this will imply that forbidding X cannot bound the path-width unless X is a forest.

How can one show that a graph—in our case, a tree—has large path-width? Let $(V_1, \dots, V_s)$ be a path-decomposition of some connected graph G, of width $\mathrm{pw}(G)$ and such that $V_1, V_s \neq \emptyset$. Pick vertices $v_1 \in V_1$ and $v_s \in V_s$, and let Q be a v_1–v_s path in G. By Lemma 12.3.2, Q meets every V_r, $r = 1, \dots, s$. Hence, the path-decomposition $(V_1 \smallsetminus V(Q), \dots, V_s \smallsetminus V(Q))$ of $G - Q$ has width at most $\mathrm{pw}(G) - 1$, so $\mathrm{pw}(G - Q) < \mathrm{pw}(G)$.

Thus every connected graph G contains a path whose deletion reduces the path-width of G. If we may further assume (e.g. by some suitable induction hypothesis) that $G - Q$ has large path-width for every path $Q \subseteq G$, then G has even larger path-width.

We now use this idea to show that trees can have arbitrarily large path-width. Let T_3^k denote the tree in which one specified vertex r has degree 3, all other vertices (except the leaves) have degree 4, and all leaves have distance k from r. If $T = T_3^{k+1}$ and Q is any path in T, then Q contains at most two of the three edges at r; hence, $T - Q$ contains a component of $T - r$, which is a copy of T_3^k. Induction on k thus shows that $\mathrm{pw}(T_3^k) \geqslant k$ for all k.

For the proof of the backward implication of Theorem 12.4.5 we need some definitions and two lemmas. Let $G = (V, E)$ be a graph. For $X \subseteq V$, we denote by ∂X the set of all vertices in X with a neighbour in $G - X$. For every integer $n \geqslant 0$ we define a set $\mathcal{B}_n = \mathcal{B}_n(G)$ of subsets of V by the following recursion:

(i) $\emptyset \in \mathcal{B}_n$;

(ii) if $X \in \mathcal{B}_n$, $X \subseteq Y \subseteq V$ and $|\partial X| + |Y \smallsetminus X| \leqslant n$, then $Y \in \mathcal{B}_n$ (Fig. 12.4.1).

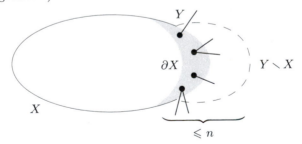

Fig. 12.4.1. If X lies in $\mathcal{B}_n$, then so does Y

Thus, a set $X \subseteq V$ lies in $\mathcal{B}_n$ if and only if there is a sequence

$$\emptyset = X_0 \subseteq \dots \subseteq X_s = X$$

such that $|\partial X_r| + |X_{r+1} \smallsetminus X_r| \leqslant n$ for all $r < s$. For example, if $(V_1, \dots, V_s)$ is a path-decomposition of G of width $< n$, then all its 'initial segments' $V_1 \cup \ldots \cup V_r$ ($r \leqslant s$) lie in $\mathcal{B}_n$, including V for $r = s$ (exercise). Conversely, we have the following:

Lemma 12.4.6. *If $V \in \mathcal{B}_n$, then* $\mathrm{pw}(G) < n$.

Proof. If $V \in \mathcal{B}_n$, then there is a sequence $\emptyset = X_0 \subseteq \ldots \subseteq X_s = V$ such that $|\partial X_r| + |X_{r+1} \smallsetminus X_r| \leqslant n$ for all $r < s$. We set

$$V_{r+1} := \partial X_r \cup (X_{r+1} \smallsetminus X_r)$$

and show that $(V_1, \ldots, V_s)$ is a path-decomposition of G (Fig. 12.4.2).

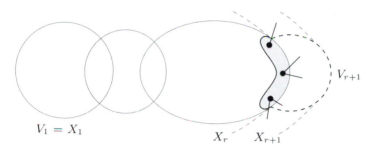

Fig. 12.4.2. Constructing a path-decomposition from $\mathcal{B}_n$

Induction on r shows that $X_r = V_1 \cup \ldots \cup V_r$ for all $r \leqslant s$; in particular, $V = X_s = V_1 \cup \ldots \cup V_s$. Hence (T1) holds. For the proof of (T2), let $xy \in E$ be given. Let $r(x)$ be minimum with $x \in X_{r(x)}$, and $r(y)$ minimum with $y \in X_{r(y)}$. We assume that $r(x) \leqslant r(y) =: r$, and show that x, like y, lies in V_r. This is clear if $r(x) = r$. Yet if $r(x) < r$, then x lies in X_{r-1}, and hence in $\partial X_{r-1} \subseteq V_r$ since $xy \in E$. For the proof of (T3), finally, let $p < q < r$ and $x \in V_p \cap V_r$ be given. Then $x \in V_p \subseteq V_1 \cup \ldots \cup V_{q-1} = X_{q-1} \subseteq X_{r-1}$, so $x \in X_{r-1} \cap V_r$. By definition of V_r this implies $x \in \partial X_{r-1}$, so $x \in \partial X_{r-1} \cap X_{q-1} \subseteq \partial X_{q-1} \subseteq V_q$. $\qquad\square$

Lemma 12.4.7. *Let $Y \in \mathcal{B}_n$ and $Z \subseteq Y$. If there is a family $(P_z)_{z \in \partial Z}$ of disjoint Z–∂Y paths in G with $z \in P_z$ for all $z \in \partial Z$, then $Z \in \mathcal{B}_n$ (Fig. 12.4.3).*

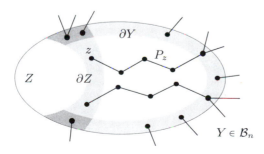

Fig. 12.4.3. Five paths P_z; three of them trivial

Proof. By definition of $\mathcal{B}_n$, there are sets $\emptyset = Y_0 \subseteq \ldots \subseteq Y_s = Y$ such that

$$|\partial Y_r| + |Y_{r+1} \smallsetminus Y_r| \leqslant n$$

Z_r

for all $r < s$. We shall deduce from this that, setting $Z_r := Y_r \cap Z$, we also have

$$|\partial Z_r| + |Z_{r+1} \smallsetminus Z_r| \leqslant n$$

for all $r < s$; then $Z = Z_s \in \mathcal{B}_n$.

Fix r. Since $Z_{r+1} \smallsetminus Z_r = Z_{r+1} \smallsetminus Y_r \subseteq Y_{r+1} \smallsetminus Y_r$, it suffices to show that $|\partial Z_r| \leqslant |\partial Y_r|$. We prove this by constructing an injective map $z \mapsto y$ from $\partial Z_r \smallsetminus \partial Y_r$ to $\partial Y_r \smallsetminus \partial Z_r$ (Fig. 12.4.4).

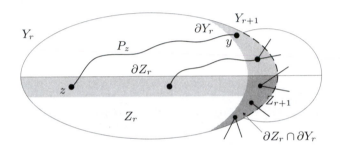

Fig. 12.4.4. An injective path linkage between $\partial Z_r \smallsetminus \partial Y_r$ and $\partial Y_r \smallsetminus \partial Z_r$

z

y

Consider a vertex $z \in \partial Z_r \smallsetminus \partial Y_r$. Then z has a neighbour in $Y_r \smallsetminus Z_r = Y_r \smallsetminus Z$, so $z \in \partial Z$. Now P_z is a path from $(Z_r \subseteq) Y_r$ to ∂Y, so P_z has a vertex y in ∂Y_r; note that $y \neq z$ by the choice of z. As z is the only vertex of P_z in Z, we have $y \in \partial Y_r \smallsetminus \partial Z_r$. Since the paths P_z are disjoint, these vertices y are distinct for different z, so $|\partial Z_r| \leqslant |\partial Y_r|$ as claimed. $\qquad \square$

(1.5.2)
(3.3.1)

Proof of Theorem 12.4.5. The forward implication of the theorem was proved earlier. For the converse, we prove the following:

n, F

If $\mathrm{pw}(G) \geqslant n \in \mathbb{N}$, then G contains every forest F with $|F| - 1 = n$ as a minor. $\qquad (*)$

Clearly, by $(*)$, if X is any forest then every graph in $\mathrm{Forb}_{\preccurlyeq}(X)$ has path-width less than $|X| - 1$.

So let $\mathrm{pw}(G) \geqslant n$, and assume without loss of generality that F is a tree. Let $(v_1, \ldots, v_{n+1})$ be an enumeration of $V(F)$ as in Corollary 1.5.2, i.e. so that v_{i+1} has exactly one neighbour in $\{v_1, \ldots, v_i\}$, for all $i \leqslant n$.

$v_1, v_2, \ldots$

For every $i = 0, \ldots, n$, we shall define a family $\mathcal{X}^i = (X^i_0, \ldots, X^i_i)$ $\mathcal{X}^i$
of disjoint subsets of V, such that $X^k_j \subseteq X^\ell_j$ whenever $j \leqslant k \leqslant \ell$ and all
X^i_j with $j > 0$ are connected in G. We then write

$$X^i := X^i_0 \cup \ldots \cup X^i_i.$$ X^i

For each i, the following three statements will hold:

(i) G contains an X^i_j–X^i_k edge whenever $1 \leqslant j < k \leqslant i$ and $v_j v_k \in$
     $E(F)$ (so $F[v_1, \ldots, v_i]$ is a minor of $G[X^i_1 \cup \ldots \cup X^i_i]$);

(ii) $|X^i_j \cap \partial X^i| = 1$ for all $1 \leqslant j \leqslant i$;

(iii) X^i is maximal in $\mathcal{B}_n$ with $|\partial X^i| \leqslant i$.

Note that (ii) and (iii) together imply $|\partial X^i| = i$.

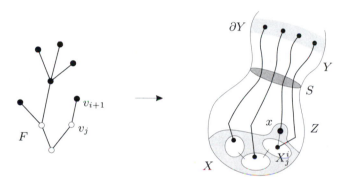

Fig. 12.4.5. Constructing an F minor in G

Let $X^0_0 \in \mathcal{B}_n$ be maximal with $|\partial X^0_0| = 0$ (possibly $X^0_0 = \emptyset$). Then X^0_0
(i)–(iii) hold for $i = 0$. Assume now that $\mathcal{X}^i$ has been defined so that
(i)–(iii) hold, for given $i \leqslant n$. If $i = 0$, let x be any vertex of $G - X^0$; x
note that $G - X^0 \neq \emptyset$, since $X^0 \in \mathcal{B}_n$ but $V \notin \mathcal{B}_n$ by Lemma 12.4.6. If
$i > 0$, consider the unique $j \leqslant i$ with $v_j v_{i+1} \in E(F)$, and let $x \in G - X^i$ x
be a neighbour of the unique vertex in $X^i_j \cap \partial X^i$; cf. (ii). Set

$$X := X^i \cup \{x\}.$$ X

If $i = n$, we have $F \preccurlyeq G[X]$ by (i) and the choice of x, so we are
done. Assume then that $i < n$. Then $X \in \mathcal{B}_n$ and $|\partial X| > i$, by (iii) and
the definition of $\mathcal{B}_n$. Since $\partial X \cap X^i \subseteq \partial X^i$, this means that

$$|\partial X| = i+1$$

and

$$\partial X = \partial X^i \cup \{x\}.$$

Let $Y \in \mathcal{B}_n$ be maximal with $X \subseteq Y$ and

$$|\partial Y| = i + 1\,;$$

this set Y will later become X^{i+1}.

By Menger's theorem (3.3.1), there exist a set $\mathcal{P}$ of disjoint X–∂Y paths in $G[Y]$ and a set $S \subseteq Y$ which separates X from ∂Y in $G[Y]$ and contains exactly one vertex from each path in $\mathcal{P}$ (but no other vertices). Let Z denote the union of S with the vertex sets of the components of $G - S$ that meet X. Clearly,

$$\partial Z \subseteq S$$

and $X^i \subsetneq X \subseteq Z$; let us show that even

$$X \subseteq Z \subseteq Y.$$

Let $z \in Z$ be given. If $z \in S$, then $z \in Y$ by the choice of S. If $z \notin S$, then z can be reached from X by a path avoiding S. If $z \notin Y$, then by $X \subseteq Y$ this path contains an X–∂Y path in $G[Y]$, contradicting the definition of S.

Thus $Z \subseteq Y \in \mathcal{B}_n$, so $Z \in \mathcal{B}_n$ by Lemma 12.4.7 applied to the Z–∂Y paths contained in the paths from $\mathcal{P}$. By (iii), $i < |\partial Z| \leqslant |S| = |\mathcal{P}|$. As every path in $\mathcal{P}$ meets ∂X, this gives $i < |\mathcal{P}| \leqslant |\partial X| = i + 1$ and hence

$$|\mathcal{P}| = i + 1\,,$$

so $\mathcal{P}$ links ∂X to ∂Y bijectively.

We now define $\mathcal{X}^{i+1}$. For $1 \leqslant k \leqslant i$ let $X_k^{i+1} := X_k^i \cup V(P_k)$, where P_k is the path in $\mathcal{P}$ containing the unique vertex of ∂X^i in X_k^i; cf. (ii). Similarly, let X_{i+1}^{i+1} be the vertex set of the path in $\mathcal{P}$ that contains x. Finally, put $X_0^{i+1} := Y \setminus (X_1^{i+1} \cup \ldots \cup X_{i+1}^{i+1})$. Clearly,

$$X^{i+1} = Y.$$

Condition (i) for $i+1$ holds by choice of x; (ii) holds by $X^{i+1} = Y$ and definition of $\mathcal{P}$; (iii) holds by $X^{i+1} = Y$ and the choice of Y, combined with $X^i \subseteq Y$ and (iii) for i.

As remarked earlier, $F \preccurlyeq G$ follows from the definition of X when $i = n$. $\square$

12.5 The minor theorem

Hereditary graph properties, those that are closed under taking minors, occur frequently in graph theory. Among the most natural examples are the properties of being embeddable in some fixed surface, such as planarity.

By Kuratowski's theorem, planarity can be expressed by forbidding the minors K^5 and $K_{3,3}$. This is a *good characterization* of planarity in the following sense. Suppose we wish to persuade someone that a certain graph is planar: this is easy (at least intuitively) if we can produce a drawing of the graph. But how do we persuade someone that a graph is non-planar? By Kuratowski's theorem, there is also an easy way to do that: we just have to exhibit an MK^5 or $MK_{3,3}$ in our graph, as an easily checked 'certificate' for non-planarity. Our simple Proposition 12.4.3 is another example of a good characterization: if a graph has tree width < 3, we can prove this by exhibiting a suitable tree-decomposition; if not, we can produce an MK^4 as evidence.

Theorems that characterize a hereditary property $\mathcal{P}$ by a set $\mathcal{X}$ of forbidden minors are no doubt among the most attractive results in graph theory. Given $\mathcal{P}$, one naturally seeks to make $\mathcal{X}$ as small as possible. This can be achieved simply by taking as $\mathcal{X}$ all the $\preccurlyeq$-minimal graphs not having $\mathcal{P}$:

Proposition 12.5.1. *A set $\mathcal{X} \subseteq \overline{\mathcal{P}}$ is minimal with $\mathcal{P} = \mathrm{Forb}_{\preccurlyeq}(\mathcal{X})$ if and only if $\mathcal{X}$ contains exactly one copy of every $\preccurlyeq$-minimal graph in $\overline{\mathcal{P}}$.* $\square$

By Proposition 12.5.1, there is essentially only one minimal set of forbidden minors for any hereditary property. Clearly, its elements are incomparable under the minor relation $\preccurlyeq$. Now the *minor theorem* of Robertson & Seymour says that any set of $\preccurlyeq$-incomparable graphs must be finite:

Theorem 12.5.2. (Minor Theorem; Robertson & Seymour 1986–97)
The finite graphs are well-quasi-ordered by the minor relation $\preccurlyeq$.

In particular, every hereditary graph property can be represented by finitely many forbidden minors:

Corollary 12.5.3. *Every graph property that is closed under taking minors can be expressed as $\mathrm{Forb}_{\preccurlyeq}(\mathcal{X})$ with finite $\mathcal{X}$.* $\square$

As a special case of Corollary 12.5.3 we have, at least in principle, a Kuratowski-type theorem for every surface:

Corollary 12.5.4. *For every surface* S *there exists a finite set of graphs* $X_1, \ldots, X_n$ *such that* $\mathrm{Forb}_\preccurlyeq(X_1, \ldots, X_n)$ *contains precisely the graphs not embeddable in* S.

The minimal set of forbidden minors has been determined explicitly only for one surface other than the sphere, for the projective plane: it contains 35 forbidden minors. It is not difficult to show that the number of forbidden minors grows rapidly with the genus of the surface (Exercise 28).

The complete proof of the minor theorem would fill a book or two. For all its complexity in detail, however, its basic idea is easy to grasp. We have to show that every infinite sequence

$$G_0, G_1, G_2, \ldots$$

of finite graphs contains a good pair: two graphs $G_i \preccurlyeq G_j$ with $i < j$. We may assume that $G_0 \npreccurlyeq G_i$ for all $i \geqslant 1$, since G_0 forms a good pair with any graph G_i of which it is a minor. Thus all the graphs $G_1, G_2, \ldots$ lie in $\mathrm{Forb}_\preccurlyeq(G_0)$, and we may use the structure common to these graphs in our search for a good pair.

We have already seen how this works when G_0 is planar: then the graphs in $\mathrm{Forb}_\preccurlyeq(G_0)$ have bounded tree-width (Theorem 12.4.4) and are therefore well-quasi-ordered by Theorem 12.3.5. In general, we need only consider the cases of $G_0 = K^n$: since $G_0 \preccurlyeq K^n$ for $n := |G_0|$, we may assume that $K^n \npreccurlyeq G_i$ for all $i \geqslant 1$.

The proof now follows the same lines as above: again the graphs in $\mathrm{Forb}_\preccurlyeq(K^n)$ can be characterized by their tree-decompositions, and again their tree structure helps, as in Kruskal's theorem, with the proof that they are well-quasi-ordered. The parts in these tree-decompositions are no longer restricted in terms of order now, but they are constrained in more subtle structural terms. Roughly speaking, for every n there exists a finite set $\mathcal{S}$ of closed surfaces such that every graph without a K^n minor has a tree-decomposition into parts each 'nearly' embedding in one of the surfaces $S \in \mathcal{S}$. (The 'nearly' hides a measure of disorderliness that depends on n but not on the graph to be embedded.) By a generalization of Theorem 12.3.5—and hence of Kruskal's theorem—it now suffices, essentially, to prove that the set of all the parts in these tree-decompositions is well-quasi-ordered: then the graphs decomposing into these parts are well-quasi-ordered, too. Since $\mathcal{S}$ is finite, every infinite sequence of such parts has an infinite subsequence whose members are all (nearly) embeddable in the same surface $S \in \mathcal{S}$. Thus all we have to show is that, given any closed surface S, all the graphs embeddable in S are well-quasi-ordered.

This is shown by induction on the genus of S (more precisely, on $2 - \chi(S)$, where $\chi(S)$ denotes the Euler characteristic of S) using the same approach as before: if $H_0, H_1, H_2, \ldots$ is an infinite sequence of graphs embeddable in S, we may assume that none of the graphs $H_1, H_2, \ldots$ contains H_0 as a minor. If $S = S^2$ we are back in the case that H_0 is planar, so the induction starts. For the induction step we now assume that $S \neq S^2$. Again, the exclusion of H_0 as a minor constrains the structure of the graphs $H_1, H_2, \ldots$, this time topologically: each H_i with $i \geqslant 1$ has an embedding in S which meets some non-contractable closed curve $C_i \subseteq S$ in no more than a bounded number of vertices (and no edges), say in $X_i \subseteq V(H_i)$. (The bound on $|X_i|$ depends on H_0, but not on H_i.) Cutting along C_i, and sewing a disc on to each of the one or two closed boundary curves arising from the cut, we obtain one or two new closed surfaces of larger Euler characteristic. If the cut produces only one new surface S_i, then our embedding of $H_i - X_i$ still counts as a near-embedding of H_i in S_i (since X_i is small). If this happens for infinitely many i, then infinitely many of the surfaces S_i are also the same, and the induction hypothesis gives us a good pair among the corresponding graphs H_i. On the other hand, if we get two surfaces S'_i and S''_i for infinitely many i (without loss of generality the same two surfaces), then H_i decomposes accordingly into subgraphs H'_i and H''_i embedded in these surfaces, with $V(H'_i \cap H''_i) = X_i$. The set of all these subgraphs taken together is again well-quasi-ordered by the induction hypothesis, and hence so are the pairs (H'_i, H''_i) by Lemma 12.1.3. Using a sharpening of the lemma that takes into account not only the graphs H'_i and H''_i themselves but also how X_i lies inside them, we finally obtain indices i, j not only with $H'_i \preccurlyeq H'_j$ and $H''_i \preccurlyeq H''_j$, but also such that these minor embeddings extend to the desired minor embedding of H_i in H_j—completing the proof of the minor theorem.

In addition to its impact on 'pure' graph theory, the minor theorem has had far-reaching algorithmic consequences. Using their tree structure theorem for the graphs in $\mathrm{Forb}_{\preccurlyeq}(K^n)$, Robertson & Seymour have shown that testing for any fixed minor is 'fast': for every graph X there is a polynomial-time algorithm[6] that decides whether or not the input graph contains X as a minor. By the minor theorem, then, every hereditary graph property $\mathcal{P}$ can be decided in polynomial (even cubic) time: if $X_1, \ldots, X_k$ are the corresponding minimal forbidden minors, then testing a graph G for membership in $\mathcal{P}$ reduces to testing the k assertions $X_i \preccurlyeq G$!

The following example gives an indication of how deeply this algorithmic corollary affects the complexity theory of graph algorithms. Let us call a graph *knotless* if it can be embedded in $\mathbb{R}^3$ so that none of its cycles forms a non-trivial knot. Before the minor theorem, it was an open problem whether knotlessness is decidable, that is, whether *any*

---

[6] indeed a cubic one—although with a typically enormous constant depending on X

algorithm exists (no matter how slow) that decides for any given graph whether or not that graph is knotless. To this day, no such algorithm is known. The property of knotlessness, however, is easily 'seen' to be hereditary: contracting an edge of a graph embedded in 3-space will not create a knot where none had been before. Hence, by the minor theorem, there *exists* an algorithm that decides knotlessness—even in polynomial (cubic) time!

However spectacular such unexpected solutions to long-standing problems may be, viewing the minor theorem merely in terms of its corollaries will not do it justice. At least as important are the techniques developed for its proof, the various ways in which minors are handled or constructed. Most of these have not even been touched upon here, yet they seem set to influence the development of graph theory for many years to come.

Exercises

1.⁻ Let $\leqslant$ be a quasi-ordering on a set X. Call two elements $x, y \in X$ *equivalent* if both $x \leqslant y$ and $y \leqslant x$. Show that this is indeed an equivalence relation on X, and that $\leqslant$ induces a partial ordering on the set of equivalence classes.

2. Let $(A, \leqslant)$ be a quasi-ordering, and assume that every descending chain $a_0 > a_1 > \ldots$ in A is finite. For subsets $X \subseteq A$ let

$$\mathrm{Forb}_{\preccurlyeq}(X) := \{\, a \in A \mid a \npreccurlyeq x \text{ for all } x \in X \,\}.$$

Show that A is a well-quasi-ordering if and only if every subset $B \subseteq A$ closed under $\geqslant$ (i.e. such that $x \leqslant y \in B \Rightarrow x \in B$) can be written as $B = \mathrm{Forb}_{\preccurlyeq}(X)$ with some finite $X \subseteq A$.

3. Find a quasi-ordering $(A, \leqslant)$, without an infinite antichain, such that *not* every subset $B \subseteq A$ closed under $\geqslant$ has the form $B = \mathrm{Forb}_{\preccurlyeq}(X)$. (Compare the previous exercise.)

4. Prove Proposition 12.1.1 and Corollary 12.1.2 directly, without using Ramsey's theorem.

5. Given a quasi-ordering $(X, \leqslant)$ and subsets $A, B \subseteq X$, write $A \leqslant' B$ if there exists an *order preserving* injection $f : A \to B$ with $a \leqslant f(a)$ for all $a \in A$. Does Lemma 12.1.3 still hold if the quasi-ordering considered for $[X]^{<\omega}$ is $\leqslant'$?

6.⁻ Show that the relation $\leqslant$ between rooted trees defined in the text is indeed a quasi-ordering.

7. Show that the finite trees are not well-quasi-ordered by the subgraph relation.

8. The last step of the proof of Kruskal's theorem considers a 'topological' embedding of T_m in T_n that maps the root of T_m to the root of T_n. Suppose we assume inductively that the trees of A_m are embedded in the trees of A_n in the same way, with roots mapped to roots. We thus seem to obtain a proof that the finite rooted trees are well-quasi-ordered by the subgraph relation, even with roots mapped to roots. Where is the error?

9. Show that the finite graphs are not well-quasi-ordered by the topological minor relation.

(Hint. Vertex degrees.)

10.+ Given $k \in \mathbb{N}$, is the class $\{ G \mid G \not\supseteq P^k \}$ well-quasi-ordered by the subgraph relation?

11. Show that a graph has tree-width at most 1 if and only if it is a forest.

12. Prove the following converse to Lemma 12.3.2: if $(T, \mathcal{V})$ satisfies condition (T1) and the statement of the lemma, then $(T, \mathcal{V})$ is a tree-decomposition of G.

13. Let $(T, (V_t)_{t \in T})$ be a tree-decomposition of a graph G. For each vertex $v \in G$, set $T_v := \{ t \in T \mid v \in V_t \}$. Show that T_v is always connected in T. More generally, for which subsets $U \subseteq V(G)$ is the set $\{ t \in T \mid V_t \cap U \neq \emptyset \}$ always connected in T (i.e. for all tree-decompositions)?

14. Let $\mathcal{H}$ be a set of graphs, and let G be constructed recursively from elements of $\mathcal{H}$ by pasting along complete subgraphs. Show that G has a tree-decomposition $(T, (V_t)_{t \in T})$ into elements of $\mathcal{H}$, such that $G[V_t \cap V_{t'}]$ is complete whenever $tt' \in E(T)$.

15. Given a tree-decomposition $(T, (V_t)_{t \in T})$ of G and $t \in T$, let H_t denote the graph obtained from $G[V_t]$ by adding all edges xy such that $x, y \in V_t \cap V_{t'}$ for some neighbour t' of t in T; the graphs H_t are called the *torsos* of this tree-decomposition. Show that G has no K^5 minor if and only if G has a tree-decomposition in which every torso is either planar, or a copy of the Wagner graph W (Fig. 8.3.1), or a subdivision of $K_{3,3}$.

(Hint. Theorem 8.3.4 and the previous exercise.)

16.+ Call a graph G *irreducible* if it is not separated by any complete subgraph. Every (finite) graph G can be decomposed into irreducible induced subgraphs, as follows. If G has a separating complete subgraph S, then decompose G into proper induced subgraphs G' and G'' with $G = G' \cup G''$ and $G' \cap G'' = S$. Then decompose G' and G'' in the same way, and so on, until all the graphs obtained are irreducible. Show that this set of irreducible graphs is uniquely determined if S is always chosen minimal. Show further that G has a tree-decomposition $(T, (V_t)_{t \in T})$ into those irreducible graphs, such that $G[V_t \cap V_{t'}]$ is complete for every edge $tt' \in E(T)$.

17. Let $\mathcal{C}$ be a set of connected vertex sets in a graph G, such that any two
     sets in $\mathcal{C}$ either meet or have an edge between them. Show that every
     tree-decomposition of G has a part that meets every set in $\mathcal{C}$.

     (Hint. Imitate the proof of Lemma 12.3.3.)

18.+ If $\mathcal{F}$ is a family of sets, then the graph G on $\mathcal{F}$ with $XY \in E(G)$ $\Leftrightarrow$
     $X \cap Y \neq \emptyset$ is called the *intersection graph* of $\mathcal{F}$. Show that a graph
     is chordal if and only if it is isomorphic to the intersection graph of a
     family of (vertex sets of) subtrees of a tree.

     (Hint. Use Proposition 12.3.8.)

19. For every $k \in \mathbb{N}$ find an $n \in \mathbb{N}$ such that the $n \times n$ grid has tree-width
     at least k.

     (Hint. By Lemma 12.3.3 and Proposition 12.4.2, it suffices to find in
     the $n \times n$ grid a minor H that contains $k + 1$ vertices pairwise linked
     by $k + 1$ independent paths in H.)

20. Reduce the (difficult) backwards implication of Theorem 12.4.4 to the
     special case that X is a grid.

21. Show that a graph has a path-decomposition into complete graphs if
     and only if it is isomorphic to an interval graph. (Interval graphs are
     defined in Ex. 36, Ch. 5.)

22. (continued)
     Prove the following analogue to Corollary 12.3.9 for path-width: every
     graph G satisfies $\mathrm{pw}(G) = \min \omega(H) - 1$, where the minimum is taken
     over all interval graphs H on $V(G)$ containing G.

23.− Show that, for any path-decomposition $(V_1, \ldots, V_s)$ of width $< n$, all
     the sets $V_1 \cup \ldots \cup V_r$ $(r \leqslant s)$ lie in $\mathcal{B}_n$.

24. Let $X \subseteq V(G)$ be maximal in $\mathcal{B}_n(G)$. Show that $|\partial X| = n$.

25.− Is the statement $(\ast)$ in the proof of Theorem 12.4.5 best possible, in
     the sense that a graph without an F minor need not have path-width
     less than $|F| - 2$?

26. Without using the minor theorem, show that the chromatic number of
     the graphs in any $\preccurlyeq$-antichain is bounded.

27. Seymour's *self-minor conjecture* asserts that 'every countably infinite
     graph is a proper minor of itself'. Make this assertion precise, and
     deduce the minor theorem from it.

     (Hint. Derive the minor theorem first for connected graphs. Then
     deduce the general version using Lemma 12.1.3.)

28. Given an orientable surface S of genus g, find a lower bound in terms
     of g for the number of forbidden minors needed to characterize embed-
     dability in S.

     (Hint. The smallest genus of an orientable surface in which a given
     graph can be embedded is called the *genus* of that graph. Use the
     theorem that the genus of a graph is equal to the sum of the genera of
     its blocks.)

Notes

Kruskal's theorem on the well-quasi-ordering of finite trees was first published in J.A. Kruskal, Well-quasi ordering, the tree theorem, and Vászonyi's conjecture, *Trans. Amer. Math. Soc.* **95** (1960), 210–225. Our proof is due to Nash-Williams, who introduced the versatile proof technique of choosing a 'minimal bad sequence'. This technique was also used in our proof of Higman's Lemma 12.1.3.

Nash-Williams generalized Kruskal's theorem to infinite graphs. This extension is much more difficult than the finite case; it is one of the deepest theorems in infinite graph theory. The general minor theorem becomes false for arbitrary infinite graphs, as shown by R. Thomas, A counterexample to 'Wagner's conjecture' for infinite graphs, *Math. Proc. Camb. Phil. Soc.* **103** (1988), 55–57. Whether or not the minor theorem extends to countable graphs remains an open problem.

The notions of tree-decomposition and tree-width were first introduced (under different names) by R. Halin, *S*-functions for graphs, *J. Geometry* **8** (1976), 171–186. Among other things, Halin showed that grids can have arbitrarily large tree-width; his proof is indicated in the hint for Exercise 19. Robertson & Seymour reintroduced the two concepts, apparently unaware of Halin's paper, with direct reference to K. Wagner, Über eine Eigenschaft der ebenen Komplexe, *Math. Ann.* **114** (1937), 570–590. (This is the classic paper containing Theorem 8.3.4; cf. Exercise 15.)

Robertson & Seymour themselves usually refer to the minor theorem as *Wagner's conjecture*. It seems that Wagner did indeed discuss this problem in the 1960s with his then students Halin and Mader. However, Wagner apparently never conjectured a positive solution; he certainly rejected any credit for the 'conjecture' when it had been proved.

Robertson & Seymour's proof of the minor theorem is given in the numbers IV–VII, IX–XII and XIV–XX of their series of over 20 papers under the common title of *Graph Minors*, which has been appearing in the *Journal of Combinatorial Theory*, Series B, since 1983. Of their theorems cited in this chapter, Theorem 12.3.5 is from Graph Minors IV, Theorem 12.4.4 from Graph Minors V, and Theorem 12.4.5 from Graph Minors I. The currently shortest available proof of Theorem 12.4.4 is given in N. Robertson, P.D. Seymour & R. Thomas, Quickly excluding a planar graph, *J. Combin. Theory B* **62** (1994), 323–348; this paper relies on the concept of *tangles* studied in Graph Minors X. Our short proof of 12.4.5 is from R. Diestel, Graph Minors I: a short proof of the path width theorem, *Combinatorics, Probability and Computing* **4** (1995), 27–30. Theorem 12.3.6 is taken from P.D. Seymour & R. Thomas, Graph searching and a min-max theorem for tree-width, *J. Combin. Theory B* **58** (1993), 22–33. Theorem 12.3.7 is due to R. Thomas, A Menger-like property of tree-width. The finite case, *J. Combin. Theory B* **48** (1990), 67–76.

The 35 minimal forbidden minors for graphs to be embedded in the projective plane were determined by D. Archdeacon, A Kuratowski theorem for the projective plane, *J. Graph Theory* **5** (1981), 243–246. An upper bound for the number of forbidden minors needed for an arbitrary closed surface is given in P.D. Seymour, A bound on the excluded minors for a surface, *J. Combin. Theory B* (to appear). B. Mohar, Embedding graphs in an arbitrary surface in linear time, *Proc. 28th Ann. ACM STOC* (Philadelphia 1996), 392–397, has developed a set of algorithms, one for each surface, that decide embeddability in that surface in linear time. As a corollary, Mohar obtains an independent and constructive proof of the 'generalized Kuratowski theorem', Corollary 12.5.4. Another independent and short proof of this corollary, which builds on Graph Minors IV & V but on no other papers of the Graph Minors series, was found by C. Thomassen, A simpler proof of the excluded minor theorem for higher surfaces, *J. Combin. Theory B* (to appear).

For every graph X, Graph Minors XIII gives an explicit algorithm that decides in cubic time for every input graph G whether $X \preccurlyeq G$. The constants in the cubic polynomials bounding the running time of these algorithms depend on X but are constructively bounded from above. For an overview of the algorithmic implications of the Graph Minors series, see Johnson's NP-completeness column in *J. Algorithms* **8** (1987), 285–303.

The concept of a 'good characterization' of a graph property was first suggested by J. Edmonds, Minimum partition of a matroid into independent subsets, *J. Research of the National Bureau of Standards (B)* **69** (1965) 67–72. In the language of complexity theory, a characterization is *good* if it specifies two assertions about a graph such that, given any graph G, the first assertion holds for G if and only if the second fails, and such that each assertion, if true for G, provides a certificate for its truth. Thus every good characterization has the corollary that the decision problem corresponding to the property it characterizes lies in NP ∩ co-NP.

Index

Page numbers in italics refer to definitions; in the case of author names, they refer to theorems due to that author. The alphabetical order ignores letters that stand as variables; for example, 'k-colouring' is listed under the letter c.

Symbol Index

The entries in this index are divided into two groups. Entries involving only mathematical symbols (i.e. no letters except variables) are listed on the first page, grouped loosely by logical function. The entry '[ ]', for example, refers to the definition of induced subgraphs $H[U]$ on page 4 as well as to the definition of face boundaries $G[f]$ on page 72.

Entries involving fixed letters as constituent parts are listed on the second page, in typographical groups ordered alphabetically by those letters. Letters standing as variables are ignored in the ordering.

Graduate Texts in Mathematics

continued from page ii

Reinhard Diestel received a PhD from the University of Cambridge, following research 1983–86 as a scholar of Trinity College under Béla Bollobás. He was a Fellow of St. John's College, Cambridge, from 1986 to 1990. Research appointments and scholarships have taken him to Bielefeld (Germany), Oxford and the US. Since 1994 he has been Professor of Mathematics at the University of Chemnitz, Germany.

Reinhard Diestel's main area of research is graph theory, especially infinite graph theory. He has published numerous papers and a research monograph, *Graph Decompositions* (Oxford 1990).